Emission and Scattering Techniques

Studies of Inorganic Molecules, Solids, and Surfaces

NATO ADVANCED STUDY INSTITUTES SERIES

Proceedings of the Advanced Study Institute Programme, which aims at the dissemination of advanced knowledge and the formation of contacts among scientists from different countries

The series is published by an international board of publishers in conjunction with NATO Scientific Affairs Division

A	Life Sciences	Plenum Publishing Corporation
B	Physics	London and New York
C	Mathematical and Physical Sciences	D. Reidel Publishing Company Dordrecht, Boston and London
D	Behavioural and Social Sciences	Sijthoff & Noordhoff International Publishers
E	Applied Sciences	Alphen aan den Rijn and Germantown U.S.A.

Series C – Mathematical and Physical Sciences

Volume 73 - Emission and Scattering Techniques

Emission and Scattering Techniques

Studies of Inorganic Molecules, Solids, and Surfaces

Proceedings of the NATO Advanced Study Institute held at Alghero, Sardinia, Italy, September 14-25, 1980

edited by

PETER DAY

Oxford University, Inorganic Chemistry Laboratory, England

D. Reidel Publishing Company

Dordrecht : Holland / Boston : U.S.A. / London : England

Published in cooperation with NATO Scientific Affairs Division

Library of Congress Cataloging in Publication Data

NATO Advanced Study Institute (1980: Alghero, Sardinia)
Emission and scattering techniques.

(NATO advanced study institutes series. Series C, Mathematical and physical sciences ; v. 73)
"Published in cooperation with NATO Scientific Affairs Division."
Includes index.
1. Emission spectroscopy–Congresses. 2. Scattering (Physics) –Congresses. I. Day, P. II. North Atlantic Treaty Organization. Division of Scientific Affairs. III. Title. IV. Series.
QD96.E46N37 1980 543'.0858 81-10681
ISBN 90-277-1317-0 AACR2

Published by D. Reidel Publishing Company
P.O. Box 17, 3300 AA Dordrecht, Holland

Sold and distributed in the U.S.A. and Canada
by Kluwer Boston Inc.,
190 Old Derby Street, Hingham, MA 02043, U.S.A.

In all other countries, sold and distributed
by Kluwer Academic Publishers Group,
P.O. Box 322, 3300 AH Dordrecht, Holland

D. Reidel Publishing Company is a member of the Kluwer Group

Printed in The Netherlands

TABLE OF CONTENTS

PREFACE

Centrally important to the progress of inorganic chemistry is the application of new physical techniques for determining crystal and molecular structures. Electronic structure, too, can now be explored by a large variety of spectroscopic techniques, most of them of quite recent origin. Realizing how essential it was to bring together experts in the techniques themselves and those who might use them for their own chemical purposes, Professor Furlani and I began in the early 1970's to organize small meetings at which this kind of interchange could take place. The first, funded by the Italian National Research Council and Ministry of Education, was at Frascati in 1971. It was followed by others at Oxford (1974) and Pugnochiuso (1977), funded under the NATO Advanced Study Institutes programme. Lectures given at the Oxford Advanced Study Institute were published by D. Reidel under the title Electronic States of Inorganic Compounds: New Experimental Techniques. A three-year interval between these Institutes has proved suitable both for introducing new generations of potential users to the methods and allowing us to incorporate advances in the methods themselves. In fact, since the last Advanced Study Institute in the series several important advances have occurred, particularly in electron, ion and neutron spectroscopies. We concentrated the course for 1980 on these newer aspects, though the more specialized lectures were prefaced with introductory material for those not familiar with the general principles. Also included are introductions to the theory of electronic states in solids and to band structure calculations of the type needed to rationalize the kind of information about bulk and surface states which the new methods provide.

The methods described in this book can be broken down into two main categories: those in which a particle (usually an electron or a photon) is emitted from the sample under bombardment by photons or electrons, and those in which a particle (electron, photon, neutron or ion) is scattered upon interaction

P. Day (ed.), Emission and Scattering Techniques, vii–viii.

with the sample. In both categories we find methods which give information about surface and bulk crystal structure and others which focus on electronic structure, both of states close to the Fermi surface and those associated with atomic cores. Accounts of all these methods will be found in this book.

It remains to thank the Scientific Affairs Division of NATO for their financial support, and my Italian and British colleagues on the organizing committee for their enthusiastic help.

Oxford,
April 1981

P. Day

ELECTRONIC PROPERTIES OF SURFACES

J.B. Pendry

Science Research Council, Daresbury Laboratory,
Warrington, WA4 4AD, UK.

The concept of a surface state is discussed in terms of trapping by repeated reflection from the crystal and surface barrier. Surface states play the role of "dangling bonds" on semiconductors but not on metals. The notion of incomplete orbitals is introduced in the context of bonding to metal surfaces. Finally an extended discussion is made of photoemission, its information content and surface sensitivity, several examples being presented.

SURFACE STATES

Surface states find many applications on semiconductor surfaces. They are the dangling bonds formed by detaching the last layer of atoms and on these surfaces they can play a role in pinning the Fermi energy within the surface state bands. On metals they play a less important role in the surface properties but on both materials they produce significant features in the photoemission spectra and they also appear in low energy diffraction experiments as resonances in the diffraction curves.

A surface state has good quantum numbers of energy and momentum parallel to the surface but it dies away exponentially into the vacuum and into the crystal. Therefore the energy of the surface state must be below the threshold for escape into the vacuum and also must be in a range of energies corresponding to a band gap of the solid. The surface state wavefunction can be expanded inside the crystal as a sum of Bloch waves with complex $\underline{k}$-vector. It follows that the band structure plays a key role in the theory of surface states. Let us begin by plotting the free electron bands for zero potential inside the

P. Day (ed.), Emission and Scattering Techniques, 1–20.

solid. In making this plot we have fixed momentum parallel to the surface and the energy of the bands is plotted as a function of momentum normal to the surface. Thus the energy is

$$E = \tfrac{1}{2}k_z^2 + \tfrac{1}{2}|\underline{k}_{/\!/} + \underline{g}|^2 \tag{P1}$$

where $\underline{g}$ is a 2-dimensional reciprocal lattice vector. There is a set of bands for every reciprocal lattice vector and the surface will couple these bands together once the potential is switched on. We can write the momentum normal to the surface as follows

$$\begin{aligned} k_z &= \pm(2E - |\underline{k}_{/\!/} + \underline{g}|^2)^{\frac{1}{2}} \\ \text{or} \quad & \pm i(|\underline{k}_{/\!/} + \underline{g}|^2 - 2E)^{\frac{1}{2}} \end{aligned} \tag{P2}$$

From equation (P2) the <u>two band rule</u> follows: for each two-dimensional reciprocal lattice vector there are two Bloch waves, either two real ones or two complex ones. This rule which is obvious for the free electron bands is a general rule.

Now let us switch on the crystal potential. The bands will distort under the influence of the potential especially where two bands cross. The real bands repel one another, but in such a way that the two band rule is always obeyed. Where gaps develop in the band structure complex bands are always to be found joining the minima and maxima across which the gap is found. In other words the complex bands leave the real band structure at the point where

$$\frac{\partial E_r}{\partial k_{zr}} = 0 \tag{P3}$$

To make these statements explicit it is possible to make a model calculation (1). Suppose we have two bands which cross at an energy of E_o and k_o in the absence of a potential coupling them together. Let the two wavefunctions be Ψ^+ and Ψ^- and we switch on a coupling,

$$V_{+-} = \langle\Psi^+|V|\Psi^-\rangle \tag{P4}$$

We shall ignore all other matrix elements. If we normalise $\Psi^{\pm}$ to unit current

$$\frac{1}{\Omega}\int_{\substack{\text{unit}\\\text{cell}}} |\Psi^{\pm}|^2 d^3\underline{r} = \left|\frac{\partial E^{\pm}}{\partial k_z}\right|^{-1} \qquad (P5)$$

then in the gap we can write for the hybridised wavefunction

$$\Psi = \Psi^{+} + e^{2i\Phi}\Psi^{-} \qquad (P6)$$

where the hybridisation must contain equal amplitudes of the two wavefunctions because no current is carried by a complex band. Another way of regarding the effect of the gap (2) is that the potential is reflecting Ψ^+ into Ψ^- with reflection coefficient exp $(2i\Phi)$. Substituting (P6) into

$$(H_o + V)\Psi = E\Psi \qquad (P7)$$

and retaining only variations of Φ and k_z

$$\begin{bmatrix} [E_o + (k_z - k_o)\dot{E}^{+} - E]\,|\dot{E}^{+}|^{-1}, & V_{+-}|\dot{E}^{+}\dot{E}^{-}|^{-\frac{1}{2}} \\ V_{-+}|\dot{E}^{+}\dot{E}^{-}|^{-\frac{1}{2}}, & [E_o + (k_z - k_o)\dot{E}^{-} - E]\,|\dot{E}^{-}|^{-1} \end{bmatrix} \begin{bmatrix} 1 \\ e^{2i\Phi} \end{bmatrix} = 0 \qquad (P8)$$

where in deriving equation (P8) we have assumed a centro-symmetric crystal so that V_{+-} is equal to V_{-+} and we have neglected all variations with energy except those of the momentum. It can be shown that there are two solutions of equation (P8) for real k_z and real E which have the form of

$$e^{2i\Phi} = -1 \qquad (\Phi = \pi/2)$$

$$e^{2i\Phi} = +1 \qquad (\Phi = \pi)$$

and these 2 solutions correspond to the band edges. Which is the lower and which the higher edge depends on the sign of V. If V is greater than 0 then we have

$$\Psi = \Psi^{+} - \Psi^{-} \qquad (P9)$$

corresponding to the bottom of the gap because this solution gives a wavefunction with minima at the atom centres and thus least interaction with the repulsive potential. On the other hand the solution

$$\Psi = \Psi^{+} + \Psi^{-} \qquad (P10)$$

corresponds to the state at the top of the gap because this leads to a wavefunction with a maximum at the atom centres. The top and bottom states are reversed if V is less than 0. Put in terms of the phase: if V is greater than 0 the phase, Φ, increases from $\pi/2$ to π, whereas if V is less than 0, Φ increases from 0 to $\pi/2$.

The condition for a surface state can easily be expressed using the concept of reflection coefficients. A surface state is formed when an electron is trapped between the barrier and the bulk of the crystal being alternately reflected from each. For this situation to come about two conditions must be satisfied.

(i) That the energy is below threshold for escape to vacuum.

(ii) That there is a band gap in the crystal preventing the electron escaping away to infinity.

Let Ψ^- be incident on the barrier.

$$\Psi^-_{tot} = (1 + e^{2i(\chi+\Phi)} + e^{4i(\chi+\Phi)} + \ldots)\Psi^-$$
$$= (1 - \exp[2i(\chi+\Phi)])^{-1}\Psi^- \qquad \text{(P11)}$$

From equation (P11) it is evident that a condition for a surface state is that

$$\chi + \Phi = n\pi \qquad \text{(P12)}$$

i.e. the barrier and the crystal play an equal role in formation of a surface state. In a given gap, the variation of Φ alone being only $\pi/2$ is not enough to ensure a surface state. If χ is relatively constant in the gap we speak of a crystal induced surface state. Thus if the phase of the barrier reflectivity χ lies between 0 and $\pi/2$ then there will be a surface state if V is less than 0. Alternatively if χ lies between $\pi/2$ and π there will be a surface state if V is greater than 0. For a high barrier located half way between atoms it is always the case than χ lies between $\pi/2$ and π. Thus the condition that V is greater than 0 is the condition for a surface state. This is known as the Shockley condition (3). It has a physical meaning which can be interpreted as follows. A positive matrix element implies covalent bonding because the valence levels at the lower edge of the gap heap up charge between atoms. This is the bond which is broken when the surface is formed and hence leads to the broken bond concept of a surface state. It is interesting to reflect why there is ever a surface state because we might expect that the matrix elements of negative potentials are generally speaking negative in sign but this neglects the concept of a pseudopotential which plays a crucial role in the concept of covalency. More of these ideas can be found in the work of Phillips and his collaborators (4).

A different sort of surface state can occur if the barrier reflectivity is more rapidly varying than the crystal reflectivity. Then we can regard the barrier as being responsible for sweeping the total phase through the condition for a surface state. For example it might be that a layer of attractive atoms is adsorbed on the surface. The attractive potential can lead to a rapidly varying phase shift which splits off a surface state below the bottom of the crystal bands. This can happen when oxygen or sulphur are adsorbed on transition metals and the p-levels very often split off below the metal bands to form a separate band of surface states. This sort of state is known as a Tamm surface state.

A curious instance of the barrier induced surface states occurs because of the coulombic potential which electrons feel outside any solid due to the image force (5), (6). Near the threshold for escape from the surface barrier the image potential dominates the reflectivity and leads to a very rapid variation of χ, sweeping through the surface state condition an infinite number of times increasingly rapidly as threshold for escape to the vacuum is achieved. A series of Rydberg surface states is formed on the material provided that the crystal has a band gap near the threshold for escape to vacuum. One instance occurs in liquid helium but other examples are to be found in a whole variety of insulators.

BONDING AT SURFACES

A very appealing concept in surface bonding is the view of surface states as dangling bonds which are ready to snap up any bonds attached to adsorbed atoms. This a useful concept but it is not always true. For the dangling bond concept to be valid we must satisfy:

(i) There must be small lateral interaction between dangling bonds, i.e. there must be small dispersion of the surface state with momentum parallel to the surface.

(ii) The surface state energy should be near the centre of the gap otherwise a small perturbation causes it to hybridise with the bulk states before any substantial bond is established.

(iii) The surface state energy should be close to the Fermi energy to allow surface bonding and antibonding levels to form below and above the Fermi energy.

These criteria are met for semiconductors such as silicon and germanium but they are not met for transition metals. In silicon and germanium the surface state is localised near surface atoms (7) and gives a good guide to adsorption sites. In contrast in the transition metals the surface states are not necessarily localised near the surface atoms and the surface state is only one amongst many factors affecting adsorption.

On metallic substrates several concepts may play a role in the bonding of an adsorbate atom. One of these concepts is the Anderson model of chemisorption. In this model the adsorbate atom is supposed to have a well localised level which can interact with the substrate and is responsible for the bonding. A little distance from the surface the adsorbate atom level is sharp. Considerations of charge neutrality mean that the adsorbate level is pinned to the Fermi energy of the metal. When the adsorbate atom moves closer to the surface the level is broadened but still remains pinned to the Fermi energy. Thus there is a net lowering of the energy of the occupied levels and hence there is a bond established between the atom and the surface. Note the contrast with the dangling bond mechanism for this current mechanism of adsorption. In bonding to a metallic substrate the key ingredient is not a specific bond available at the surface but the general presence of continuum levels in the metal which are capable of broadening the adsorbate level. The Anderson model makes a specific prediction for photoemission experiments. It is that an adsorbate bonding according to this mechanism must produce valence levels due to the adsorbate in the vicinity of the Fermi surface, some below and some above.

There is another mechanism of bonding to metals which I gave the title to some years ago of 'incomplete orbitals' (8). Whereas the Anderson scheme is appropriate for very well defined adsorbate levels such as occur when the adsorbate has a d like character, the incomplete orbital scheme is more appropriate when the adsorbate levels have p type character and therefore are less tightly bound. An essential ingredient of the Anderson scheme is that the adsorbate level is pinned to the Fermi energy or close to the Fermi energy because in this way charge neutrality is achieved by appropriate floating of the level relative to the Fermi energy. Very often when p type orbitals are found at surfaces the adsorbate level is not found near the Fermi energy and a new mechanism has to be found. In the incomplete orbital scheme as the adsorbate level approaches to the surface it broadens as in the Anderson scheme but a significant new effect takes place. Because the angular momentum barrier confining a p orbital to the adsorbate scribes a metal with a single plasmon mode typically with f the order of 15 eV, then there are 2 cases of refraction urface. In the first case the electric field is perpen- to the plane of incidence: the case of s polarisation. d inside the solid, E_2, is given by the following equation

$$= \frac{2\cos\Theta_o}{\cos\Theta_o + \left(\cos^2\Theta_o - \frac{w_p^2}{w^2}\right)^{\frac{1}{2}}} \qquad (P27)$$

$_o$ is the incident field amplitude, and Θ_o the polar angle dence. This equation leads to a linear increase of the nside the solid from zero up to a maximum of twice the l field. On the other hand the case of the electric field l to the plane of incidence, p polarisation, can give rise re drastic correction to the internal field which is given following equation

$$\frac{\;_2}{\;_o} = \frac{2\cos\Theta_o\left(1 - \frac{w_p^2}{w^2}\right)^{\frac{1}{2}}}{\cos\Theta_o\left(1 - \frac{w_p^2}{w^2}\right) + \left(\cos^2\Theta_o - \frac{w_p^2}{w^2}\right)^{\frac{1}{2}}} \qquad (P28)$$

are 2 singularities in this second equation - one at w_p, ther at $w_p/\cos\Theta_o$. At the first singularity the internal goes to zero and at the second it reaches a peak. Another tant point to note about the photon field inside the solid at the decay length is always much greater than the electron length and therefore the surface sensitivity is always ated by the electronic properties of the medium.

emission Intensities

If we assume a linear response of the solid to the photon d then the photocurrent is described by first order perturb- n theory. Application of the golden rule gives the following ula for the photon current at a specified value of momentum llel to the surface of the escaping electron and of energy .

$$I(\underline{k}_{/\!/}, E + w) = \frac{L}{k_z}\left|\left\langle \underline{k}_{/\!/}, E + w\left|\frac{a.P}{2c}\right|E\right\rangle\right|^2 \rho(E) \qquad (P29)$$

electron escapes from the system in a time reversed LEED te. The operator is simply the photon field which we ountered in equation (P26) whereas the state from which the

atom is rather low the broadening of the level means that the top half of the now resonant state can flow over the angular momentum barrier and no longer constitutes a well defined bound state. Only the bottom half of the level persists as a true bound state, and the rest dissolves into the metal. This qualitative picture can be confirmed by numerical calculations of scattering properties of atoms on the surface and in the bulk of metals. Charge neutrality of the adsorbate is now achieved not by the level floating at the Fermi energy but rather the level floats near the top of its own angular momentum barrier and loses the right fraction of the orbital to preserve charge neutrality.

The top half of the broadened level is dissolved by the metal leaving the bottom half of the resonance intact and occupied if the whole level is below the Fermi energy, and because of this preservation of occupied states with lowest energy the incomplete orbital scheme leads to a bonding of the adsorbate atom to the surface. The incomplete orbital scheme has some features of covalency and some features of ionicity in its character. It is covalent to the extent that the metal bands perform the role of dissolving the top half of the adsorbate resonance but it is ionic to the extent that the remaining occupied levels are localised near the adsorbate atom not half way between adsorbate atoms and substrate atoms as they would be in true covalency. This localisation does not involve necessarily a charging of the adsorbate level because due to its incompleteness the resonance need not hold an integral number of electrons. The model accounts for the numerous systems in which the adsorbate level is observed well below the Fermi energy as though it had ionic character and yet where dipole moment measurements indicate very little net charge associated with the adsorbate atom. Again this particular mechamism makes a specific prediction for the photoemission experiments. An essential ingredient of the model is that the adsorbate valence levels are pinned not at the Fermi energy but just above the angular momentum barrier for escape into the metal. This usually occurs very close to the bottom of the metal sp band. Instances of 'incomplete orbital' bonding are found in chalcogen bonding to transition metals where the p levels of the chalcogens are found about 6 eV below the Fermi energy of the metal near the bottom of the metal sp band.

It is an interesting speculation that the incompleteness mechanism may be an important ingredient for lowering barriers to certain reactions. The ability for an orbital at the surface to "go incomplete" considerably reduces the electrostatic energy for a given configuration which would otherwise be highly charged.

PHOTOEMISSION

In a photoemission experiment (9) photons incident on a crystal are absorbed by the valence or conduction electrons of the solid. The electrons thus acquire enough energy to escape from the crystal and if all the valence bands are equally sampled, energy analysing the escaping electrons gives us an idea of the density of states in the occupied bands of the crystal, i.e. an energy level diagram of the crystal is displayed. It gives us information about:

(i) Surface states.

(ii) Adsorbate levels.

(iii) Deep valence band structure.

All these quantities are relevant to bonding.

If the photon energy is in the X-ray region electrons escaping from the crystal have quite strong penetrating properties and we see essentially bulk properties of the crystal. The electrons coming from an X-ray photoemission experiment have energies of around 1keV and resolution poses some difficulty in this range. It is also difficult to resolve the momentum of the electrons because of the narrow angular aperture required, with consequent sacrifice of count rates. In contrast in an ultraviolet photoemission experiment the electrons have at most a few tens of eV kinetic energy, are much less penetrating, and give greater surface sensitivity. It is also possible to get better energy and momentum resolution.

Perhaps the most powerful photoemission experiments are those in which the angle and therefore the momentum of the escaping electron is specified. The importance of these experiments can be understood if we make two naive assumptions. They are

$$\underline{k}_f = \underline{k}_i + \underline{q} \qquad (P18)$$

$$E_f = E_i + \hbar w \qquad (P19)$$

These equations are not strictly true and certainly equation (P18) is not remotely true for X-ray photoemission spectroscopy, but for ultraviolet photoemission they are sufficiently true that angle resolved photoemission experiments define all the good quantum numbers of an electron in a crystal: the energy, the momentum parallel to the surface and to a lesser extent the momentum normal to the surface. Thus we can get a detailed band by band plotting of the energy levels inside and at the surface of the solid

because of this complete definition
Angle resolved photoemission is the u
provides us with experimental access t
tially theoretical quantities.

The Photon Field

The electron Hamiltonian is descr

$$H_o = \tfrac{1}{2}P^2 + V$$

We now have to introduce a photon field

$$\underline{A} = \underline{a}\cos(\underline{q}.\underline{r} - wt)$$

$$\underline{B} = \text{curl } \underline{A}$$

The radiation field can be described ent
we dispose of the choice of gauge as foll

$$\text{div } \underline{A} = 0$$

$$\text{Therefore } \underline{q}.\underline{a} = 0$$

In the presence of this photon field the H
to become

$$H = \tfrac{1}{2}|\underline{P} + \tfrac{1}{c}\underline{A}|^2 + V$$

which results in a perturbation term of the

$$\frac{1}{2c}\left[\underline{P}.\underline{A} + \underline{A}.\underline{P}\right]$$

where we are neglecting effects of second or
field. That is to say we are assuming linea

The dielectric properties of the solid
simple picture. Quantum corrections for whi
-zero at the surface produce terms which are
with in the usual formulation of photoemissio
them in the following. Classical corrections
handle. They simply give rise to refraction
the solid and can be described by the Fresnel
assume a simple form for the dielectric consta

$$\varepsilon = 1 - w_p^2/w^2,$$

which de
energy
at the
dicular
The fie

where
of inc
field
extern
parall
to a m
by the

atom is rather low the broadening of the level means that the top half of the now resonant state can flow over the angular momentum barrier and no longer constitutes a well defined bound state. Only the bottom half of the level persists as a true bound state, and the rest dissolves into the metal. This qualitative picture can be confirmed by numerical calculations of scattering properties of atoms on the surface and in the bulk of metals. Charge neutrality of the adsorbate is now achieved not by the level floating at the Fermi energy but rather the level floats near the top of its own angular momentum barrier and loses the right fraction of the orbital to preserve charge neutrality.

The top half of the broadened level is dissolved by the metal leaving the bottom half of the resonance intact and occupied if the whole level is below the Fermi energy, and because of this preservation of occupied states with lowest energy the incomplete orbital scheme leads to a bonding of the adsorbate atom to the surface. The incomplete orbital scheme has some features of covalency and some features of ionicity in its character. It is covalent to the extent that the metal bands perform the role of dissolving the top half of the adsorbate resonance but it is ionic to the extent that the remaining occupied levels are localised near the adsorbate atom not half way between adsorbate atoms and substrate atoms as they would be in true covalency. This localisation does not involve necessarily a charging of the adsorbate level because due to its incompleteness the resonance need not hold an integral number of electrons. The model accounts for the numerous systems in which the adsorbate level is observed well below the Fermi energy as though it had ionic character and yet where dipole moment measurements indicate very little net charge associated with the adsorbate atom. Again this particular mechamism makes a specific prediction for the photoemission experiments. An essential ingredient of the model is that the adsorbate valence levels are pinned not at the Fermi energy but just above the angular momentum barrier for escape into the metal. This usually occurs very close to the bottom of the metal sp band. Instances of 'incomplete orbital' bonding are found in chalcogen bonding to transition metals where the p levels of the chalcogens are found about 6 eV below the Fermi energy of the metal near the bottom of the metal sp band.

It is an interesting speculation that the incompleteness mechanism may be an important ingredient for lowering barriers to certain reactions. The ability for an orbital at the surface to "go incomplete" considerably reduces the electrostatic energy for a given configuration which would otherwise be highly charged.

PHOTOEMISSION

In a photoemission experiment (9) photons incident on a crystal are absorbed by the valence or conduction electrons of the solid. The electrons thus acquire enough energy to escape from the crystal and if all the valence bands are equally sampled, energy analysing the escaping electrons gives us an idea of the density of states in the occupied bands of the crystal, i.e. an energy level diagram of the crystal is displayed. It gives us information about:

(i) Surface states.

(ii) Adsorbate levels.

(iii) Deep valence band structure.

All these quantities are relevant to bonding.

If the photon energy is in the X-ray region electrons escaping from the crystal have quite strong penetrating properties and we see essentially bulk properties of the crystal. The electrons coming from an X-ray photoemission experiment have energies of around 1keV and resolution poses some difficulty in this range. It is also difficult to resolve the momentum of the electrons because of the narrow angular aperture required, with consequent sacrifice of count rates. In contrast in an ultraviolet photoemission experiment the electrons have at most a few tens of eV kinetic energy, are much less penetrating, and give greater surface sensitivity. It is also possible to get better energy and momentum resolution.

Perhaps the most powerful photoemission experiments are those in which the angle and therefore the momentum of the escaping electron is specified. The importance of these experiments can be understood if we make two naive assumptions. They are

$$\underline{k}_f = \underline{k}_i + \underline{q} \tag{P18}$$

$$E_f = E_i + \hbar w \tag{P19}$$

These equations are not strictly true and certainly equation (P18) is not remotely true for X-ray photoemission spectroscopy, but for ultraviolet photoemission they are sufficiently true that angle resolved photoemission experiments define all the good quantum numbers of an electron in a crystal: the energy, the momentum parallel to the surface and to a lesser extent the momentum normal to the surface. Thus we can get a detailed band by band plotting of the energy levels inside and at the surface of the solid

because of this complete definition of the quantum numbers. Angle resolved photoemission is the ultimate experiment. It provides us with experimental access to what used to be essentially theoretical quantities.

The Photon Field

The electron Hamiltonian is described by

$$H_o = \tfrac{1}{2}P^2 + V \qquad (P20)$$

We now have to introduce a photon field of the form

$$\underline{A} = \underline{a}\cos(\underline{q}.\underline{r} - wt) \qquad (P21)$$

$$\underline{B} = \text{curl}\ \underline{A} \qquad (P22)$$

The radiation field can be described entirely by an $\underline{A}$ vector and we dispose of the choice of gauge as follows

$$\text{div}\ \underline{A} = 0 \qquad (P23)$$

$$\text{Therefore}\quad \underline{q}.\underline{a} = 0 \qquad (P24)$$

In the presence of this photon field the Hamiltonian is modified to become

$$H = \tfrac{1}{2}\left|\underline{P} + \frac{1}{c}\underline{A}\right|^2 + V \qquad (P25)$$

which results in a perturbation term of the form

$$\frac{1}{2c}\left[\underline{P}.\underline{A} + \underline{A}.\underline{P}\right] \qquad (P26)$$

where we are neglecting effects of second order in the photon field. That is to say we are assuming linear response.

The dielectric properties of the solid will complicate this simple picture. Quantum corrections for which grad $\underline{A}$ is none -zero at the surface produce terms which are difficult to deal with in the usual formulation of photoemission. We shall neglect them in the following. Classical corrections are more easy to handle. They simply give rise to refraction at the surface of the solid and can be described by the Fresnel equations. If we assume a simple form for the dielectric constant

$$\varepsilon = 1 - w_p^{\,2}/w^2 ,$$

which describes a metal with a single plasmon mode typically with energy of the order of 15 eV, then there are 2 cases of refraction at the surface. In the first case the electric field is perpendicular to the plane of incidence: the case of s polarisation. The field inside the solid, E_2, is given by the following equation

$$\frac{E_2}{E_o} = \frac{2\cos\Theta_o}{\cos\Theta_o + \left(\cos^2\Theta_o - \frac{w_p^2}{w^2}\right)^{\frac{1}{2}}} \tag{P27}$$

where E_o is the incident field amplitude, and Θ_o the polar angle of incidence. This equation leads to a linear increase of the field inside the solid from zero up to a maximum of twice the external field. On the other hand the case of the electric field parallel to the plane of incidence, p polarisation, can give rise to a more drastic correction to the internal field which is given by the following equation

$$\frac{E_2}{E_o} = \frac{2\cos\Theta_o\left(1 - \frac{w_p^2}{w^2}\right)^{\frac{1}{2}}}{\cos\Theta_o\left(1 - \frac{w_p^2}{w^2}\right) + \left(\cos^2\Theta_o - \frac{w_p^2}{w^2}\right)^{\frac{1}{2}}} \tag{P28}$$

There are 2 singularities in this second equation - one at w_p, the other at $w_p/\cos\Theta_o$. At the first singularity the internal field goes to zero and at the second it reaches a peak. Another important point to note about the photon field inside the solid is that the decay length is always much greater than the electron decay length and therefore the surface sensitivity is always dominated by the electronic properties of the medium.

Photoemission Intensities

If we assume a linear response of the solid to the photon field then the photocurrent is described by first order perturbation theory. Application of the golden rule gives the following formula for the photon current at a specified value of momentum parallel to the surface of the escaping electron and of energy (10).

$$I(\underline{k}_{/\!/}, E + w) = \frac{L}{k_z} \left|\langle \underline{k}_{/\!/}, E + w \left| \frac{a.P}{2c} \right| E \rangle\right|^2 \rho(E) \tag{P29}$$

The electron escapes from the system in a time reversed LEED state. The operator is simply the photon field which we encountered in equation (P26) whereas the state from which the

electron is excited is a valence state of the crystal. The final term in the expression is a density of valence states. This apparently innocuous formula hides a wealth of complications. No one of the terms described above is easy to calculate and it is quite a tour de force to put them all together. However it is possible to make these calculations and programs are currently available for doing this, at least for simple materials (11).

In evaluating photoemission matrix elements there is a useful and important transformation which can be made. As we have derived it the matrix element is in what is known as the velocity form described by equation (P30) below.

$$M = \langle E_i | \frac{\underline{A}.\underline{P}}{c} | E_i + w \rangle \qquad (P30)$$

However by commuting the photon field operator with the zeroth order Hamiltonian a new expression can be derived for the matrix element

$$M = \frac{1}{wc} \langle E_i | \underline{A}.\underline{P}H_0 - H_0\underline{A}.\underline{P} | E_i + w \rangle$$

$$= \frac{-i}{wc} \langle E_i | \underline{A}.\underline{\nabla V} | E_i + w \rangle \qquad (P31)$$

and because the gradient of the potential is involved in this new expression it is known as the acceleration form. It is also possible by considering other commutators with the Hamiltonian to derive yet another form of the matrix element

$$M = w \langle E_i | \frac{\underline{A}.\underline{r}}{c} | E_i + w \rangle \qquad (P32)$$

and this version is known as the length form of the matrix element. All three expressions are equal if the states between which the matrix element is taken are exact eigenstates of the zeroth order Hamiltonian, but they are not equal if exact eigenstates are not used. Each stresses the accuracy of the eigenstates in different regions of space. In fact the length form actually diverges in extended systems unless exact eigenstates are used.

We must now discuss the elements which go into our model Hamiltonian. The first and most important ingredient is the potential which dictates the behaviour of both the valence electrons and of the escaping high energy electron. It is usually calculated using local density techniques to approximate exchange and correlation effects and it is desirable to have a self-consistent calculation of the charge density for the system. The potentials are usually regarded as the input to a photo-emission calculation.

Another important ingredient is the lifetime of the escaping electron. In fact the decay of the escaping electron can be expressed either as a lifetime or as a path length. It is usual to define the decay in terms of an inverse lifetime

$$T_2^{-1} = -2V_{20i} \tag{P33}$$

and in metals the quantity V_{20i} takes a value between -1 and -4 eV. In insulators V_{20i} can be very small especially when the photon frequency is less than twice the band gap. Under these circumstances, the direct transition model gives nonsense for the results of a photoemission experiment. The key effect on spectra of lifetimes of the escaping electron is a smearing of the momentum normal to the surface of these states and thus a blurring of the z-momentum conservation rule.

Not only does the escaping electron have a finite lifetime in a solid but the hole it leaves behind also decays after a finite length of time. The lifetimes of holes in solids vary enormously and have not been much studied in the past. Essentially they must be determined from photoemission experiments. If we define the inverse lifetime by

$$T_1^{-1} = -2V_{10i}$$

then in metals

$$V_{10i} \ \alpha \ (E - E_F)^2 \tag{P34}$$

and typically in metals V_{10i} is less than about 0.5 eV but in insulators it can be very small indeed, sometimes essentially zero. Important points to note here are that the energy of the hole is almost always well defined in solids despite its finite lifetime which has the effect of smearing of the resolution of the energy bands.

Further complications which occur are those processes in which the electron escaping from the crystal interacts yet another time with a solid before escaping from the crystal, creating a secondary electron spectrum. These processes are very hard to describe accurately but fortunately the secondary spectrum is usually weak and well separated in energy from the primary spectrum. Another sort of complication is that after the

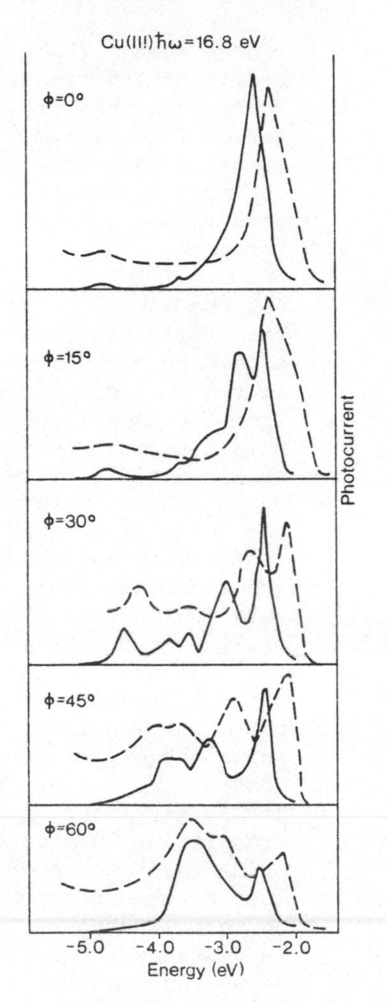

Fig. 1
Photoemission spectra for a copper (111) surface. Unpolarized radiation, $h\omega = 16.8$ eV is incident normally on the surface and the electrons are emitted at $\Theta = 45^{\circ}$ to the normal. The angle Φ is measured relative to the $(\overline{2}11)$ azimuth (——) theory, (----) experiment. The energy is measured relative to E_F and the imaginary part of the hole self energy is given by $\Sigma_1 = 0.054$ eV.

initial excitation the electron and the hole which it leaves behind in the crystal can subsequently interact - an extreme case being the creation of an exciton in a semiconductor. Again these processes are difficult to describe but they are usually not of much importance in a photoemission experiment because the energy range of the electron is not suitable for creating excitons in most materials.

We now pass to the calculations. In figure 1 we see some calculations of the photoemission spectrum for a copper 111 surface and compared with experiments taken by Nilsson (12) the first point to note is that the calculation shows all the correct features with more or less the correct relative intensities but the energies at which these features occur (representing the energies of bands in the solid) are a little in error. Therefore the potential we have used is not as accurate as we would wish. On the other hand the more serious difficulties which we feared, that is to say a breakdown of the model, do not seem to be evident in the spectra. For example the secondary electron spectrum does not seem to be creating any problems. No new peaks appear in the experiment that are not predicted by the theory. Although the experiments do not measure absolute intensities nevertheless separate experiments confirm that the absolute scale in the calculations seems to be about right. Most of the peaks can be ascribed to interband transitions but in some cases, though not in this particular example, surface states

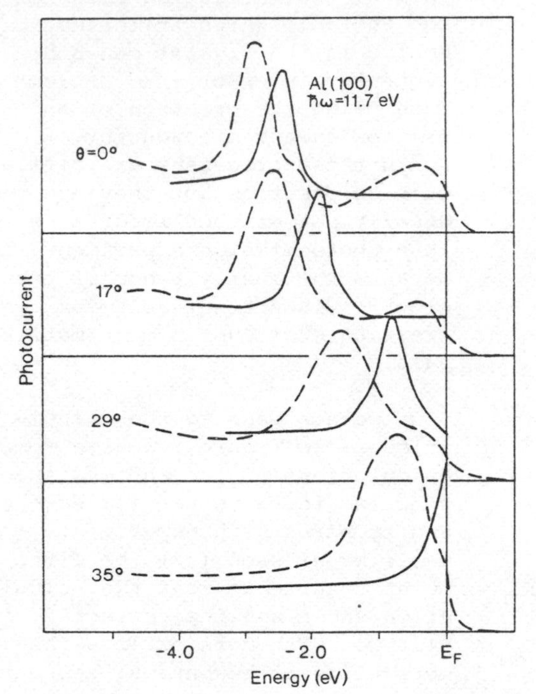

Fig. 2
Photoemission spectra for an aluminium (001) surface. Unpolarized radiation, hω = 11.7 eV, is incident at 45° to the normal in the ($\overline{1}\overline{1}0$) azimuth and the electrons are emitted at various polar angles, Θ, in the (110) azimuth. (——) theory, (----) experiment. $\Sigma_1 = -0.27$ eV.

can be seen. The main origin of the transitions comes from the core region in materials like copper. That is where the matrix element gathers most of its strength.

By way of contrast the matrix elements in aluminium are not dominated by contributions from the atom cores. In this material the strongest potential is the surface barrier retaining the conduction and valence electrons inside the solid and is is from this region that the matrix elements receive their strongest contribution. For this reason photoemission from aluminium has extreme surface sensitivity. There are hardly any interband transitions in strong evidence in the spectrum, most of the spectrum being dominated by surface states. Figure 2 shows the spectrum for aluminium - both theory and experiment (13) - and it is evident that the correct disperson of the surface state is being obtained by the theory. Figure 3 shows the dispersion plotted as a function of the momentum parallel to the surface as determined from the angle of emission. The plot shows a quadratic dispersion as a function of $\underline{k}_{//}$, with an effective mass of almost exactly unity. Despite the electron being bound normal to the surface it is free to move parallel to the surface and aluminium has a free-electron-like surface just as for the bulk.

The most common sort of photoemission experiment keeps the photon energy constant whilst the energy of the emitted electrons is analysed. Yet another experiment is one in which the photon energy and the energy detected by the analyser are moved in phase so that electrons are detected always from the same initial state

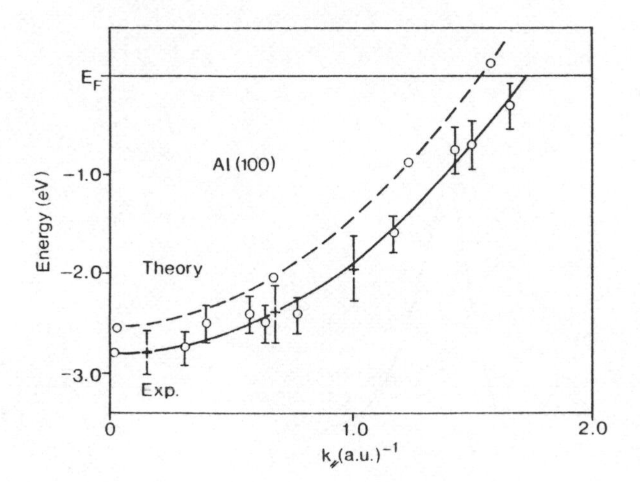

Fig. 3
Experimental and theoretical dispersion of the surface state peak in an aluminium (001) surface.

but at different final state energies. Changes in the spectra are due to band structure in the final state. Figure 4 shows band structure relevant to such a series of transitions on the copper 001 surface. The vertical line marks the k-value of the initial state selected by the constant initial state spectroscopy. Since the light is incident normally on the (001) surface peaks in the normal emission spectrum are expected only in the bands of Δ_1 symmetry. The initial state below E_F has Δ_5 symmetry. The dashed lines show other final states that do not couple. The initial d band is drawn in and as the photon and electron analyser energies are changed in step the direct transitions sweep through various final state bands. Figure 5 shows a calculation of the photoemission spectrum seen under these constant initial state conditions. The arrows show where we expect interband transitions to occur and are seen to predict the peak positions. Note that because of the mode in which the spectra are taken the peaks are much broader than those taken in the conventional mode. For constant initial state spectroscopy widths are essentially determined by the electron lifetime in the final

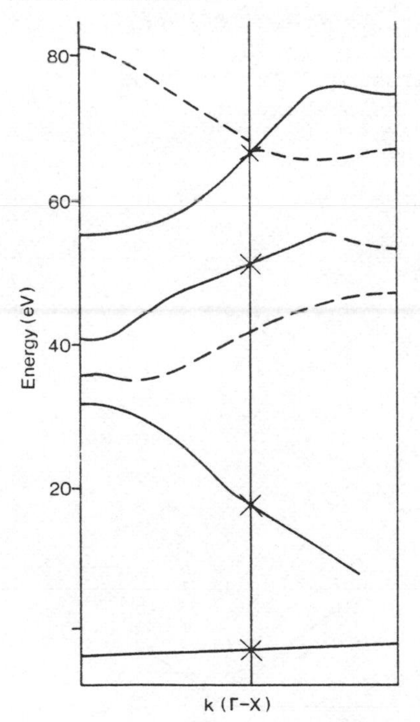

Fig. 4
Band structure along the (001) (Γ - X) direction for copper

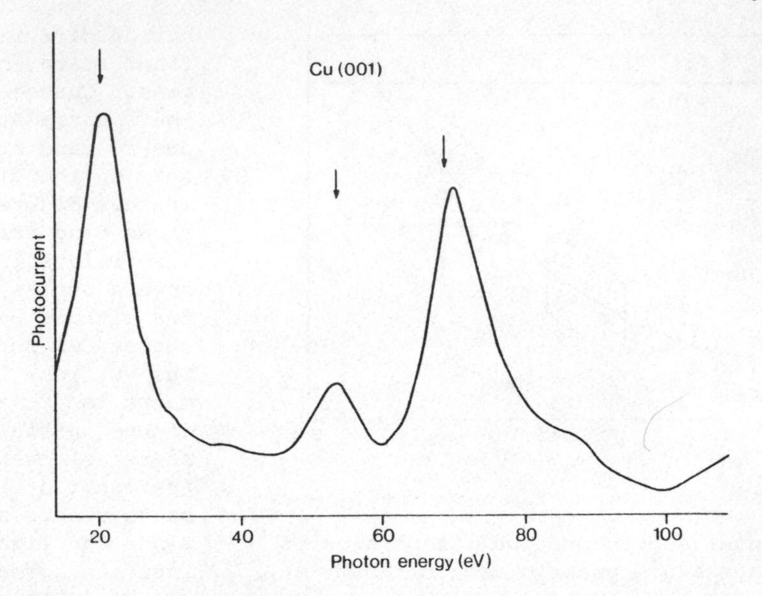

Fig. 5
Photoemission spectrum calculated for a copper (001) surface. Unpolarised radiation is incident normally on the sample and electrons are collected in normal emission. Arrows mark the energies at which interband transitions are expected.

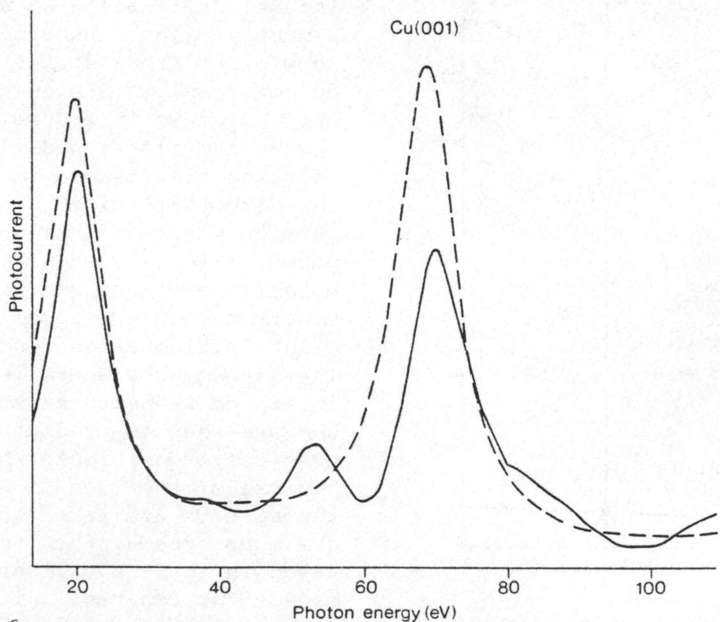

Fig. 6
The same spectrum shown in figure 5 (————) compared with a calculation in which final state multiple scattering is omitted (----------).

state which is much shorter than that in the initial state, hence the broader peaks!

This calculation can also be used to illustrate the effect of scattering in the final state on the photoemission spectrum. In figure 6 two calculations are shown: one in which the final state has had all the scattering correctly included and another calculation in which a free electron wave is chosen as the final state. It is clear that some of the structure is incorrectly described when the final state scattering is omitted and therefore it is important to include this scattering for an accurate interpretation. This is especially true when photoemission intensities are plotted as a function of angle of emission because the final state scattering often acts so as to redistribute the flux from one angle to another.

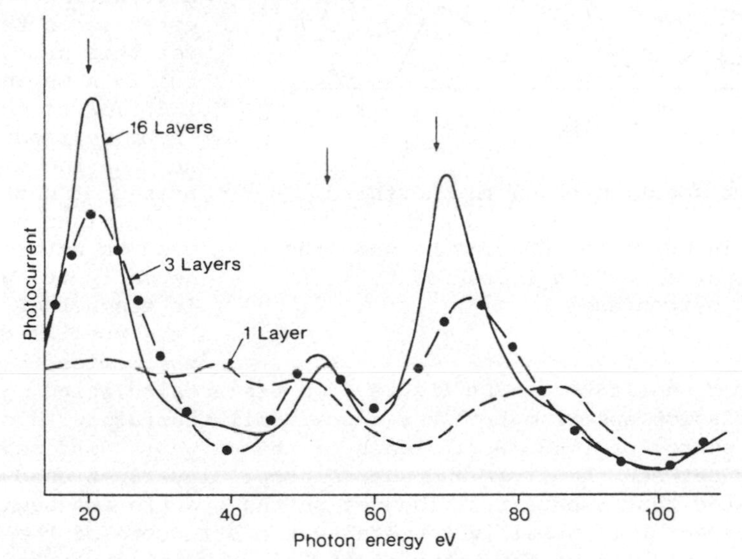

Fig. 7
The same spectrum shown in figure 5 (full line) compared with calculations in which the photon field is restricted to the first layer (----) and to the first three layers (-·-)

Yet another demonstration that can be made from this series of calculations is the surface sensitivity of photoemission. In figure 7 we show the same calculation together with several other calculations in which the photon field has artificially been confined to the first few layers of the crystal. When 16 layers have received the full photon flux essentially the correct photon spectrum is reproduced because the electron cannot escape from much deeper than about 8 layers in the solid. One layer is not enough to describe the calculation but by 3 layers we are already

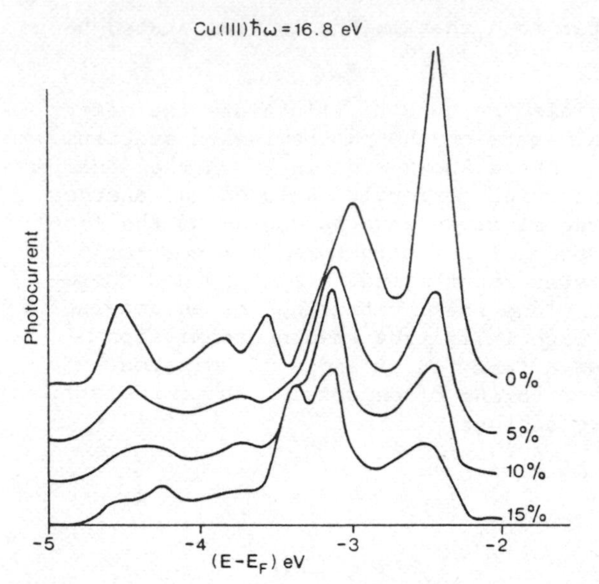

Fig. 8
The same spectrum shown in the third panel of figure 1 compared with calculations in which the spacing between the two copper layers is increased by various percentages

seeing the photoemission spectrum well on the way towards convergence. The photoemission spectrum is dominated by the first 3 layers of the crystal.

One of the quantities to be determined for the surface is the geometry. Low energy electron diffraction is the established technique for this procedure but in some instances LEED leaves something to be desired in accuracy and the question arises of whether photoemission can be used to determine geometry at surfaces. It seems that in certain special cases photoemission may give good sensitivity. In figure 8 we see a calculation for the photoemission spectrum from the copper (111) surface with constant photon energy of 16.8 eV. In each of the curves a different spacing between the first and second copper layers is used and it is noticed that substantial changes in the spectra are seen. We might expect this sensitivity because the structure of the d bands is very sensitive to the inter atomic spacing in copper and by altering the spacing of the last layer we alter the coupling of the d bands in the last layer to the bulk. We expect that on pulling away the top layer the coupling will decrease and therefore the d band width in the last layer will tend to decrease. The position of the bands will not alter because they are determined by the interaction of several layers but we expect to see in the spectrum a heaping of the weight towards the centre of the d bands and this is what happens to the curves. In figure 9 two of the calculated curves are compared with experiment and it is evident from this comparison that the dilation of the top layer of the copper (111) surface is certainly less than 5%.

These calculations show some of the potential for photoemission experiments which is beginning to be exploited. The field is characterised by strong interplay between theory and experiment and holds out the possibilities of much mutual benefit in these areas.

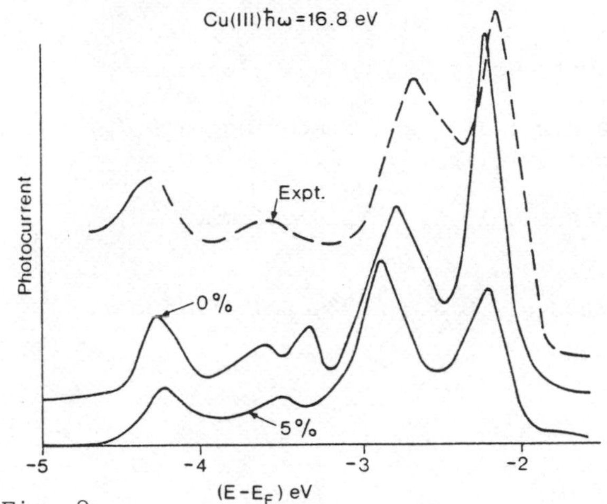

Fig. 9
Spectra from figure 8 for 0 and 5% increased spacing between the last two copper layers compared with experiment (---)

REFERENCES

(1) Pendry J.B. and Gurman S.J., 1975, Surf. Sci., 49, pp. 87-105.

(2) Appelbaum J.A. and Blount E.I., 1973, Phys. Rev., B8, pp. 483-491.

(3) Shockley W., 1939, Phys. Rev., 56, pp. 317-

(4) Phillips J.C., 1973, "Bonds and Bands in Semiconductors" (New York: Academic).

(5) McRae E.G. and Caldwell C.W., 1976, Surf. Sci., 57, pp. 63-92.

(6) Echenique P.M. and Pendry, J.B., 1978, J. Phys. C., 11, pp. 2065-2075.

(7) Appelbaum J.A., Baraff G.A. and Hamann D.R., 1975, Phys. Rev., B11, pp. 3822-3831.

(8) Pendry J.B., 1977, J. Phys. C., 10, pp. 809-824.

(9) Feuerbacher B., Fitton B. and Willis R.F., 1978, "Photoemission and the Electronic Properties of Surfaces" (Chichester: Wiley).

(10) Pendry J.B., 1976, Surf. Sci., 57, pp. 679-705.

(11) Hopkinson, J.F.L., Pendry, J.B. and Titterington D.J., 1980, Comp. Phys. Comm., 18, pp. 69-92.

(12) Ilver L. and Nilsson P.O., 1976, Sol. St. Comm., 18, pp. 677-680.

(13) Gartland P.O. and Slagsvold B.J., 1978, Sol. St. Comm., 25, pp. 489-492.

VALENCE BAND PHOTOELECTRON SPECTROSCOPY OF SOLIDS

P A Cox

University of Oxford, Inorganic Chemistry Laboratory,
South Parks Road, Oxford, UK

Photoelectron spectroscopy may be used to provide information about the valence electronic levels in solids. Applications to a range of simple solids, as well as to more complex examples of transition metal compounds, are decribed here.

INTRODUCTION

In recent years, photoelectron spectroscopy has proved to be one of the most powerful techniques for investigating the filled valence electronic levels in solids. The principal advantage over more traditional forms of absorption spectroscopy is that in PES the electron is removed altogether from the solid, and so to a certain approximation the features of the spectrum are determined solely by the filled levels, and unaffected by details of empty states. In the next section we shall examine this assumption and show under what conditions it can be fulfilled. After a brief discussion of the problem of calibration of spectra and reference levels - which is essential if measured spectra are to be related to theoretical predictions of binding energies - we shall use selected PE spectra to illustrate the nature of valence levels in solids.

Some of the most interesting recent developments have been concerned with aspects such as the angular dispersion of photoelectron spectra and studies of surface electronic levels, rather than the bulk levels considered in this paper. Such developments are discussed in detail in later contributions to this volume. The present account, by contrast, is concerned with rather more "traditional" aspects of PES. As such, it is intended as an introduction to these later contributions, both as an illustration of the uses of PES and as an account of some basic notions of the electronic structure of solids.

P. Day (ed.), Emission and Scattering Techniques, 21–60.

THE THREE-STEP MODEL

A fully satisfactory theory of PES should describe the ionisation process in a solid in terms of a unified many-electron framework. Such theories have been developed (1), and in some circumstances their use may be necessary. For most practical purposes, however, and especially as a conceptual framework to help us think about the the PES experiment, "one-step" theories of photoionisation are too complicated, and it is convenient to divide the photoionisation event into three distinct steps (2):

(i) An electronic transition in the solid from a filled (bulk) level into an empty level.

(ii) Transport of the electron to the surface of the solid.

(iii) Escape from the solid surface into the vacuum.

Making many approximations, it is possible to use the three-step concept to derive an expression for the photoelectron current $j(E,h\nu)$, at electron energy E and photon energy $h\nu$, in the following form:

$$j(E,h\nu) \propto \int d^3\underline{k} \; |M(E,E_i,\underline{k})|^2 \; n(E_i,\underline{k}) \; n(E,\underline{k}) \; T(E,\underline{k}) \qquad (1)$$

In this expression:

$E_i = E - h\nu$ is the initial energy of the electron (= - binding energy, I);

$\underline{k}$ is the wave-vector of the initial and final electronic levels in the solid;

$M(E,E_i,\underline{k})$ is a matrix element for the radiative transitions between the two levels;

$n(E_i,\underline{k})$ is a function derived from the valence band structure, giving the number of electrons in filled states at energy E_i and wavenumber $\underline{k}$;

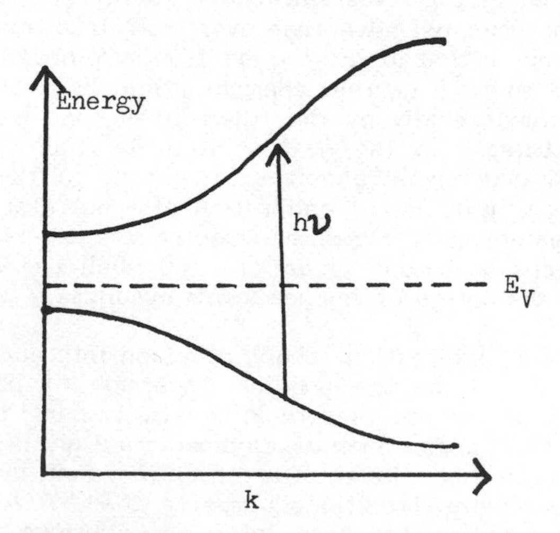

Figure 1. A direct (k conserving) transition.

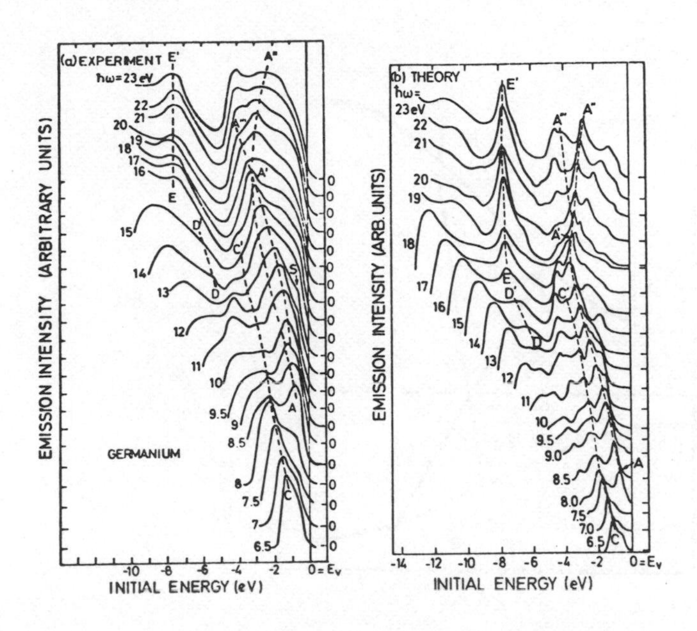

Figure 2. PE spectra of germanium at different photon energies.(a) experimental spectra; (b) spectra calculated with direct transition effects.

n(E,k) is a similar function for empty levels at energy E;

T(E,k) is a transmission function which incorporates steps (ii) and (iii), and represents the probability that an electron in its final bulk level (E,k) can escape into the vacuum and be detected in the photoelectron spectrometer.

Equation (1) incorporates the normal selection rule for electronic transitions in a solid, that is that k should not change (3). In effect the photon carries negligible momentum, so that the momentum of the electron must be conserved. This so-called "direct transition" effect can have an important influence on the PE spectrum of a solid. Figure 1 illustrates k conservation in a solid with a single filled valence band and a single empty band above the vacuum level E_V. Although the photon has enough energy to ionise electrons from anywhere in the valence band, the combined effect of energy and k conservation in this case is to allow only the single sharp transition shown. Thus the PE spectrum would not show the complete valence band, but only this sharp signal. Some such direct transition effects are evident in the series of PE spectra of germanium taken at different photon energies (4) and shown in Figure 2. If all the structure in these spectra represented peaks in the filled valence band density of states, it should appear at the same initial energy in all spectra. In fact many of the peaks move, particularly with lower photon

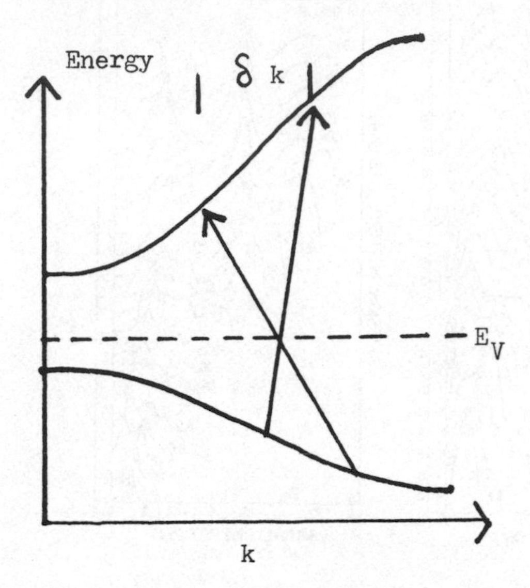

Figure 3. The breakdown of k conservation.

energies. The theoretical spectra on the right incorporate final state densities and k conservation. Studies such as these, with variable photon energies usually provided from a synchrotron source, have been exploited recently for the purposes of "mapping" initial and final state E against k curves in solids (5).

In practice there are many factors which tend to reduce the importance of the k selection rule in PES. One of these is already present in equation (1) as the integral over all wave-vectors k. Once the electron emerges from the solid, k can be measured as the photoelectron momentum, obtainable for example from its kinetic energy and direction of travel. Detailed information about k, and hence about direct transitions, can be obtained by measuring the angular dependence of the PES signals from single crystal surfaces. It will be seen on the other hand that equation (1) implies an integration of the PES over all angles; since the spectrometer may have a limited angular acceptance of photoelectrons, this is best achieved by measuring non-oriented polycrystalline samples. Information about k is then lost and direct transition effects become less apparent.

Another important limitation on direct transitions is provided by the finite path length of the electron in its final state, as we shall discuss below (6). A mean free path of λ effectively locates the electron in space, and hence through the uncertainty principle, gives an uncertainty to k of the order:

$$\delta k \simeq 1/\lambda \qquad (2)$$

If $\underline{k}$ is not well defined, "non-vertical" transitions become possible, as illustrated in figure 3. Another factor to be considered is that of lattice vibrations (phonons) which themselves carry a wave-vector $\underline{k}$. If phonons are created or destroyed during an electronic transition, only the total $\underline{k}$ (electronic plus vibrational) needs to be conserved (3). The proportion of purely electronic, vibrationless, transitions is given by a Debye-Waller factor like that in X-ray diffraction or Mössbauer spectroscopy; this factor decreases with increasing temperature and with increasing photon energy (7).

At low photon energies, in the ultra-violet (UPS) range the factors mentioned above lead to some smearing of the structure which results from k conservation (8). At higher photon energies, the final states form a close mesh in E - $\underline{k}$ space, and within the limits of definition of $\underline{k}$, the final state function n(E,k) appearing in equation (1) is practically constant. Under these conditions k conservation can by ignored (9). If we further assume that the transition function T(E,$\underline{k}$) is also constant over the energy range of interest, then the equation for the photoelectron current can be greatly simplified. In place of equation (1), we have:

$$j(E,h\nu) \propto |M(E,E_i)|^2 N(E_i) \qquad (3)$$

where, as before:

$E_i = E - h\nu = -I$;

$M(E,E_i)$ is the matrix element averaged over the k vector;

$N(E_i)$ is the total density of states for electrons in the valence band at E_i.

Under these conditions then, the PE spectrum shows the valence band density of states (DOS) weighted by the squared matrix element which gives the ionisation cross-section of the orbitals concerned. Since orbitals with different atomic composition may have very different cross-sections which vary in a known way with photon energy (10), both the factors in equation (3) contain useful information.

The PE spectra of copper metal taken at different photon energies (11) are shown in Figure 4. The spectum changes with photon energy, showing the importance of final state and k conservation effects. However, at the highest energies shown in Figure 4, the PE spectrum agrees quite well with a smoothed version of the calculated density of states. This illustrates the way in which final state and k conservation effects become less important at higher photon energy.

So far, we have only discussed step (i) in the three-step model. Much less is known in detail about steps (ii) and (iii). It is essential to recognise, however, that electrons are scattered quite strongly by matter and so travel only a short distance before losing energy in inelastic collisions (6). The mean free path for electrons in a variety of solids has

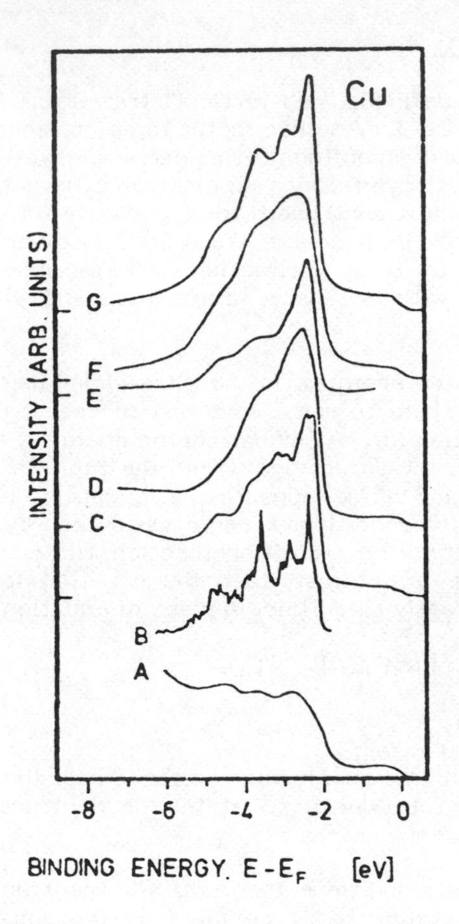

Figure 4. PE spectra of copper at several photon energies. Curves A,C,D and E: 11.4, 21.2, 40.8 and 48.4 eV; F: 1487 eV; G: deconvolution of F; B: theoretical density of states (11).

been found to vary in a fairly systematic manner with energy (12), as shown in Figure 5. Although such a plot has become known as the "universal curve", it needs to be treated with some degree of reservation; it is becoming recognised that, especially at lower energies, electron path lengths may vary quite widely between different types of solid (13). Nevertheless, Figure 5 gives a useful guide to the degree of surface sensitivity of PES, as the mean free path obviously determines the escape depth for photoelectrons from the solid.

Since it is quite clear that PES probes only a thin surface region of a solid, it is necessary to ask whether the density of states observed is really characteristic of the bulk of the solid. The answer seems to be that

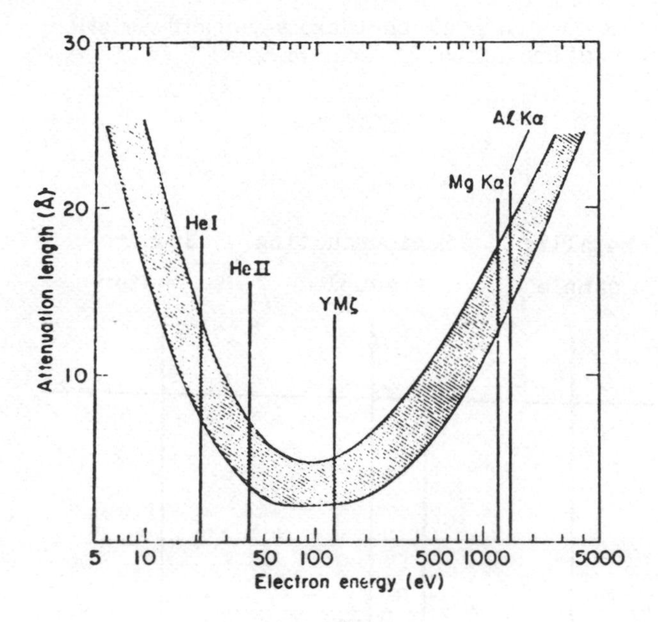

Figure 5. A version of the "universal curve" (12).

provided the surface is clean and representative of the bulk in composition and structure, surface effects are largely confined to the top atomic layer, and do not contribute strongly to angular integrated spectra, especially from polycrystalline materials. The requirement of surface purity is essential, and necessitates careful cleaning and preparation techniques and the best possible UHV conditions (14). The use of PES to study surface electronic structure is described in later contributions to this volume.

Apart from the surface sensitivity, one important consequence of the inelastic scattering of electrons is that it contributes a low-kinetic-energy background to the PE spectrum. It may be necessary to subtract this background before a satisfactory comparison with a theoretical DOS is possible.

It may be useful to end this section with a summary of the conditions under which PES can give direct information about the bulk DOS of a solid.

(i) Clean surface, representative of the bulk in composition and structure;

(ii) Angular integration of the spectrum, preferably from a polycrystalline sample;

(iii) Sufficiently high photon energy to avoid final state and/or direct transition effects;

(iv) Allowance for matrix element (cross-section) variation;

(v) Subraction of background if necessary.

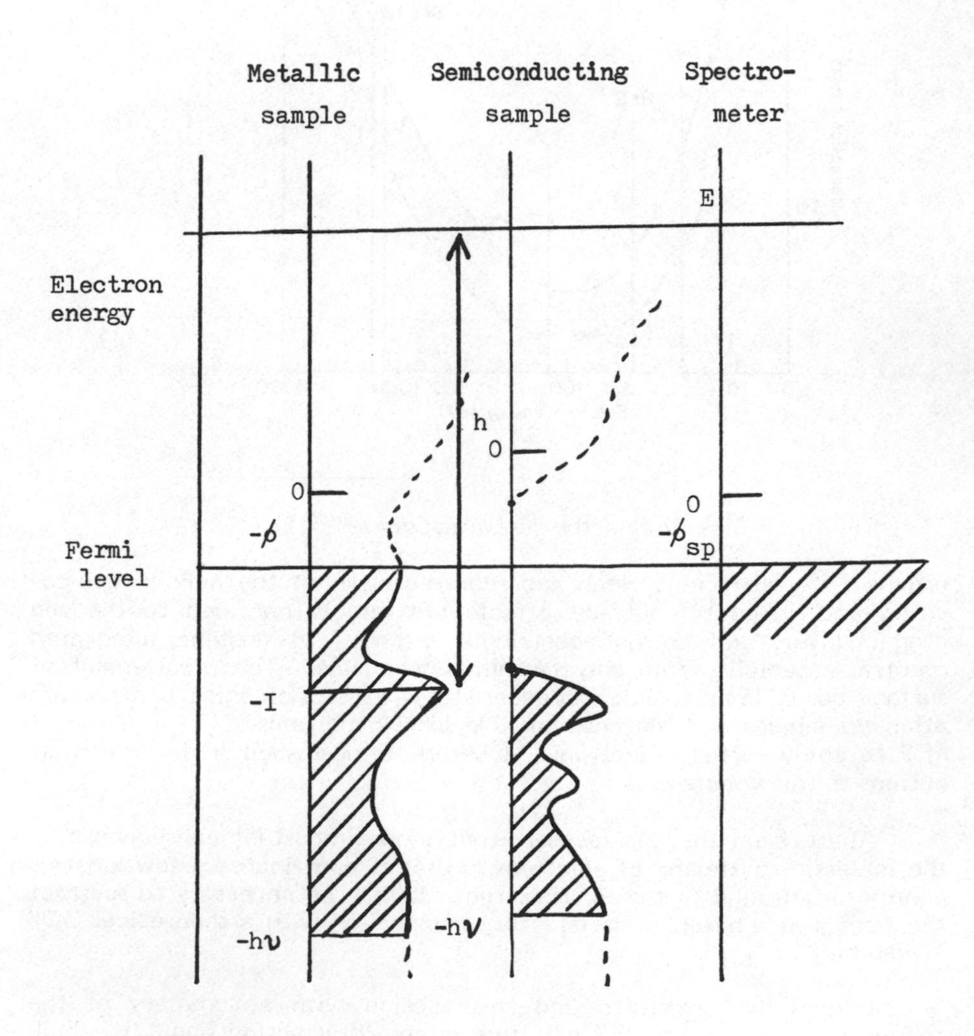

Figure 6. Reference levels in photoelectron spectroscopy. Note the different zero (vacuum level) in each case. E: measured vacuum level; ϕ: sample work function; ϕ_{sp}: spectrometer work function.

CALIBRATION AND REFERENCE LEVELS

In gas phase PES, the ionisation energy of a molecule defined by

$$I = h\nu - E \qquad (4)$$

(where E is the measured electron kinetic energy) refers directly to the vacuum level, which represents an electron at rest far from any electric charge. Calculations of energy levels of electrons in solids also refer generally to the vacuum level, but unfortunately measured IEs do not, and therefore need to be corrected before comparison with theory (15). The ideal situation is illustrated in Figure 6.

A solid is characterised by a Fermi level, which in a metal at absolute zero is simply the boundary between the top filled level and the lowest empty level (16). When two metallic specimens are placed in contact, the Fermi levels must coincide at equilibrium, as otherwise one would have empty energy levels on one side of the junction below filled ones on the other side. The difference between the Fermi level and the vacuum level is the work function of the metal, and if the work functions for the two metals in contact are not equal, then their vacuum levels will be different. In a photoelectron spectrometer, the electron kinetic energy is measured with respect to the vacuum level of the spectrometer. If the work functions ϕ for a metal sample, and ϕ_{sp} for the spectrometer, are known,the IE, corrected to the vacuum level of the sample, can be found (17) from the equation

$$I = h\nu - E + \phi - \phi_{sp} \qquad (5)$$

With an insulating sample, the situation is more complicated. One still defines the Fermi level as a thermodynamic level for electrons in a solid, somewhere between the highest filled and lowest empty energy levels, and for samples in contact at equilibrium, the Fermi levels coincide. In a pure solid, the Fermi level is very nearly midway in the band gap, but in real solids it is likely to be fixed by impurities, defects, or surface states, and is therefore difficult to locate a priori. More serious, however, is the fact that under the conditions of measurement of a PE spectrum, an insulator may be subject to serious charging effects which make equilibrium criteria quite irrelevant (18).

A rather direct way of referring measured spectra to the vacuum level of the sample has been recommended (19). The maximum IE which it is possible to measure from a sample must correspond to electrons which can just escape from the solid, and thus with a final energy of zero with respect to the vacuum level. Thus the equation:

$$I_{max} = h\nu \qquad (6)$$

allows an absolute calibration of the spectrum. There may not be any filled levels at $-I_{max}$ in the solid, but signals equivalent to these come from inelastically scattered electrons ionised from other levels. Thus in practice I_{max} often represents the cut-off in the inelastic background which we have seen is always present in a PE spectrum.

Faced with the difficulties involved in absolute calibrations of PE spectra, many spectroscopists adopt the simple procedure of referring to the ionisation threshold as the energy zero. In a metal, such a zero represents the Fermi level of the sample; in an insulator, it refers to the top of the valence band. Figures 2 and 4 show examples where this form of reference is adopted. It is worth mentioning here that alternative conventions also exist in another important aspect of the presentation of PE spectra: that is the direction in which the energy scale runs. The more usual convention is that measured electron kinetic energy increases from left to right, so that IE increases from right to left. However, some PE spectrometers generate plots running in the other direction. Figures 12 and 14 in the next section show examples where IE increases from left to right.

VALENCE LEVELS IN SIMPLE SOLIDS

From an elementary point of view solids can be divided into four main classes:

(i) molecular
(ii) ionic
(iii) covalant
(iv) metallic

In this section we shall examine the PES of some examples taken from each of these classes.

(i) Molecular solids.

In a molecular solid such as benzene, the molecules retain their identity, and are held together by relatively weak van der Waals forces. We would expect the valence levels to resemble closely those observed in isolated molecules in the gas phase, and this prediction is borne out experimentally (20) in spectra such as those in Figure 7. The gas phase PES show ionisation from the filled molecular orbitals, in many cases with superimposed vibrational excitations which may broaden the lines (21). Spectra from solid samples show some additional broadening, and also a shift to lower binding energy. This shift represents a very important aspect of the PES of solids (22). Even in atoms and molecules, Koopmans' theorem -which predicts that IEs should equal the negative of orbital energies - breaks down because of relaxation and correlation effects (23). In solids additional relaxation effects are contributed by the surrounding molecules. A simple model of this is provided by considering what happens when a sphere of radius R carrying an electric charge Q is moved from empty space into a medium of relative dielectric constant D (24) : because of the dielectric polarisation of the medium, there is an energy change of:

$$\Delta E = - Q^2 (1 - 1/D)/(8\pi \varepsilon_0 R) \qquad (7)$$

Such polarisation always lowers the energy of a charged object in the solid relative to free space, and thus acts to lower the IE in a solid from the gas phase value.

A molecular solid with considerably stronger interaction than benzene is TTF-TCNQ (25), the 1:1 compound formed between tetrathiafulvenene (TTF) and tetracyanoquinodimethane (TCNQ), the molecular

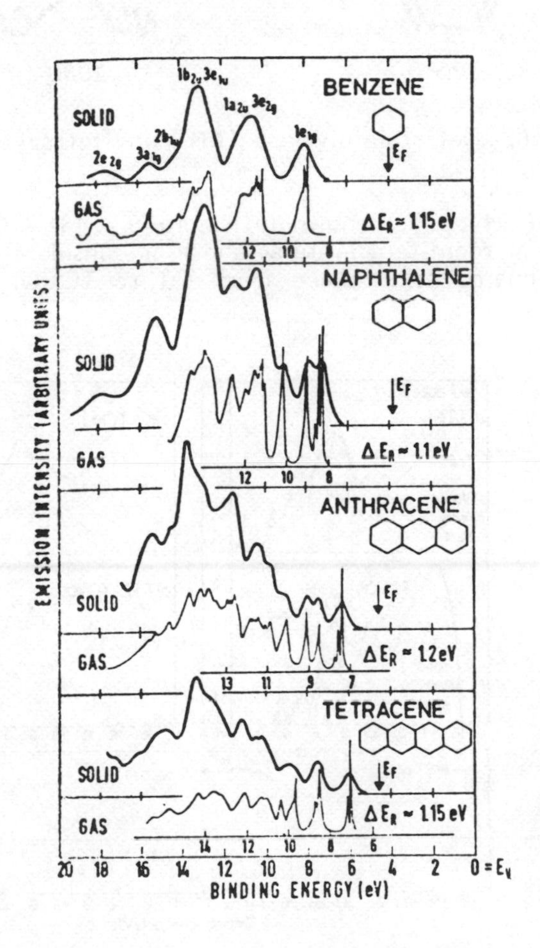

Figure 7. Solid state and gas phase UPS of benzene and other acenes. Note the shift of the energy scales of the gas phase relative to the solid state spectra (20).

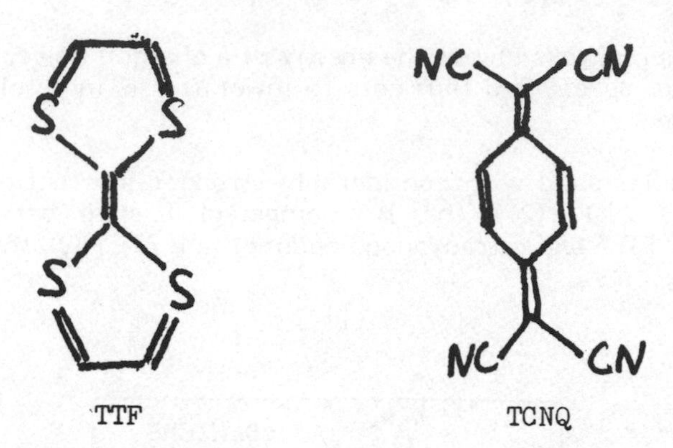

Figure 8. Tetrathiafulvalene (TTF) and tetracyanoquinodimethane (TCNQ).

structures of which are shown in Figure 8. TTF-TCNQ has metallic conductivity at room temperature, and it is considered likely that some transfer of charge takes place from TTF to TCNQ, giving a partially

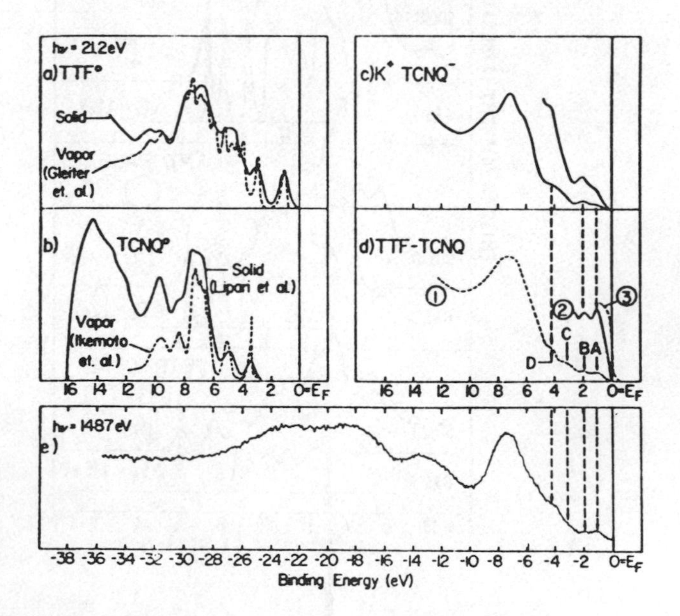

Figure 9. Spectra of TTF-TCNQ (23). (a) and (b) He-(I) UPS of TTF and TCNQ; (c) and (d) UPS of K^+TCNQ^- and TTF-TCNQ; (e) XPS of TTF-TCNQ.

empty band of TTF levels, and a partially filled band of TCNQ orbitals. Some evidence for charge transfer is indeed provided by the PES (26) shown in figure 9. Spectra of pure TTF and TCNQ solids show a resemblance to gas phase spectra similar to that found in the previous example. The potassium salt of TCNQ, which is known to contain the $TCNQ^-$ anion, shows different structure in the low IE region of its PES. Such structure is also observed in the UPS of TTF-TCNQ (spectrum d in the Figure), and apparently also in the XPS (spectrum e).

(ii) Ionic solids

The previous example might better be considered as an ionic solid. In simpler examples of this class, such as sodium chloride, NaCl, an electron is transferred from one atom, which becomes the cation (Na^+) to another which becomes an anion (Cl^-). Bonding is provided by the electrostatic interaction between charged ions, but otherwise the interaction between valence levels in the different ions is not considered to be important (27). We can use such a simple chemical picture to derive an electronic energy level diagram for NaCl without any elaborate calculation. The highest filled level, from which electrons are most easily removed, is the top filled level of the anions, chlorine 3p in this case. At rather higher binding energy is the chlorine 3s orbital. Similarly, the lowest empty band is the lowest empty cation level, in this case the sodium 3s orbital from which an electron was removed to form the ionic compound. A schematic diagram of the density of states is shown in Figure 10.

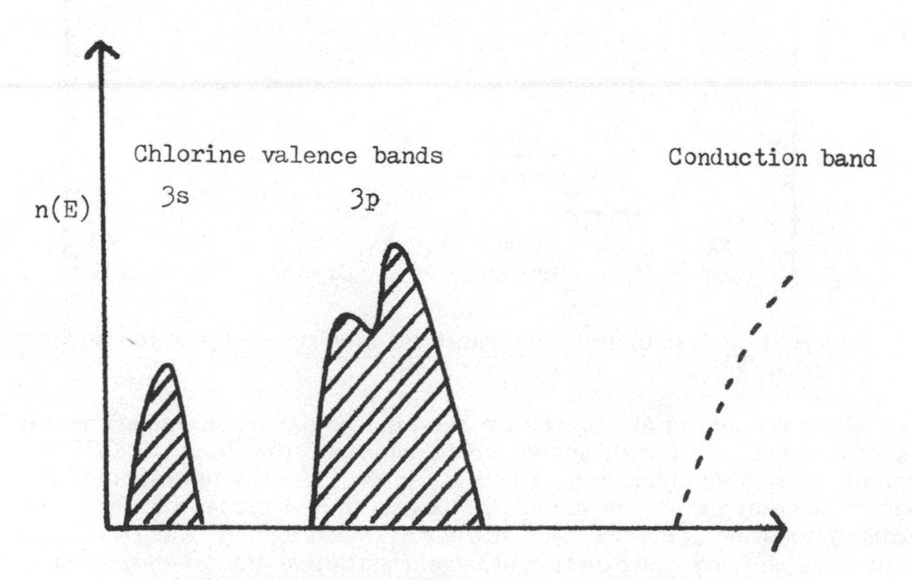

Figure 10. Schematic density of states for sodium chloride.

It is straightforward to use the ionic model quantitatively to calculate IEs (28). For the lowest IE, corresponding to the top filled anion level, we have:

$$I = A + V_m - E_r \quad (8)$$

where

A is the electron affinity of the non-metal atom (or the IE of the anion in free space);

V_m is the electrostatic Madelung potential due to other ions in the crystal; this is a positive term because the nearest ions are cations and have a stabilising influence;

E_r is the relaxation or polarisation energy which we discussed above.

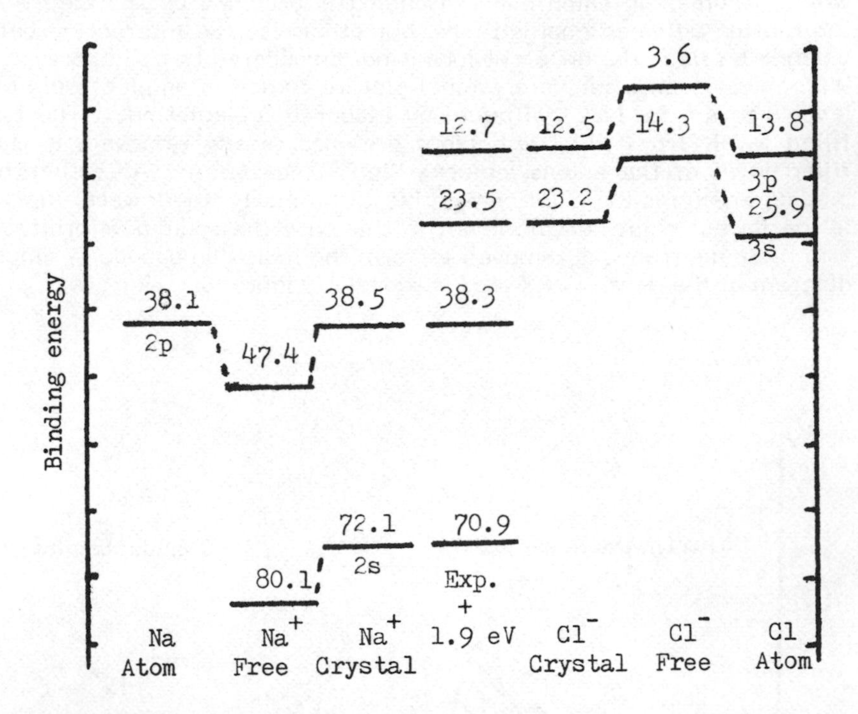

Figure 11. Calculated and measured binding energies for sodium chloride (28).

Quite successful attempts have been made to compare experimental IEs from ionic solids with simple calculations of this kind (28). The example of sodium chloride is shown in Figure 11. The data include the sodium core levels. The calculated levels only include the first two terms of equation (8); the effect of the relaxation term E_r was taken into account by shifting the experimental values to give the best agreement with the theory. It is remarkable that the difference between the 3p energy level in a neutral chlorine atom and in a free chloride ion, as shown in figure 11, is almost cancelled by the difference between the free ion and the ion in the crystal. This must be due to the similarity between the one-centre electronic repulsion in the ion, and the attractive potential of surrounding ions, so that the chloride IE in the compound is much closer to that of a chlorine atom than of a free chloride ion. Thus because of the Madelung potential, the binding energy in a solid is much less sensitive to charge distribution than one would expect at first sight (29).

The model discussed so far is one of point ions which have only an electrostatic interaction with each other. In fact valence levels in ionic solids have an appreciable bandwidth and this may often come from overlap between anions (30). Such overlap gives rise to off-diagonal Hamiltonian matrix elements or transfer integrals (t) between atomic orbitals on neighbouring ions. The simplest tight-binding theory (31) then predicts a band-width:

$$W = 2z|t| \tag{9}$$

where z is the number of nearest neighbours of an ion.

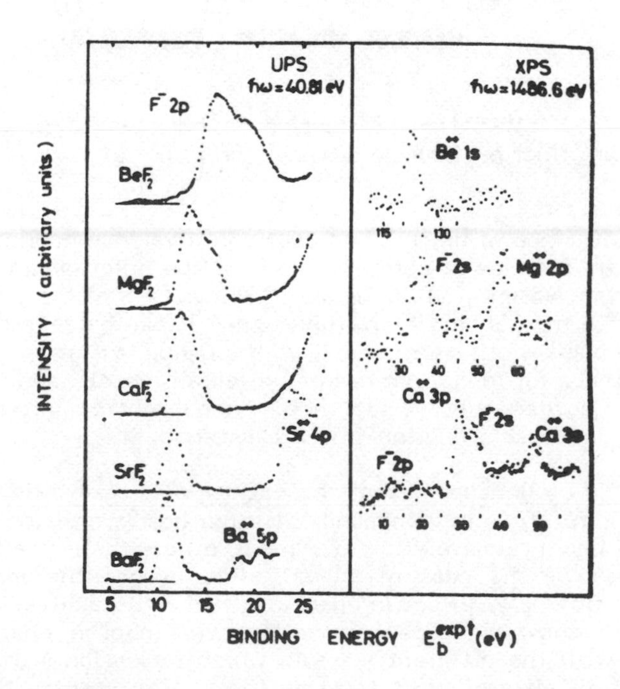

Figure 12. PE spectra of alkaline earth halides, showing the fluorine 2p band (30).

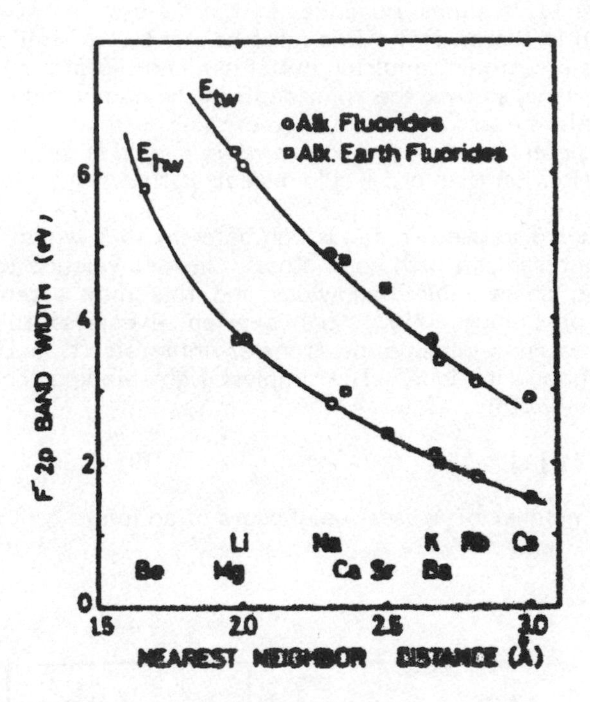

Figure 13. Variation of fluorine 2p band width in ionic fluorides with the fluorine-fluorine distance (30).

The influence of anion-anion overlap on valence band width is nicely illustrated by a series of compounds where the anion-anion distance, and hence overlap, varies (30). Figure 12 shows UPS of the fluorides of the alkaline earth metals beryllium to barium. It can be seen how much the fluorine 2p band-width decreases down the group. Figure 13 shows these data and others for monofluorides of the alkali metals, and illustrates the monotonic decrease of overlap, and hence transfer integral and band width, with increasing fluorine-fluorine distance.

Sometimes it happens that the cations also have filled levels in the valence region. Thus in compounds of some post-transition elements the (n-1)d shell has a relatively low IE, and in others there is a filled ns shell. Figure 14 shows the case of silver iodide, where the iodine 5p level expected at low IE overlaps in energy with the filled silver 4d band (32). The spectra show significant variation with photon energy, which is consistent with the different ways in which ionisation cross-sections for 4d and 5p orbitals are known to vary (33). The spectra thus allow the total density of states to be broken down into silver 4d and iodine 5p components, as illustrated on the right hand side of Figure 14.

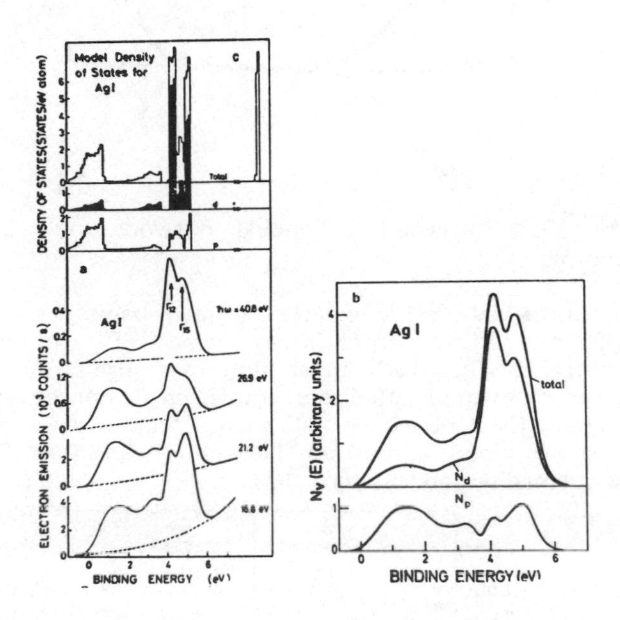

Figure 14. PE spectra of silver iodide at different photon energies (left), with breakdown of density of states into silver 4d and iodine 5p components (33).

(ii) Covalent solids.

In a covalent solid such as diamond, chemical bonding occurs through the sharing of electrons in localised bonds between adjacent atoms. These bonds are essentially similar to carbon-carbon single bonds in a saturated hydrocarbon, and the simplest view of the band structure of a covalent solid is that the highest filled (valence) band is composed of bonding orbitals, while the lowest empty (conduction) band is made up of orbitals which are antibonding between nearest neighbour atoms. The width of the band in this case results from the interaction between neighbouring bonds, around the same or nearby atoms. Indeed it is possible to get quite a good approximation to the valence band structure of a covalent semiconductor from a simple model starting with a network of bonds as shown in figure 15. Figure 16 shows the resulting E-$\underline{k}$ curves, both for the simplest model in which only nearest neighbour bonds interact, and for the next level of approximation where next-nearest-neighbour interactions are also considered (34). This latter calculation gives a reasonable estimate of the valence band DOS, as shown by the comparison in Figure 16 with a much more sophisticated (orthogonalised plane wave) calculation (35).

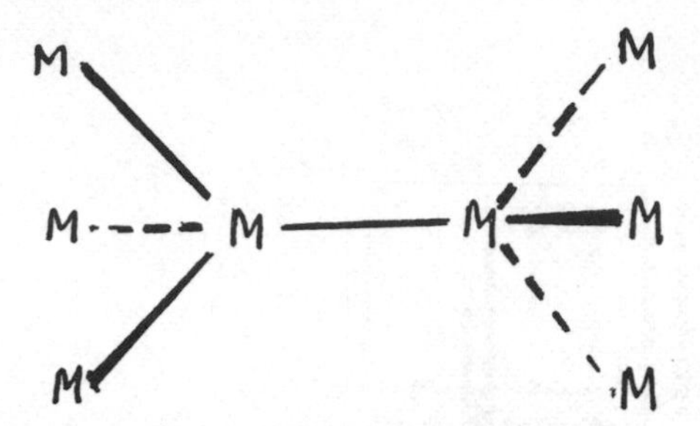

Figure 15. Tetrahedral bonding network in a covalent semiconductor.

Some quite detailed comparisons have been made between calculated DOS and experimental X-ray or UV-PE spectra of covalent semiconductors. The study illustrated in figure 2 has already been referred to as an example of direct transition (k conservation) effect in the UPS range. By contrast, in XPS such effects are not significant, and Figure 17 shows good agreement between calculated DOS and XPE spectra for germanium and silicon (36).

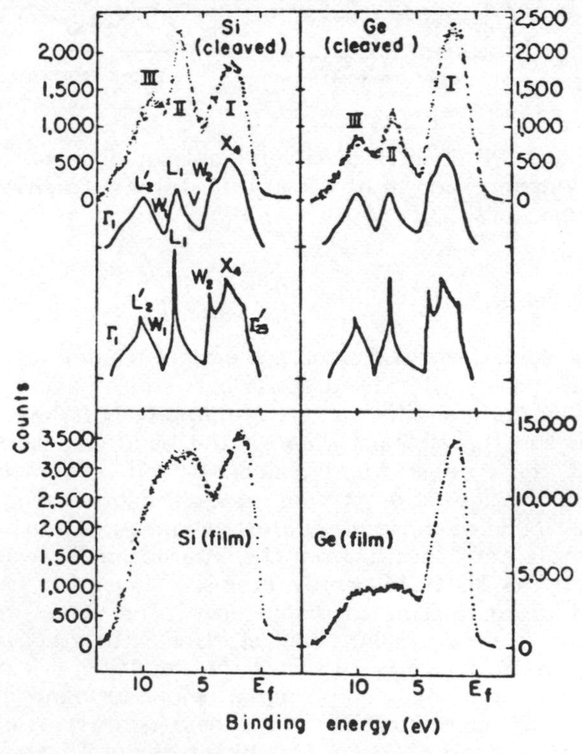

Figure 17. XPS of silicon and germanium compared with theoretical structures of amorphous semiconductors (37). The lower spectra are from amorphous films.

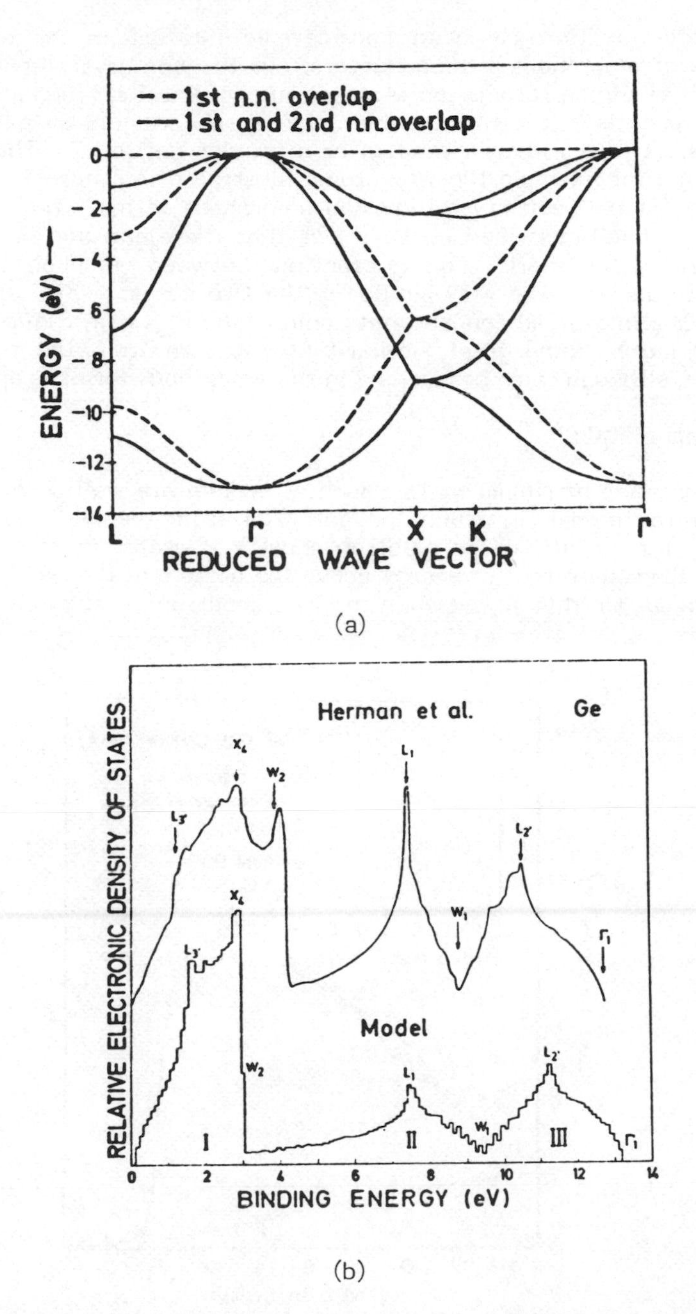

Figure 16. Germanium band structure calculated with bond-orbital model (34): (a) E-k curves; (b) DOS compared with OPW calculation of Herman et al. (35).

There has recently been considerable interest in the electronic structure of amorphous semiconductors. In the non-crystalline state, a solid such as silicon retains the tetrahedral coordination found in crystals; the difference is in the way in which the tetrahedra join up over longer distances, to give either a random or a regular network. Thus over a short range, for example the few atoms illustrated in Figure 15, there is no difference between crystalline and amorphous forms. The model for the valence band described above shows that the electronic structure is dominated by the short range interactions between neighbouring bonds, and should therefore be very similar in the two cases. XPE spectra of amorphous films of silicon and germanium are shown in Figure 17, and these do indeed show great similarity to spectra from the crystalline materials, although some broadening in the amorphous forms is apparent.

(iv) Simple metals.

Many properties of simple metals such as sodium are well described by a free electron model, in which the potential from the Na^+ ion cores is ignored. The resulting free electron density of states is simply proportional to the square root of energy above the bottom of the band (38). A form similar to this is shown in the XPE spectrum of aluminium (39) in

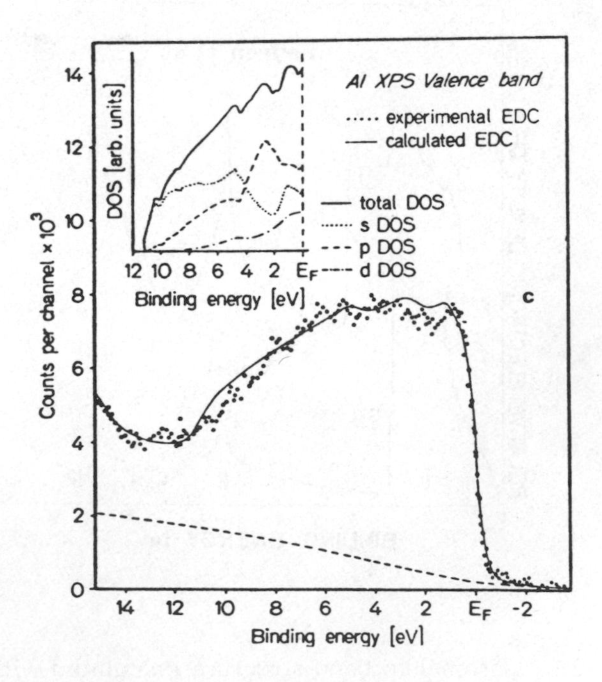

Figure 18. XPS of aluminium, with (inset) breakdown of total DOS into s,p and d partial waves (39).

Figure 18. Also characteristic of a metal is the Fermi edge - the boundary between the top filled and lowest empty levels, where the PE signal falls sharply to zero. (Thermal excitation at the Fermi surface at room temperatures will contribute a broadening of the order of 0.05 eV to this edge; in the spectra of Figure 18 and 19 this is insignificant compared with the experimental broadening provided by the width of the Al K_α exciting radiation - around 1 eV). An accurate fit of the XPS of simple metals has proved quite difficult, however (39). To reproduce correctly the band shape in aluminium, it was necessary to take account of the variation of ionisation cross-section with over the band. This was done by breaking the total DOS down into contributions from s, p and d partial waves, as shown in inset in Figure 18. Each partial wave was then assigned a different XPS cross-section to achieve the fit shown in this Figure.

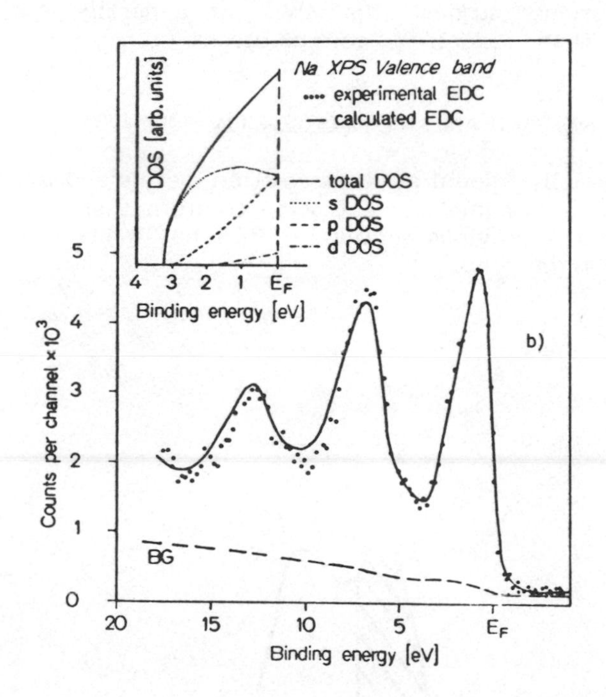

Figure 19. XPS of sodium, showing plasmon loss features (40).

At higher binding energies the valence band spectra of metals can show quite strong satellite structure (40). This is illustrated in the sodium spectrum in Figure 19. Similar structure in the core PES of metals is ascribed to plasmon loss events (41). The potential of the core hole or of the out-going photoelectron acts on the valence electrons to excite collective density waves, or plasma oscillations. Quantisation of these waves gives plasmons which have energies depending on the electron density, generally in the range 5 - 15 eV. The spectrum in Figure 19 shows a valence band peak closely below the Fermi level, and then two peaks apparently at higher IE; these correspond to energy loss from the photoelectron of one or two plasmon excitations. To give a theoretical fit to this spectrum, the partial density of states analysis was performed in the same way as for aluminium, and the resulting band was convoluted with the plasmon loss structure appearing in the XPS of (sharp) core lines in sodium. It is remarkable that the response of valence electrons to ionisation of another valence electron is apparently identical to that from ionising a highly localised core orbital.

The PE spectrum of sodium - a solid which in terms of its band structure must be regarded as very simple - is a salutory reminder that a simple one-electron picture cannot always be used. Satellite peaks, which dominate Figure 19, may be present in PE spectra of any solid, although experience suggests that they are generally less apparent in valence bands than in spectra of core electrons (42).

TRANSITION METALS AND METALLIC COMPOUNDS

With their partially filled d shells, transition metals and their compounds often show a more complex electronic structure than the simple solids considered in the previous section. PES has made some significant contributions in this area.

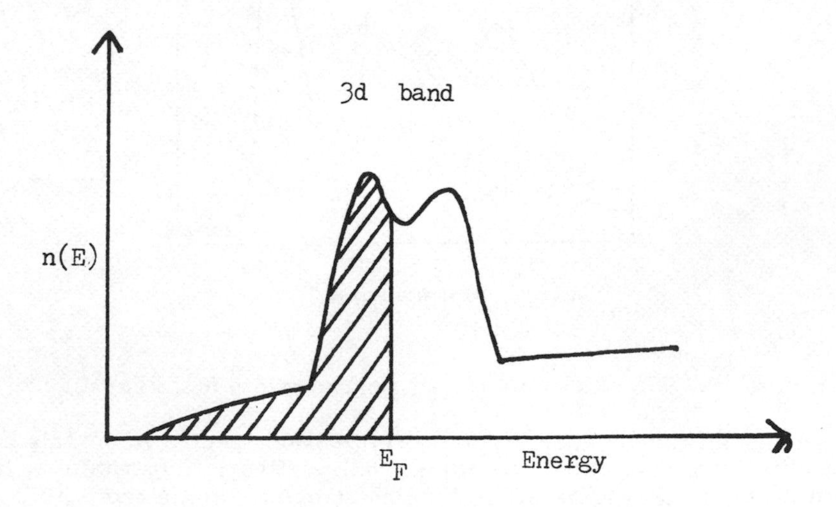

Figure 20. Schematic density of states for a transition metal.

For an element in the first transition series, the valence atomic orbitals are 3d, 4s and at higher energy the 4p. The 3d orbitals are fairly contracted relative to the normal interatomic spacings, and so in the elemental metals and their compounds they give rise to rather narrow bands. By contrast the 4s and 4p orbitals are spatially extended and overlap strongly with surrounding atoms. In the metals, this gives rise to a broad, free-electron-like band, overlapping the d band (43). A schematic picture of the resulting DOS for an element near the middle of the transition series is shown in Figure 20. As the transition series is traversed, the increasing nuclear charge moves the d band to lower energy relative to the s-p band. In addition, a contraction of the d orbitals reduces the overlap and makes the d band narrower. XPE spectra of the metals magnanese to copper are shown in Figure 21, where these effects can be seen (44). It is interesting that the spectrum of nickel shows a

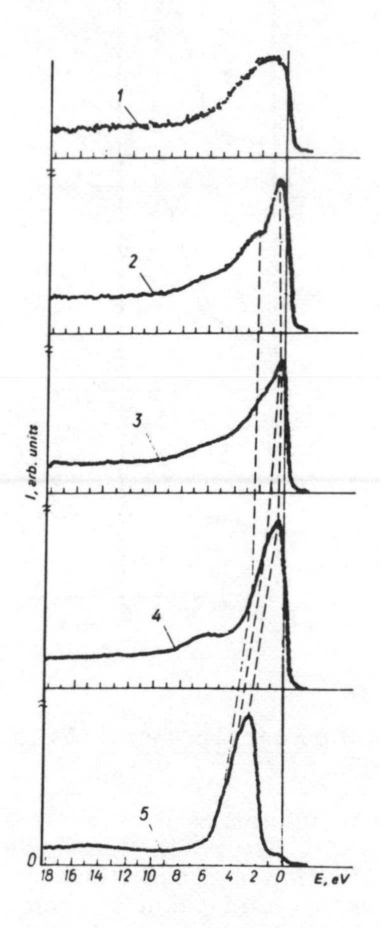

Figure 21. XPS of transition metals: (1) manganese; (2) iron; (3) cobalt; (4) nickel; (5) copper (44).

high DOS at the Fermi level where the states are predominantly 3d in character. By the time copper is reached the d band is clearly full and the Fermi level lies in the free-electron-like s-p band.

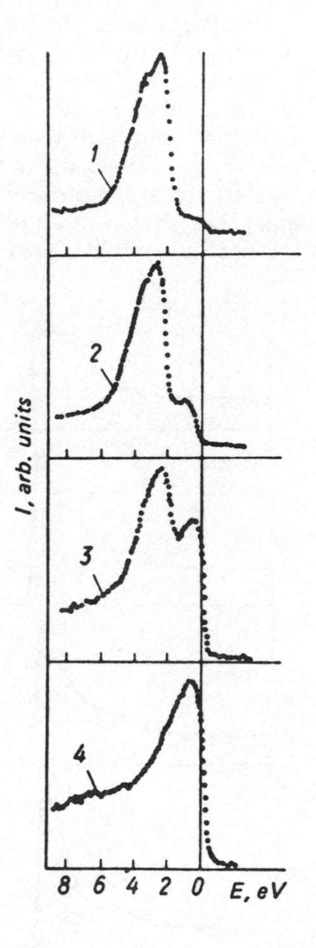

Figure 22. XPS of copper (1); nickel (2); and alloys $Ni_{10}Cu_{90}$ (3); $Ni_{53}Cu_{47}$ (4); (45).

XPS has shed some interesting light on the electronic structures of transition metal alloys. Figure 22 shows again copper and nickel, together with alloys of different composition (45). The simplest model of such alloys would suggest a d band at an energy intermediate between those of the elements. On the other hand, the spectra show that the d bands of the constituent atoms retain their identity in the alloys. Thus the d levels of the later transition elements must be quite localised.

Spectra such as these have given a considerable stimulus to the "coherent potential approximation", which is the simplest theory of alloys to treat correctly such localisation (46).

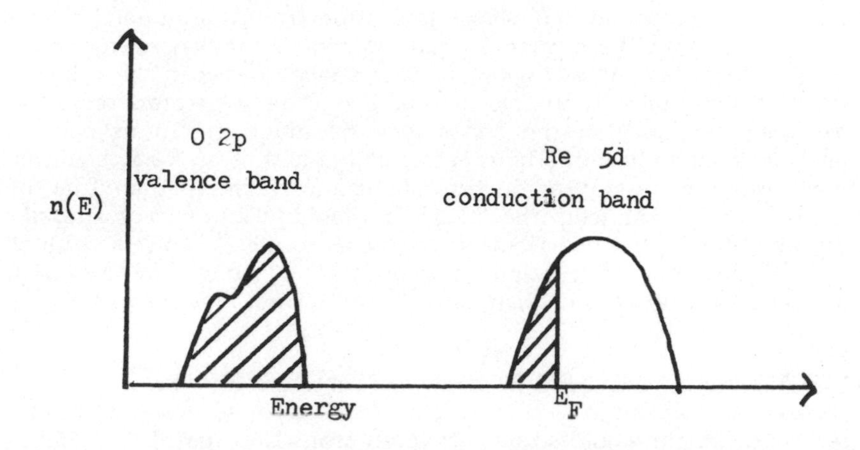

Figure 23. Schematic density of states for rhenium trioxide.

In compounds with more electronegative elements like oxygen, we expect the ionic model to give a first approximation to the electronic structure. As above, we predict a filled valence band based on the oxygen 2p atomic orbitals, and at higher energy, bands made up from cation orbitals. The important difference between sodium chloride and a

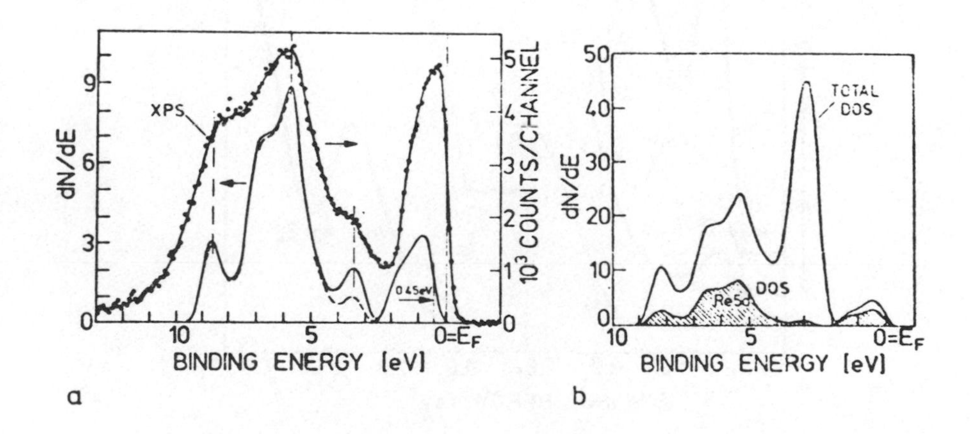

Figure 24. XPS of rhenium trioxide (48). (a) experimental and calculated spectra; (b) total DOS and rhenium 5d contribution.

transition metal compound is that in the latter, the cation orbitals may be partially occupied (47). Thus we would treat ReO_3 formally as a compound of Re^{6+}, with one remaining 5d electron per rhenium atom. Overlap in the solid forms a band from the 5d levels, and this is partially occupied in ReO_3, as we show schematically in Figure 23. The XPE spectrum of this compound (48) shows ionisation from the 5d band at the Fermi level, and from the oxygen 2p band at higher binding energy: see Figure 24. In fact there is an important discrepancy between the relative intensities of the peaks in this spectrum, and those expected from the DOS and the occupancy of the bands (one 5d electron, 18 oxygen 2p electrons per rhenium atom). The reason for this is that at X-ray energies the ionisation cross-section of 5d orbitals is much larger than that of oxygen 2p. The signal from the "O 2p" region in Figure 24 is actually dominated by the Re 5d contribution to these levels. This is a clear indication of covalent interaction between orbitals on different atoms, which reduces the real atomic charges from the formal ionic values.

NON-METALLIC COMPOUNDS OF TRANSITION ELEMENTS.

The previous example would suggest that all transition metal compounds with partially filled d levels should be metallic, yet any chemist knows that this is not the case: the great majority of halides, and a large number of oxides and other compounds are non-metals. Many of these compounds certainly have partially filled d levels, but the spectroscopic and magnetic properties are characteristic of localised configurations of d

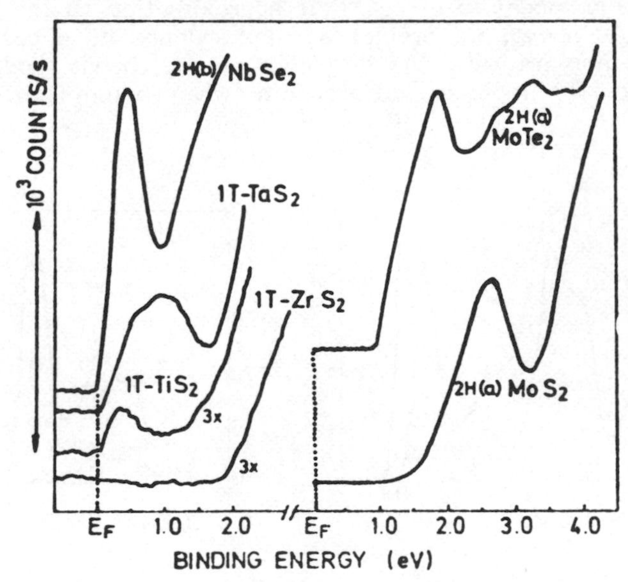

Figure 25. UPS of transition metal dichalcogenides, showing the metal d band and the edge of the chalcogen p band (50).

electrons. To a first approximation, the ions appear to be isolated from one another, rather than forming the bands of overlapping levels which we have discussed above. There are several factors which can contribute to such non-metallic behaviour, and the examples in this section demonstrate how PES can help to throw some light of these phenomena.

(i) Band splitting.

In a solid, the local environment of a transition metal atom provides a ligand field which splits the five d orbitals into groups with different energies (49). It is thus to be expected that the d band will split into two or more sub-bands. If the splitting is sufficient, the sub-bands may not overlap in energy, and appropriate electron configurations will give an insulator with a band gap. A fairly straightforward example is provided by molybdenum disulphide, MoS_2. Figure 25 shows UPS of this and of some similar chalcogenides (50). The d electron signal is apparent at lower IE than the chalcogenide p levels; it increases in intensity from the formally d^0 (Ti and Zr) to the d^1 (Nb and Ta) and d^2 (Mo) compounds. Some of these compounds have the cadmium iodide structure, but in the similar layer structure of MoS_2, the nearest-neighbour sulphur atoms provide a trigonal prismatic (D_{3h}) coordination for each molybdenum (51). This geometry, and the resulting ligand field splitting of the d orbitals, is shown in Figure 26. In the solid, interaction between neighbouring

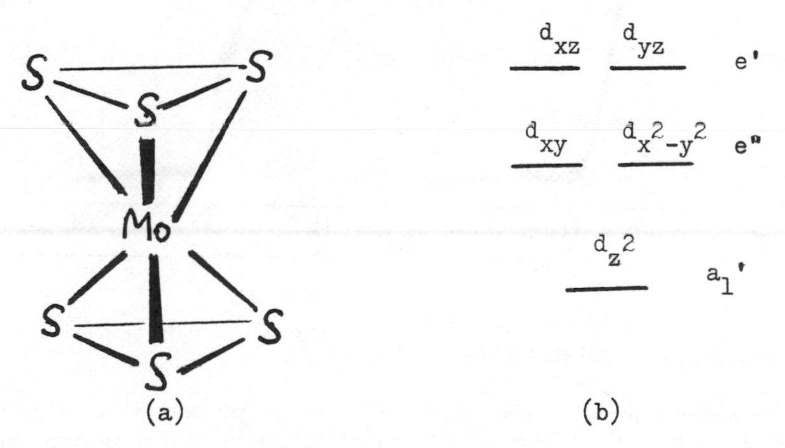

Figure 26. (a) Portion of the MoS_2 structure, showing trigonal prismatic coordination of molybdenum. (b) Ligand field splitting of Mo 4d levels.

molybdenum atoms within a layer broadens the levels into bands; there nevethertheless remains a gap between the lowest band from the d_{z^2} orbitals and the higher ones. In MoS_2 this band is full and the result is a semiconductor; by contrast the d^1 compounds are metallic.

More complex cases of band splitting appear to result from

interaction between one transition metal atom and another, rather than from simple ligand field effects. Such a mechanism has been proposed for the transition metal dioxides which have structures based on that of rutile, TiO_2 (52). Some of these oxides are metallic, some are not. The d^1 compounds VO_2 and NbO_2 even have a temperature-dependent metal-insulator transition. Figure 27 shows UPS of VO_2 at temperatures above and below the transition (53). Again the d band appears at lower IE than the oxygen 2p band, and at higher temperatures shows the sharp Fermi edge expected for the metallic phase. This edge disappears in the non-metallic phase, as shown by the spectrum recorded at 293 K.

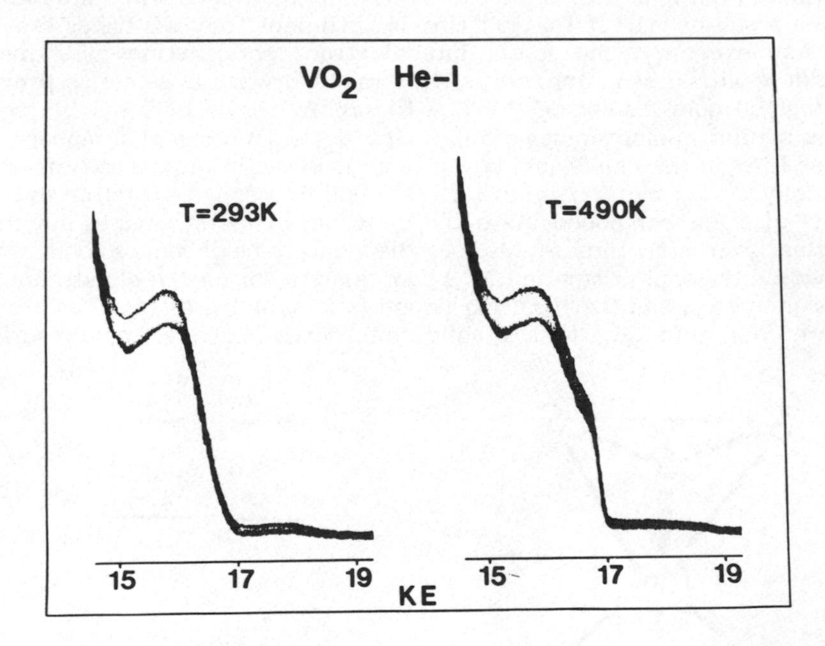

Figure 27. UPS of the d band of VO_2 (53), with the Fermi edge visible in the metallic phase at 490 K.

An explanation for the varied electronic properties of the rutile dioxides is based on the fact that the structure has chains of edge-shared metal-oxygen octahedra (52). As shown in Figure 28, these allow fairly close interaction between metal atoms in the direction of the _c_-axis. A schematic representation of the electronic energy levels is shown in Figure 29. The levels labelled M-Oσ and M-Oπ come from oxygen 2p with some covalent admixture of metal d orbitals. These shaded levels show the filled bands of the d^0 compound TiO_2. Particularly important for understanding the d^1 compounds is the next band, marked M-Mσ. It results from the metal d orbitals which point along the c-axis, and so have appreciable overlap between neighbouring metal atoms. In the regular form of the rutile structure, this band should hold two electrons per atom, even if it does not overlap in energy with the next higher level. Indeed,

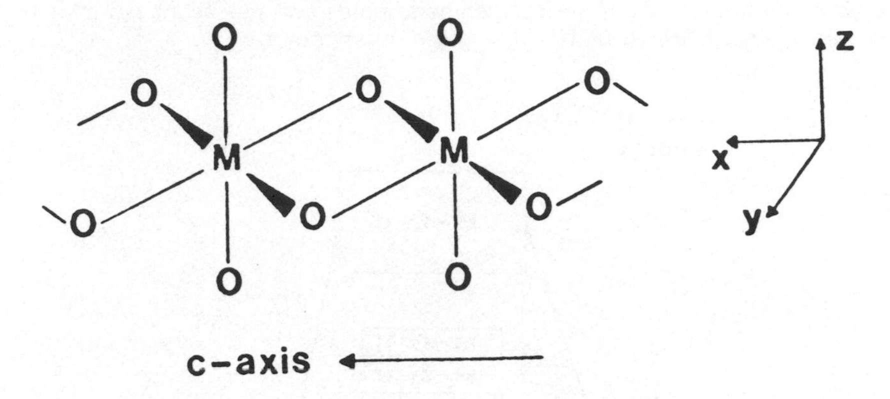

Figure 28. A portion of the rutile structure, showing edge-sharing of metal-oxygen octahedra along the c-axis.

the high temperature, metallic, form of VO_2 does have the regular structure. In the non-metallic form a structural distortion occurs which results in an alternation of the V-V distances, and thus in a kind of pairing of vanadium atoms along the c-axis. The result of this is that the M-M band, as shown in Figure 30, splits into a lower M-M bonding level and an

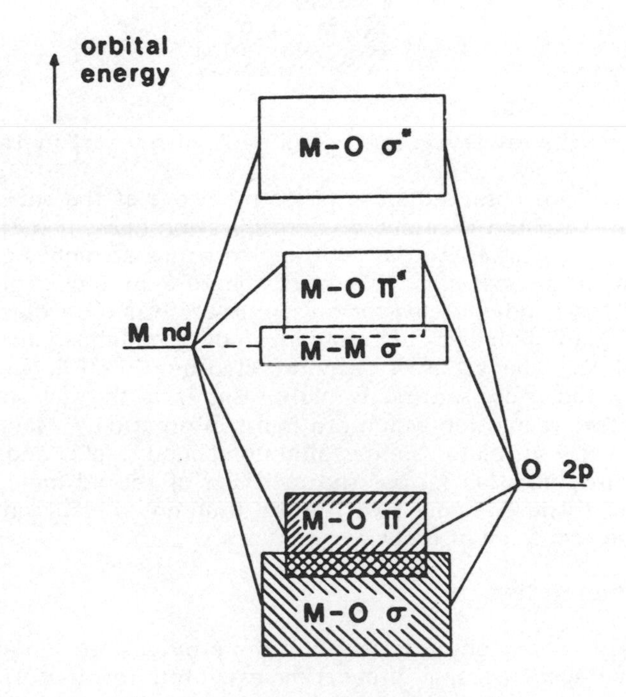

Figure 29. Energy levels for a rutile dioxide (see text).

upper antibonding one. The former band holds two electrons per pair of vanadium atoms, and is thus filled in the d^1 compound.

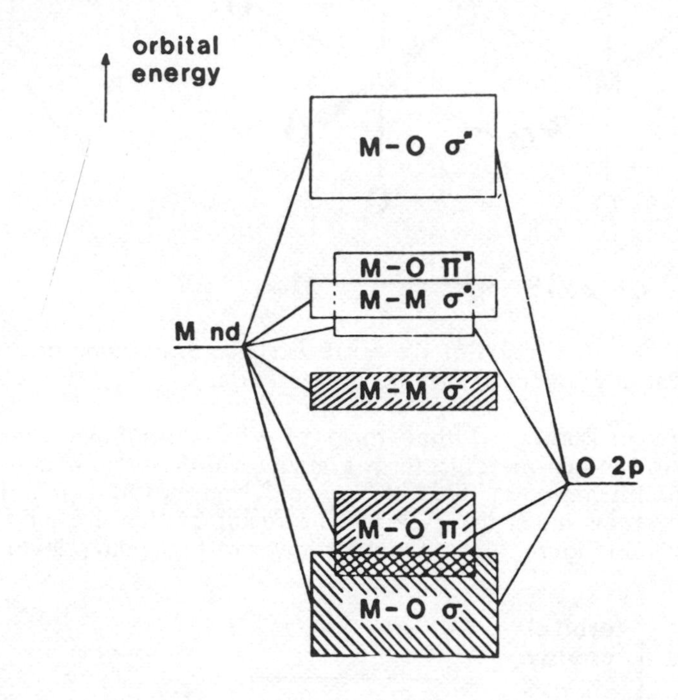

Figure 30. Energy levels for a d^1 dioxide with metal-metal pairing.

Distortions are observed some other members of the series, such as MoO_2 although with the d^2 electron configuration here, one electron per molybdenum atom must enter the overlapping bands at higher energy, and so this compound is metallic. Indeed, the PES of MoO_2 gives strong evidence that the band-splitting model is correct (54). Figure 31 shows XPS and UPS of dioxides of niobium ($4d^1$), molybdenum ($4d^2$) and ruthenium ($4d^4$). The 4d level appears stronger in XPS, since at this photon energy the cross-section is high. But it is the UV spectra with their much better resolution which are more interesting. Not only is the Fermi edge clearly visible in the metallic compounds MoO_2 and RuO_2, and absent in the insulator NbO_2, but the splitting of the 4d band, predicted for a distorted rutile compound with more than one d electron, is clearly apparent in the spectrum of MoO_2.

(ii) Electron correlation.

The non-metallic compounds discussed in the previous section are diamagnetic and have spectroscopic properties expected for materials with a band gap. Many transition metal compounds, however, are paramagnetic and show the spectroscopic properties characteristic of localised

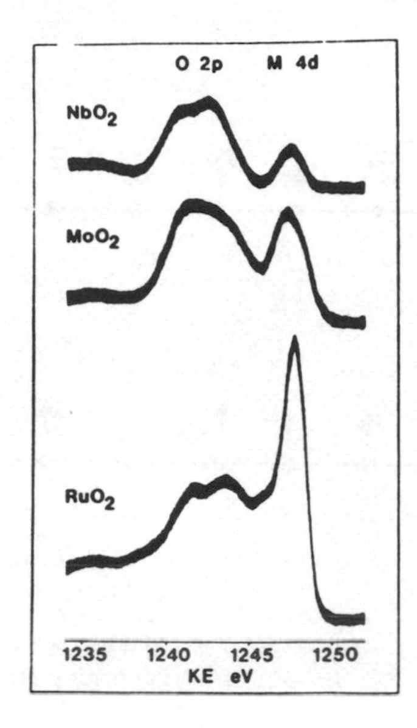

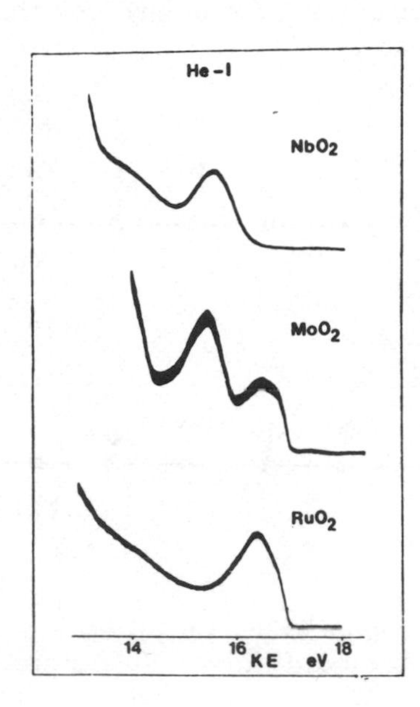

Figure 31. XPS (left) and UPS (right) of dioxides of niobium, molybdenum and ruthenium (54).

configurations of d electrons (55). Thus Ni^{2+} in nickel oxide, NiO, seems to be essentially the same as in isolated octahedral complexes such as $Ni(H_2O)_6^{2+}$. In compounds such as NiO the band theory of solids breaks down, and this seems to happen because of repulsion between electrons (56). In band theory, electron-electron repulsion is averaged out, and merely contributes to the periodic potential of the lattice. Sometimes this orbital approximation is seriously misleading. A qualitative guide to when this is likely to happen is provided by the Hubbard model (57):

Consider a one-dimensional chain of atoms, each having a single non-degenerate valence orbital and one electron. The interaction between atoms gives rise to a transfer integral t and a consequent band-width, W. If W is small, the ground-state electron configuration will consist of one electron localised on each atom, as shown in Figure 32 (a). This is a non-metallic state, since motion of electrons in a conduction band must involve configurations such as that in Figure 32 (b), where one orbital is doubly occupied. Such double occupancy incurs an energy penalty, through the extra inter-electron repulsion U for two electrons on one atom. If $U \gg W$, we have the localised, non-metallic, ground state of Figure 32. If the band-width W is much larger than U, on the other hand, electron repulsion becomes unimportant and we recover normal band theory, which predicts a metal in this case. Intermediate cases - which must include many transition metal compounds - can show complex behaviour (55), but in any case the simple Hubbard model is not adequate for these.

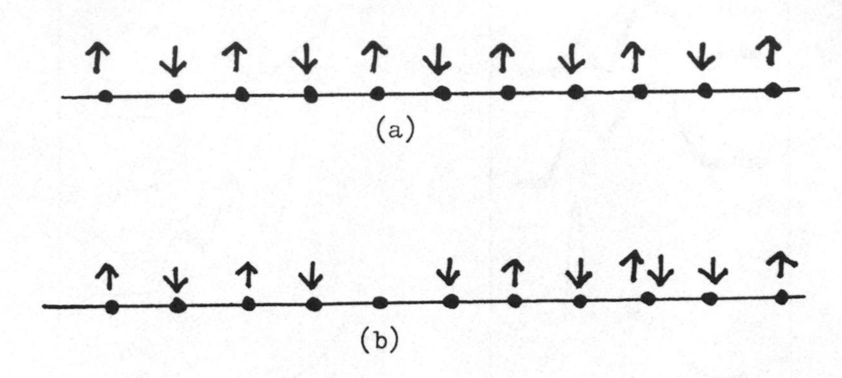

Figure 32. Localised electrons. (a) Ground state with singly occupied atoms; (b) state with double occupancy and consequent extra electron-electron repulsion.

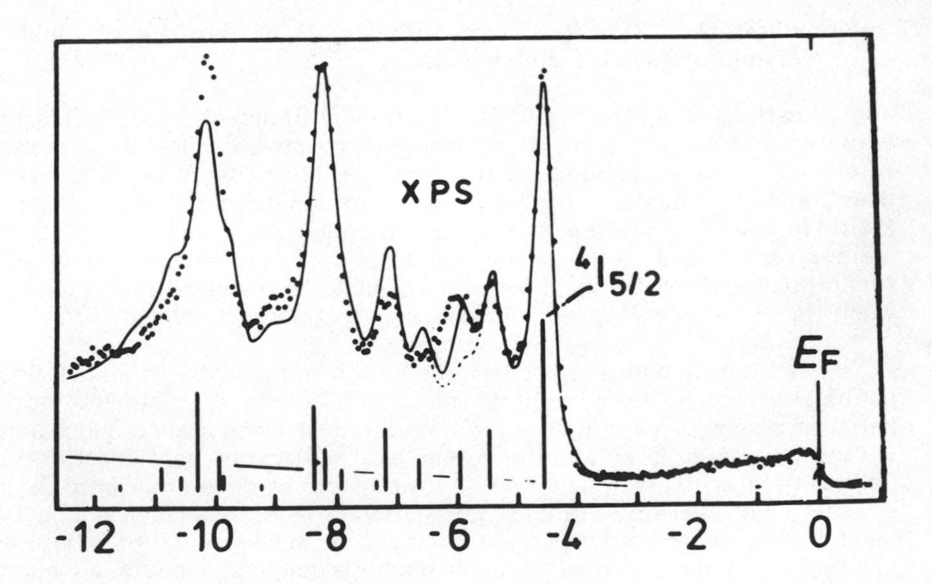

Figure 33. XPS of thulium metal, with theoretical fit of states from 4f ionisation (59)

It appears that electron localisation is likely when valence atomic orbitals are small relative to the interatomic spacings, because then band-widths will be small, and single-site electron repulsions large. This often happens with 3d valence orbitals of elements in the first transition series (55). The most dramatic examples, however, come from rare-earth elements and compounds with partially filled 4f orbitals (58). Although they have energies comparable to valence orbitals, the 4f are core orbitals in terms of their size. 4f band-widths may be around 0.01 eV, whereas electron repulsion integrals are in the range 5-7 eV. (The contribution by Baer to this volume describes the measurement of U in rare-earth metals by combining data from XPS and another form of electron spectroscopy, Bremsstrahlung Isochromat Spectroscopy). Localised 4f electrons coexist in rare-earth metals with the conduction electrons, and give rise to complex multiplet structure in XPS. Figure 33, for example, shows the spectrum of thulium (59), which has the

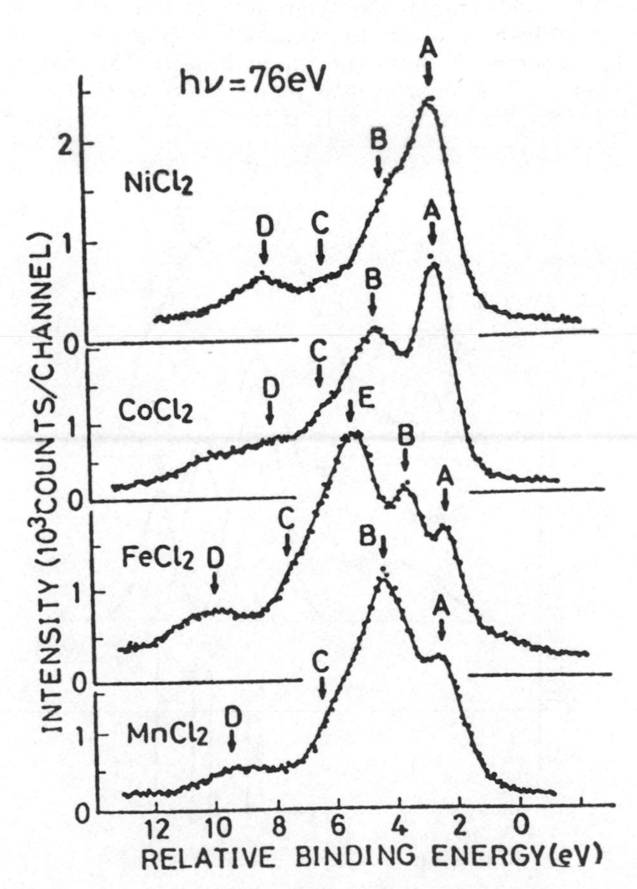

Figure 34. PES of dichlorides of manganese, iron, cobalt and nickel excited with synchrotron radiation at 76 eV (62).

electron configuration $4f^{12}$. Since the 4f orbitals are so strongly localised, the interpretation of this spectrum must come from atomic, rather than solid-state, physics (60). Assuming Russell-Saunders coupling, the twelve 4f electrons couple to give a $^3H_6$ ground state. Ionisation from this can give many possible states of the $4f^{11}$ configuration. The relative energies of these are known from optical and UV spectroscopy of other systems with $4f^{11}$, for example erbium compounds containing Er^{3+}. The intensities of the different multiplet states in the PE spectrum can be calculated using the coefficients of fractional parentage which arise in atomic theory (60). The resulting agreement between theory and experiment is very satisfactory, and confirms the picture of fully localised 4f levels (61).

Multiplet structure also appears in the PES of transition metal compounds with localised d-electron configurations. Figure 34, for example, shows spectra of some transition metal dichlorides (62). The bands labelled C and D in each spectrum are attributed to ionisation from the chlorine 3p orbitals, the remaining bands A, B and E to 3d states. The case of $FeCl_2$ is shown in more detail in Figure 35, together with the predicted levels. The appropriate model is now that of ligand field theory, which treats the energy levels of localised d^n configurations in

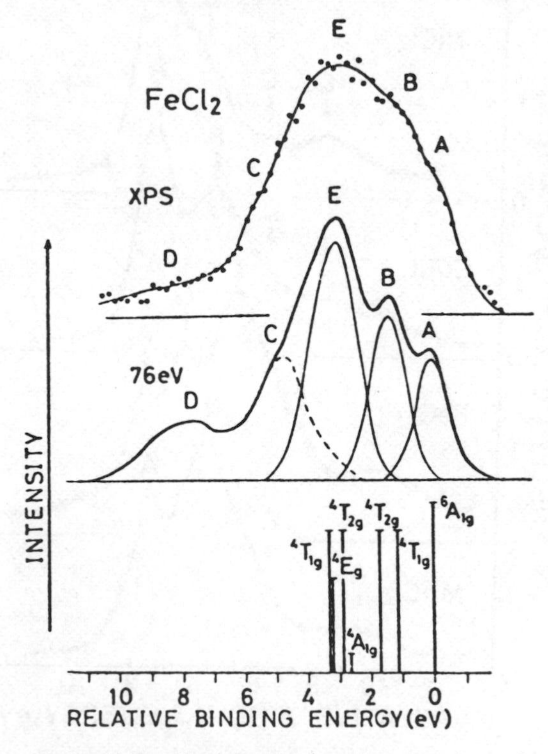

Figure 35. PES of $FeCl_2$, together with predicted multiplet states from the iron 3d levels (62).

crystal or molecular environments (49). The dichlorides adopt the cadmium chloride structure in which the metal atom occupies a site with approximately cubic symmetry. The resulting ground state configuration for Fe^{2+} is $t_{2g}^{4}e_{g}^{2}$ ($^5T_{2g}$). Ionisation from this can give a variety of states from the $t_{2g}^{3}e_{g}^{2}$ and $t_{2g}^{4}e_{g}^{1}$ configurations, and the energies and intensities of these states are shown in Table 1. The agreement between theory and experiment in Figure 34 is good. It remains true, however, that the available data do not allow such a stringent test of the multiplet theory for d electrons in transition metal compounds, compared with the f electrons in rare earths (63).

Table 1. Multiplet states formed in ionisation from $t_{2g}^{4}e_{g}^{2}$ ($^5T_{2g}$). B and C are Racah electrostatic pârameters; Δ is the ligand field splitting (60).

Configuration	State	Relative energy	Relative intensity
$t_{2g}^{3}e_{g}^{2}$	$^6A_{1g}$	Δ - 35B	6/5
	$^4A_{1g}$	Δ - 25B + 5C	2/15
	$^4E_{g}$	Δ - 16B + 8C	2/3
	$^4T_{1g}$	Δ - 16B + 7C	1
	$^4T_{2g}$	Δ - 22B + 5C	1
$t_{2g}^{4}e_{g}^{1}$	$^4T_{1g}$	- 25B + 6C	1
	$^4t_{2g}$	- 17B + 6C	1

(iii) Structural disorder

In the normal development of the theory of solids, Bloch's theorem shows how the periodicity of a crystal creates orbitals which are delocalised throughout the solid (64). When lattice periodicity breaks down, localised electronic states may develop even without the assistance of electron correlation. In all real solids, lattice periodicity is broken at the surfaces and at defects within the solid, and it is well known that both situations may have localised electronic levels associated with them (65,66). The investigation of states either at clean surfaces or in adsorbate layers is a very important aspect of current PES work, and is described in other contributions to this volume.

We shall mention here the case of solids with extensive structural disorder, which have been the subject of much recent interest (67). If a high proportion of lattice sites are defective, it may no longer be possible to think of states as localised at individual defects. Nevertheless it appears that extensive disorder can give rise to states which are spatially localised (68). A rather crude picture of how this can happen is presented in Figure 36. The potential is shown to vary randomly from site to site, so

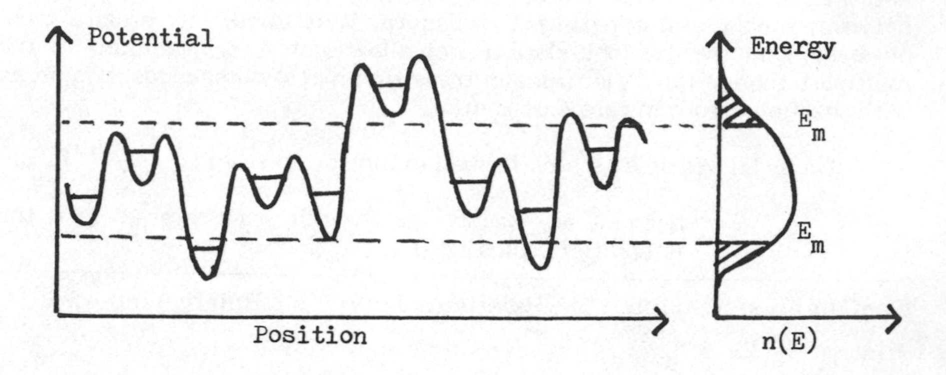

Figure 36. Schematic representation of the energy levels for electrons in a disordered solids. The DOS curve (right) shows localised states (shaded) separated from extended states by mobility edges E_m.

that the electronic levels within each potential well also fluctuate. Levels near the middle of the energy range are likely to have neighbours with similar energy, so that by overlap between adjacent wells a state delocalised throughout the solid can be formed. However, levels at particularly high or low energies are unlikely to have any similar neighbours, and these will remain localised. The density of states is shown on the right of Figure 36, and marked on this are the two mobility edges which separate localised states in the wings of the band from extended states in the centre (68). The existence of mobility edges suggests the possibility of a metal-insulator transition, known as the Anderson transition, which depends on electron concentration and disorder. If the electron concentration is such that the highest filled levels are extended states, a metal will result. If, by changing the electron concentration or the disorder, the highest filled levels can be made to move into the localised region, the solid should become non-metallic.

The mixed oxide lanthanum strontium vanadate, $La_{1-x}Sr_xVO_3$ shows a transition which is supposed to be of the Anderson type (69). With x=0 the trivalent vanadium atoms have a d^2 electron configuration localised by electron correlation. As some lanthanum is replaced by strontium which has one less valence electron, a corresponding number of vanadium atoms are oxidised to the V^{4+} state. At low x the compound remains insulating, but at x=0.24 a transition to a metal takes place without any major structural change. UPS of compounds either side of this transition (70) are shown in Figure 37. The vanandium 3d band appears at low IE, and for metallic samples a sharp Fermi edge can be seen. The naive picture presented above suggests that this edge should persist in insulating samples, since no discontinuous change in DOS takes place at the transition point. Nevertheless, Fermi edges are absent in the insulators. This may be partly due to sample charging, but it seems also that the independent electron picture of the Anderson transition is too simple, and that the effects of electron correlation may be very important. It is likely that PES can contribute significant information about the valence levels in disordered solids, but at the present moment it is doubtful whether the theory exists which can give an adequate interpretation of these spectra.

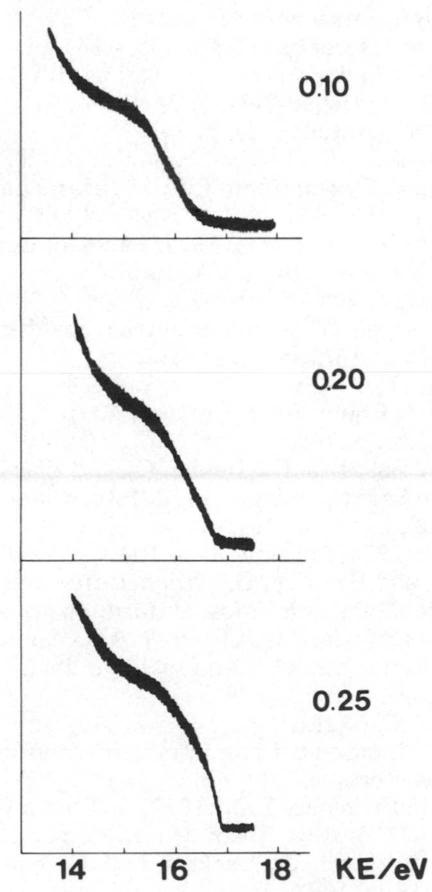

Figure 37. UPS showng the d band of lanthanum strontium vanadate, $La_{1-x}Sr_xVO_3$, for different values of x (70).

REFERENCES

1. Schaich, W.L. in "Photoemission in Solids: I" (eds. M. Cardona and L. Ley), 1978, Springer-Verlag, Berlin, pp. 105 - 134.
2. Berglund, C.N. and Spicer, W.E.: 1964, Phys. Rev. 136A, pp.1030, 1044.
3. Kittel, C. "Introduction to Solid State Physics" (third edition), 1966, John Wiley, New York, pp. 302 - 3.
4. Grobman, W.D., Eastman, D.E. and Freeouf, J.L.: 1975, Phys. Rev. B12, p. 4405.
5. Wehner, P.S., Williams, R.S., Kevan, S.D., Denley, D. and Shirley, D.A.: 1979, Phys. Rev. B19, P.6164.
6. Brundle, C.R.: 1974, Surface Sci. 48, p. 99.
7. Reference (3), pp. 68 - 70.
8. Feibelman, P.J. and Eastman, D.E.: 1974, Phys. Rev. B10, p. 4932.
9. See discussion in references (4) and (8).
10. Manson, S.T. in reference (1), pp. 135 - 163.
11. Hüfner, S. in "Photoemission in Solids: II" (eds. L. Ley and M. Cardona), 1979, Springer-Verlag, Berlin, p. 175,
12. Shirley, D.A. in reference (1), p. 193.
13. (a). Eastman, D.E.: 1973, Phys. Rev. B8, p. 6027.
    (b). Grobman, W.D. and Koch, E.E. in reference (11), p.264.
14. Fuggle, J.C. in "Handbook of X-ray and Ultra-violet Photoelectron Spectroscopy", (ed. D. Briggs) 1977, Heyden, London, pp. 273 - 312.
15. Evans, S. in reference (14), pp. 121 - 151.
16. Reference (3), pp. 204 - 5.
17. Siegbahn, K. et al. "Electron Spectroscopy for Chemical Analysis", 1967, Almqvist and Wiksell, Uppsala.
18. See reference (15).
19. Evans, S.: 1973, Chem. Phys. Letters, 23, p. 134.
20. Reference (13 b), p. 269.
21. Turner, D.W., Baker, A.D., Baker, C. and Brundle, C.R. "Molecular Photoelectron Spectroscopy", 1970, Interscience, London.
22. Reference (12), pp. 176 - 181.
23. Koopmans, T.: 1934, Physica, 1, p. 104.
24. Bleaney, B.I. and Bleaney, B. "Electricity and Magnetism" (second edition), 1965, Clarendon Press, Oxford, pp. 16 - 28.
25. Heeger, A.J. amd Garito, A.F in "NATO Summer Institute on Low Dimensional Conductors" (ed. H.J. Keller), 1975, Plenum Press, New York, p. 89.
26. Reference (13 b), p. 281.
27. Huheey, J.E. "Inorganic Chemistry", (second edition), 1978, Harper and Row, New York, pp. 51 - 65.
28. Citrin, P.A. and Thomas, T.D.: 1972, J. Chem. Phys. 57, p. 4446.
29. Shirley, D.A.: 1973, Adv. Chem. Phys. 23, pp. 123 - 6.
30. Poole, R.T., Szajman, J., Leckey, R.C.G. and Liesegang, J.: 1975, Phys. Rev. B12, p. 5872.
31. Jones, H. "Theory of Brillouin Zones and Electronic States in Crystals", (second edition), 1975, North Holland, Amsterdam, pp. 221 - 234.

32. Goldman, A., Tejeda, J. Schevchik, N.J. and Cardona, M.: 1974, Phys. Rev. B10, p. 4388.
33. Hagenau, H.J., Gudat, W. and Kunz, C.: 1975, J. Opt. Soc. Am. 65, p. 742.
34. Schevchik, N.J. Tejeda, J. and Cardona, M.: 1974, Phys. Rev. B9, p. 2627.
35. Herman, F., Kortum, R.L., Kuglin, C.D. and Van Dyken, J.P.: 1968, "Methods in Computational Physics", Academic Press, New York, vol. 8, p. 193.
36. Ley, L., Kowalczyk, S.P., Pollak, R. and Shirley, D.A.: 1972, Phys. Rev. Letters, 29, p. 1088.
37. Ziman, J.M. "Models of Disorder", University Press, Cambridge, 1979, pp. 447 - 459.
38. Reference (3), p. 210.
39. Steiner, P., Höchst, H. and Hüfner, S. in reference (11), p. 369.
40. Reference (39), p. 368.
41. Pardee, W.J., Mahan, G.D., Eastman, D.E., Pollak, R.A., Ley, L., McFeely, F.R., Kowalzcyk, S.P. and Shirley, D.A.: 1975, Phys, Rev. B11, p. 3164.
42. Reference (12), pp. 181 - 9.
43. Mattheis, L.F.: 1964, Phys. Rev. 134, pp. 477 - 386
44. Hüfner, S. amd Wertheim, G.K.: 1974, Phys. Letters, 47A, P. 349.
45. Hüfner, S., Wertheim G.K. and Wernik J.H.: 1973, Phys. Rev. B8, p. 4511.
46. Soven, P.: 1967, Phys. Rev. 156, p. 809; 1969, Phys. Rev. 178, p. 1136.
47. Harrison, W.A. "Electronic Structure and the Properties of Solids", 1980, W.H. Freeman, San Francisco, pp. 431 - 476.
48. Wertheim, G.K., Mattheis, L.F., Campagna, M. and Pearsall, T.P.: 1974, Phys. Rev. Letters, 32, p. 997.
49. Reference (27), pp. 348 - 390.
50. Williams, P.M. and Shepherd, F.R.: 1973, J. Phys. C, 6, p. L36.
51. Wilson, J.A. and Yoffe, A.D.: 1969, Adv. Phys. 18, p. 193.
52. Goodenough, J.B. in "Progress in Inorganic Chemistry" (ed. H. Reiss), Pergamon,New York, 1971, vol. 5, pp. 344 - 364.
53. Beatham, N., Fragala, I.L., Orchard, A.F. and Thornton G.: 1980, J. Chem. Soc. Faraday II, 76, p. 929.
54. Beatham N. and Orchard, A.F.: 1979, J. Elec. Spectr. 16, p. 77.
55. Reference (52), pp. 145 - 399.
56. Mott, N.F.: 1949, Proc. Phys. Soc. London, Sect. A 62, p. 416.
57. Hubbard, J.: 1963, Proc. Roy. Soc. A176, p. 328.
58. Dieke, G.H. "Spectra and Energy Levels of Rare Earth Ions in Crystals", 1969, John Wiley, New York.
59. Lang, J., Baer, Y. and Cox, P.A.: 1981, J. Phys. F 11, p. 121.
60. Cox, P.A.: 1975, Structure and Bonding, 24, p. 59.
61. Campagna, M., Wertheim, G.K. and Baer, Y. in reference (11), pp. 217 - 260.
62. Ishii, T., Kono, S., Suzuki, S., Nagakura, I., Sagawa, T., Kato, R., Watanabe, M. and Sato, S.: 1975, Phys. Rev. B12, p. 4320.
63. Hüfner, S. in reference (11), pp. 173 - 216.

64. Reference (3), p. 259.
65. Appelbaum, J.A. and Hamann, D.R.: 1976, Revs. Mod. Phys. 48, pp. 479 - 496.
66. Reference (3), pp. 571 - 5.
67. See reference (37).
68. Anderson, P.W.: 1958, Phys. Rev. 109, p. 1492.
69. Sayer, M., Chent, R., Flethcer, R. and Masingh, A.: 1975, J. Phys. C 8, p. 2059.
70. Egdell, R.G and Wall, G., unpublished results.

CORE LEVEL PHOTOELECTRON SPECTROSCOPY

G. K. Wertheim

Bell Laboratories
Murray Hill, New Jersey 07974

I. INTRODUCTION

The electronic states of solids can be divided, with little ambiguity, into core and valence orbitals. The core electrons comprise the more tightly bound inner electrons whose wave functions lie well within the atomic radius. Such orbitals do not overlap those of neighboring atoms and therefore do not form energy bands. Generally speaking they are fully occupied and do not participate in chemical bonding. The major exceptions are the 4f states of the rare earths which are core-like in spatial extent, but valence-like in binding energy.

At first sight it would appear that core electrons are entirely atomic in character, i.e. that they can be fully characterized by measurements on free atoms, and that solid state effects should be vanishingly small. This is true only to first order. In fact, it was the recognition that chemical effects on core electrons are readily measurable, that provided one of the early motivations for the development of x-ray photoelectron spectroscopy. The acronym ESCA, Electron Spectroscopy for Chemical Analysis, was coined by K. Siegbahn to emphasize the wide range of applications of this technique in chemistry which he and his collaborators had demonstrated (1).

A. Binding Energy

The most fundamental property of a core electron is its binding energy, i.e. the energy required to raise the electron to the vacuum level (2). To first approximation one might expect this energy to be equal to the eigen-energy of the electron in the atom in its initial state, a concept often called the frozen orbital approximation or Koopmans' theorem. For core levels it gives an answer which is too large by 3-30 eV. This may not seem like a serious error for the deeper core levels with binding energies measured in keV, but it points to an important omission. The removal of a core electron perturbs the other electrons in the atom especially those in orbitals with

P. Day (ed.), Emission and Scattering Techniques, 61–74.

greater radial extent. These will relax while the photoelectron is being emitted, so that the final hole-state atom has a ground state energy significantly lower than the original eigen energy of the photo-emitted electron. However, relaxation does not always populate the ground state. In fact it can be shown that excited states are populated to such an extent that the centroid of the final states lies at the Koopmans' energy, see Fig. 1. This serves to give the Koopmans' energy an operational reality,

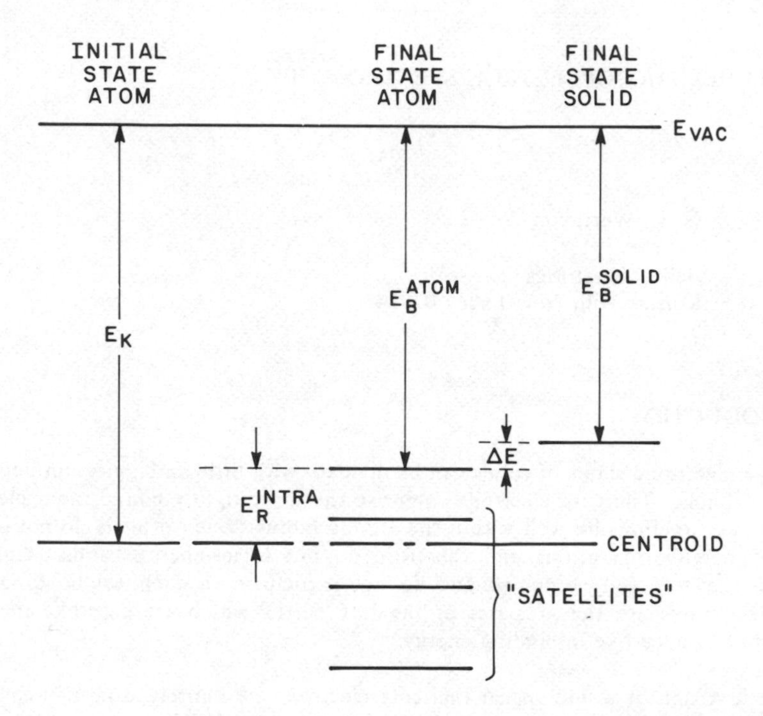

Figure 1 Core electron binding energies in the initial and final state of an atom and in the final state of a solid. ΔE incorporates both initial state changes during the formation of the solid and extra-atomic final state effects.

but it has not proved possible to determine it experimentally by this method because extrinsic loss structure is superimposed on the spectrum of final excited states. When individual excited states have large intensity they are usually referred to as satellites. Measured binding energies always refer to the ground state line (2).

Electrons in orbitals which lie outside of that of the photoemitted electron suddenly see a core potential screened by one less electron. This observation motivates the so-called 'equivalent cores' approach to the treatment of the final-state in which the core-ionized Z atom is replaced by a Z+1 atom. In *solids* the response to photoionization is not limited to the atom itself, but may extend to its neighbor and beyond. This subject will be considered further in Section II, below. More detailed exposition can be found in the items cited in the bibliography.

B. Line Width

The width of a core level is determined by the lifetime of its hole state. The dominant decay channels are x-ray emission and two-electron processes. For all but the innermost electrons of the heavier atoms the latter dominate. In de-excitation one electron drops into the original hole state and the second is emitted with the excess energy of the initial one-hole state over that of the final two-hole state. If both electrons come from a shell with principal quantum number greater than that of initial state hole the process is called Auger effect; if one comes from the same shell it is referred to as Coster-Kronig effect. The latter has greater probability when energetically allowed, because of the greater overlap between initial and final state wave functions. As a result, the most weakly bound electron state of a given shell always has the narrowest linewidth because it is not subject to Coster-Kronig decay. Since core-core transitions dominate the decay process, chemical effects on core-hole lifetime are generally negligible. The exceptions are states with small binding energy which can decay only via valence states. The best examples are provided by the 4f electrons of the rare earths (3).

C. Level Splitting

Core electron states are well known to be split by the spin-orbit interaction which was recognized in early studies of x-ray spectra. In a sense this is also a final state effect reflecting the energy difference between hole states with spin and orbital angular momentum parallel and antiparallel. Core s-electron states, which have no orbital angular momentum, have no spin-orbit splitting. All core states can, however, be split by coupling to outer incomplete shells, a phenomenon generally termed multiplet coupling. The most striking examples are found among the 4s and 4d electrons of the rare earths and the 3s electrons of the transition metals. Simple splitting are found only when a core s-hole and the incomplete shell have different principal quantum numbers. If they are the same, a more complex many-electron treatment must be employed, and a large number of final states may be resolved experimentally.

This manifestation of the ubiquitous many-electron aspects of core electron photoemission will not be considered in this chapter. For detailed exposition see the bibliography. An overview is given in Ref. 4.

II. BINDING ENERGIES IN SOLIDS

The discussion of core electron binding energies in solids generally takes the free atom binding energy as its starting point. Intra-atomic effects can then be left out of the discussion. The basic question is how does the excitation potential change when atoms are assembled into a solid? The relevant phenomena may be usefully classified into initial and final state effects. The former take place when the solid is created, the latter represent phenomena that happen as a result of photoionization.

The most important initial state effects are due to changes in electronic configuration, e.g. in the solid the atom may exist as an ion, or charge may redistribute itself among the outer orbitals in response to the formation of energy bands. Furthermore the charge on the other ions in the solid will produce an electrostatic potential at the site the atom in question, the Madelung potential, which directly modifies all binding energies.

Final state effects collectively denote the response of the solid to the sudden creation of a core hole. In metals the dominant effect is the screening response of the conduction electrons, which typically lower the energy of the final state by an energy of the order of 3 eV. In insulators a similar reduction in energy is obtained by the polarization of the neighboring atoms. There will also be a vibrational response as phonons are excited by a Frank-Condon mechanism.

It is worth emphasizing at this point that binding energy comparisons require a common reference level. By virtue of the use of the free atom as the reference substance this reference level is the vacuum level. Measurements in solids must always be corrected to the vacuum level before a meaningful discussion of binding energy shifts can be given.

We shall now consider some illustrative examples.

A. Metals

In metals the dominant extra-atomic final state effect is due to the screening response of the conduction electrons. It may be estimated by comparing the binding energy of a free atom with that in the corresponding metal. There may, however, be an initial state contribution of comparable magnitude from a change of configuration between atom and metal. This is particularly important in the noble and transition metals in which charge may flow between s, p, and d-states (5,6).

As a particularly simple example we consider the behavior of surface atoms on metallic gold (7). Experiment has shown that the binding energy of these atoms is 0.4 eV smaller than that of the bulk atoms, Fig. 2. It is reasonably clear that there is no net charge flow from or to the surface. Furthermore, the core electron line shape of bulk and surface atom is similar suggesting (see III) that the screening response is not altered significantly. The major source of the change in binding energy is then to be found in the change in band structure at the surface which will cause a redistribution of charge between s, p and d-states. On the basis of the smaller coordination number at the surface we expect to find narrower bands. This in turn implies less hybridization and less empty d-character above the Fermi energy. In other words we expect gold surface atoms to be more nearly $5d^{10}$ than those in the bulks. It is then easy to see that there will be a reduction in core electron binding energy because more charge has been put into the more compact d-orbitals. Similar surface binding energy changes have now also been seen in W, (8) and Ir (9).

Slightly more complex is the analysis of core electron shifts in alloys and intermetallic compounds. part of the problem is that there is no common reference level as in the case of the surface atoms. Interpretation of shifts requires correction of all binding energies to the vacuum level. This can be accomplished even when the work function of the alloy is not known, by eliminating unknown work function shifts between equations for the individual component, see Ref. 10. The resulting equation for the charge transfer in an equiatomic compound has been used to obtain an estimate of charge flow in CsCℓ-structure intermetallic compounds of Au.

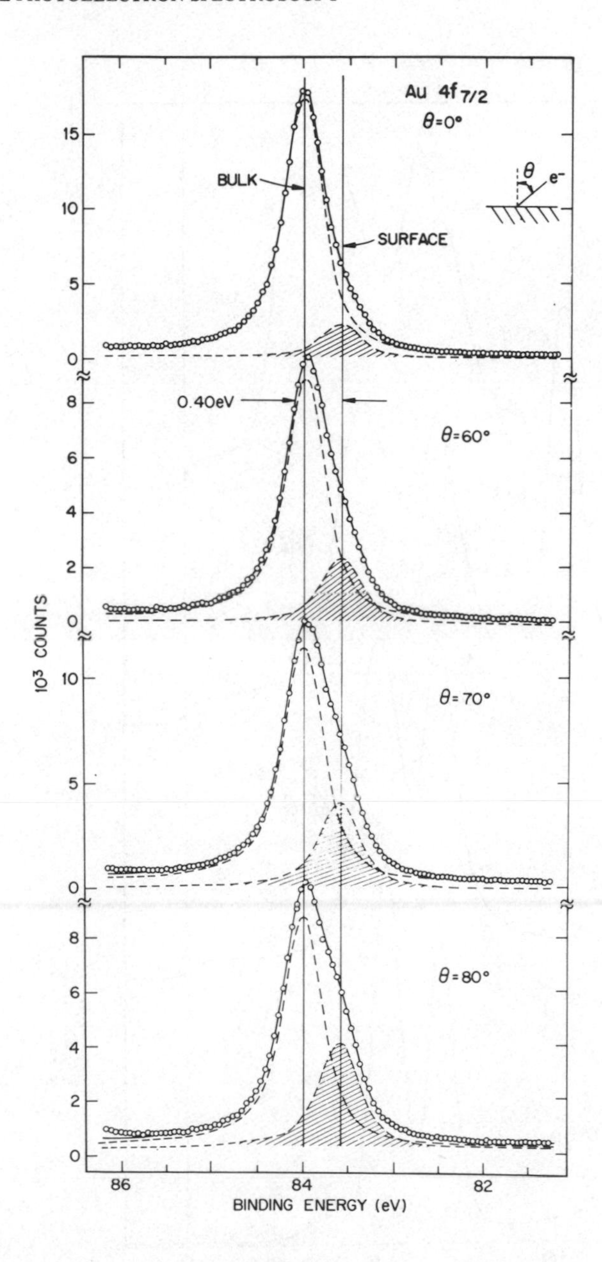

Figure 2 Take-off angle dependent data for the $4f_{7/2}$ line of a clean gold surface. The least squares fits show the bulk component at 84.0 eV and the surface component at 83.6 eV.

Valence band data for some of these compounds, Fig. 3, illustrate the narrowing of the Au 5d band as we proceed from metallic fcc gold to ionic CsAu where it

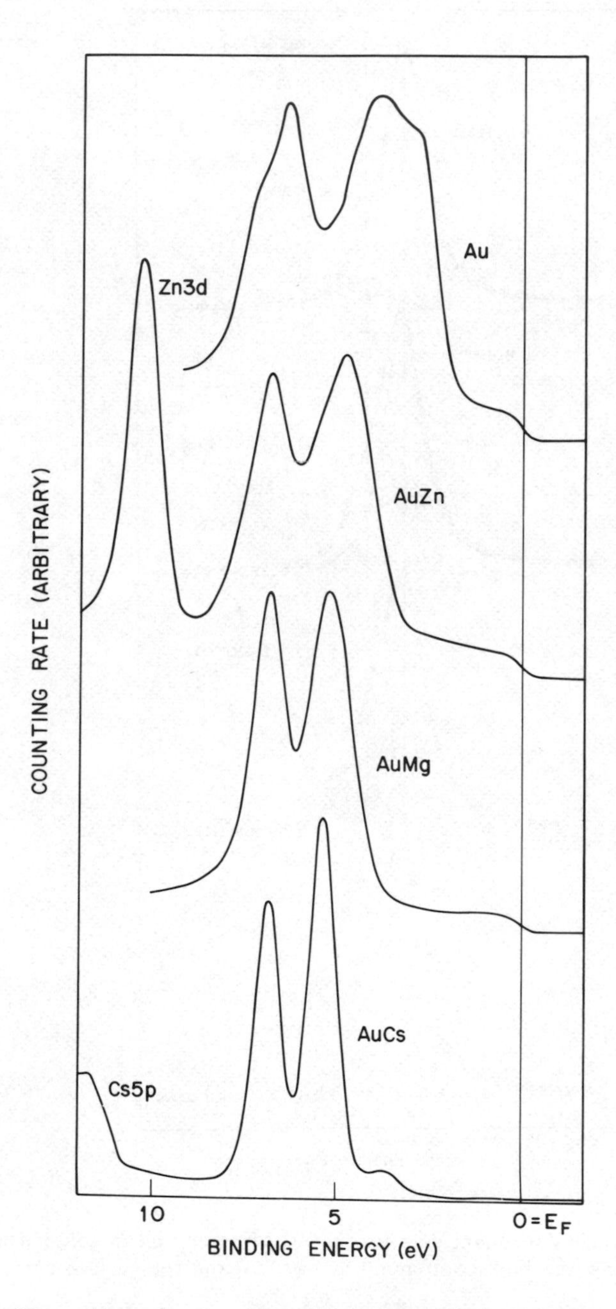

Figure 3 The valence bands of metallic gold and three CsCl-structure gold compounds. With increasing ionic character the gold 5d band assumes the character of a spin-orbit doublet with the free ion splitting of 1.5 eV.

approaches the character of the free-ion $5d^{10}$ spin orbit doublet. According to the analysis of Ref. 10 the redistribution of charge between gold 6s, 6p and 5d states has an effect on the binding energy comparable to that due to charge flow to the gold from the other element in the compound. The limiting case of CsAu is considered in detail below.

As a final example consider the graphite intercalation compound LiC_6. In this case (11) the Fermi level in the compound is known because the band structure of the two-dimensional graphite layers is largely retained when the compound is formed. Moreover the shift of E_F in LiC_6 relative to graphite, obtained from valence band spectra and from a band structure calculation, are in good agreement. Core electron binding energy measurements for C ls and Li ls show that both binding energies increase when the Fermi level is used as reference. Correction to vacuum, see Fig. 4, shows that the Fermi level shifts is actually larger than the core electron shift. The net result is that the C ls binding energy decreases while that of Li ls increases, as we would expect for Li-to-C electron donation. Quantitatively the size of shifts are also in line with the 6:1 ratio based on the composition and the estimated coulomb integrals.

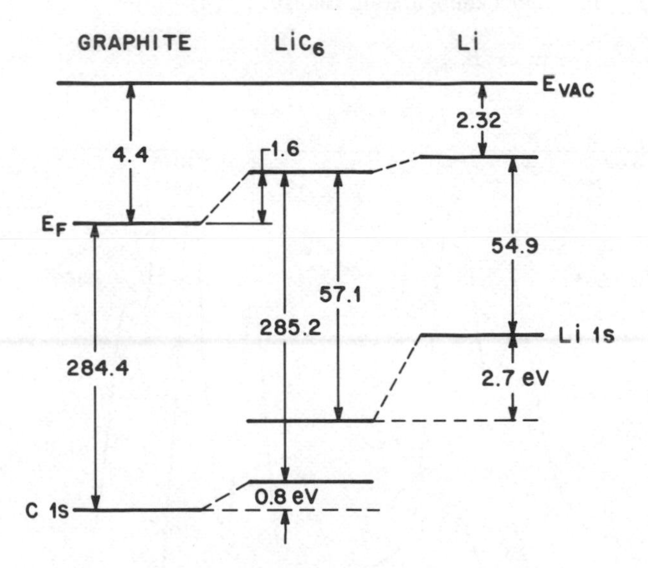

Figure 4 Energy level diagram for graphite, LiC_6 and Li showing shifts of the Fermi level and Cls and Lils core levels relative to the vacuum level.

B. Ionic Compounds

Core electron binding energies in ionic compounds can be derived from atomic binding energies by the following stratagem. First calculate the change produced by converting a free atom into a free ion. The shift is approximately equal to the potential inside a spherical shell with radius equal to that of the valence orbitals occupied by the ionic charge, i.e. ~1Ry. Then assemble a set of positive and negative ions into a crystal. Each ionic site now experiences an additional potential due to all the other ions, the Madelung potential. It will largely cancel the single ion potential, leaving a net shift of a few eV. Note, that since the Madelung potential depends on the lattice constant, the binding energy will be temperature dependent (12). On photoionization final state effects again come into play. Most important is the polarization of the neighboring ions which serves to screen the core hole. In addition there is the possibility of exciting lattice vibrations, i.e. emitting phonons, because a coulombic force appears between the photoionized atom and its neighbor when the photoelectron departs. These final state effects must be considered if meaningful information is to be extracted from experiment.

As an example consider the case of CsAu, an ionic insulator with 2.5 eV bandgap, Fig. 5. The data (13) show that the Fermi level in the compound is pinned to the bottom of the conduction band so that we can use the electron affinity to establish the zero of energy. In order to extract the charge transfer from the comparison between metallic Au and CsAu we must refer the Au binding energies in both substances back to atomic gold. For metallic gold the major corrections are due to electronic screening, E_{scr}, and a configuration change, E_{conf},

$$E_{metal} = E_{atom} - E_{scr} - E_{conf}.$$

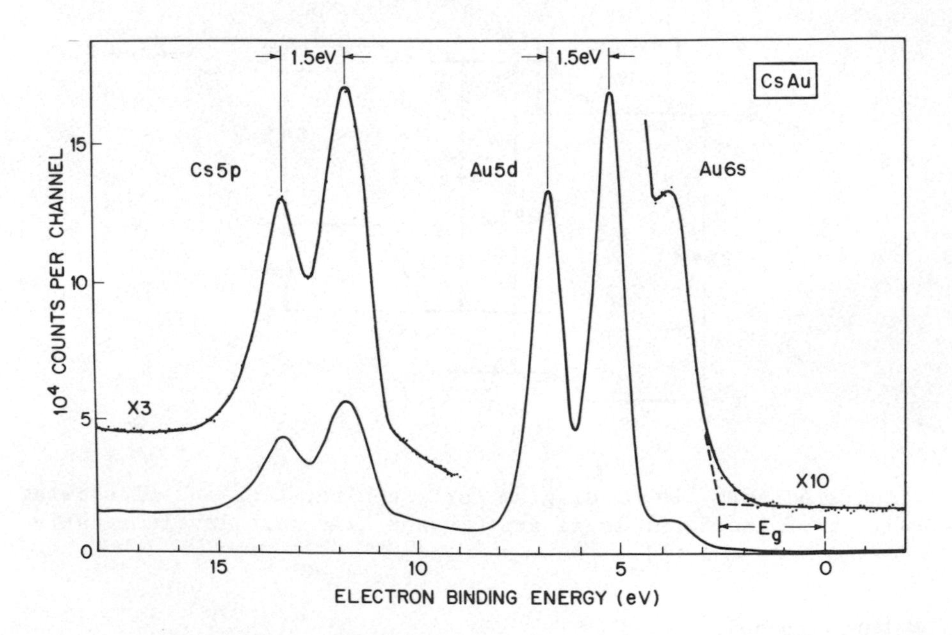

Figure 5 The extended valence band spectrum of CsAu. Note that the compound is an insulator with a 2.5 eV band gap. The Fermi level is pinned to the bottom of the conduction band.

For CsAu the dominant effects are due to the ionic nature of the compound, i.e. we must consider the effect of the ionic charge, $F_o\delta z$, and the Madelung potential, $M\delta z$, as well as due to the polarization of the neighboring Cs ions,

$$E_{CsAu} = E_{atom} + (F_o - M)\delta z - E_{pol}.$$

Eliminating E_{atom} and solving for δz yields

$$\delta z = \frac{E_{CsAu} - E_{Au} - E_{scr} - E_{conf} + E_{pol}}{F_o - M}$$

In this case the three correction terms are readily estimated. E_{scr} (2.7 eV) is based on measurements on atomic and metallic mercury (14) in which E_{conf} is negligible. E_{conf} (-1.6 eV) is estimated on the basis of results (5) for the similar metal copper. E_{pol} (1.9 eV) is taken from work (15) on CsI. F_o (9.38) is approximated by scaling the calculated value for the neutral gold atom according to the ionic radius. M (6.88) is calculated for the known lattice constant of CsAu. The value of $E_{CsAu} - E_{Au}$ (-2.3 eV) is obtained from the measurements as illustrated in Fig. 6, making corrections for the work function of Au (5.1 eV) and the electron affinity of CsAu (1.5 eV).

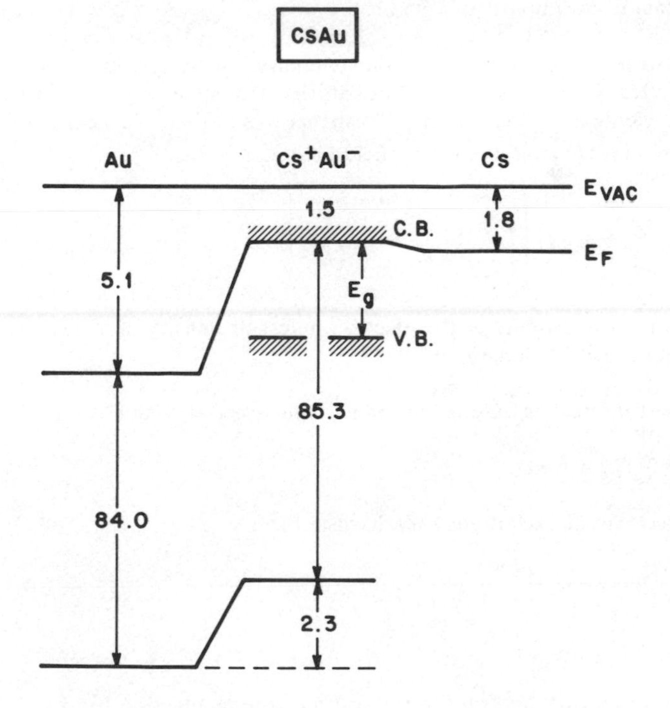

Figure 6 Energy level diagram for Au, CsAu and Cs. Note that although the Au 4f binding energy measured from E_F increases from Au to CsAu, correction to E_{vac} leads to 2.3 eV decrease.

The salient observation is that all the terms that enter into the numerator are of comparable magnitude. The poorly known correction terms are as large as the measured shifts. As a result the reliability of calculated charge transfer (0.6 electrons to the gold) is not high.

The major purpose of this exercise is to delineate the physical processes that define a binding energy, and to emphasize that the difference in binding energies measured relative to the Fermi energy by itself conveys no information about charge transfer.

IV. MANY-ELECTRON EFFECTS

In the interpretation of core electron binding energies, the fact that conduction electron screening reduces the energy of the core hole state in metals is central to the analysis. However, just as in the case of intra-atomic relaxation in which relaxation does not necessarily lead to the ground state, so here as well it becomes necessary to inquire into the spectrum of final states. A major difficulty is that the required final states are states of the atom with a core hole in a solid which has itself responded to the formation of the core hole. A proper treatment must incorporate the fact that the hole-state atom is an "impurity" in its host.

We will for the moment ignore this complication and assume that we know the spectrum of electron-hole pair excitations in the final state as well as the corresponding matrix element. The spectral response can then be calculated using a generalization of a formula due to Hopfield (16)

$$I(\omega) = \int_{-\infty}^{\infty} e^{i\omega t} dt \, \exp\left[\int_{0}^{\omega_c} \frac{g(\omega')}{\omega'^2} \left(e^{-\omega' t} - 1\right) d\omega\right]$$

in which $g(\omega)$ is the product of the electron-hole pair density of states and the square of the excitation matrix element.

The most interesting case is that of a simple metal in which

$$g(\omega) = N_o^2 V_o^2 \omega = \alpha\omega$$

The spectrum of excitations then has the form

$$i(\omega) = \left[\xi/\hbar(\omega - \omega_o)\right]^{1-\alpha}$$

$$\omega \geq \omega_o$$

where ξ is an energy of the order of the width of the conduction band.

The constant, α, is also expressible (17) in terms of the Friedel phase shift, δ, obtained in the discussion of the screening of an impurity atom in a metal;

$$\alpha = \sum_{l=0}^{\infty} 2(2\ell+1)(\delta_\ell/\pi)^2,$$

The connection is clear since the core-ionized atom is, in effect, an impurity with unit charge (18).

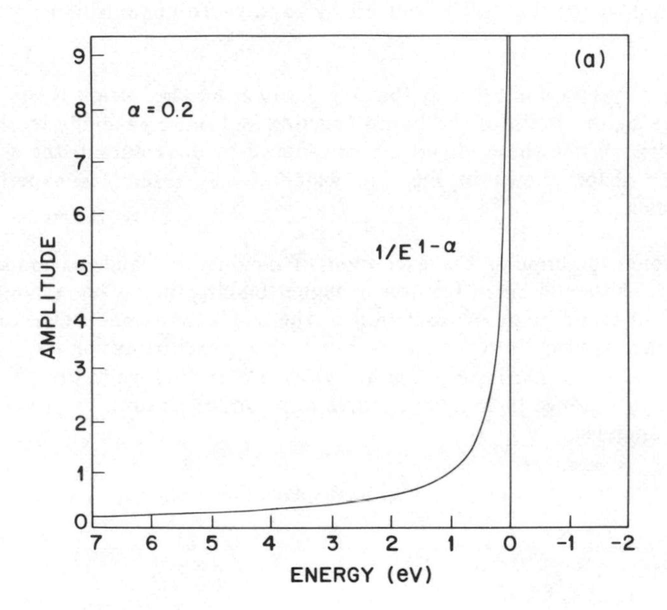

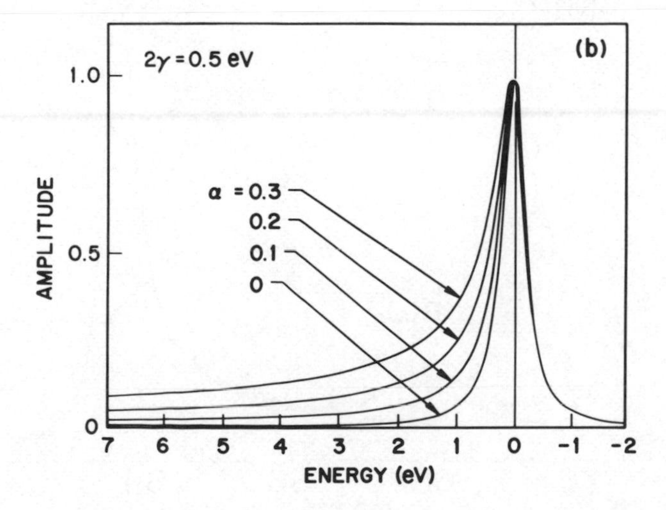

Figure 7 Theoretical core electron line shape in a simple metal. (a) the many-body component reflecting the effect of electron-hole pair excitations at E_F. (b) the result of adding the Lorentzian lifetime width by convolution.

The line shape I(ω), illustrated in Fig. 7a, is singular but integrable at $\hbar\omega_o$. This is the energy of the optimally screened hole state, screened by an infinity of infinitesimal electron excitations. The power law tail to greater binding energy contains finite excitations and is closely related to the satellites in the core electron spectrum of insulators. This part of the theoretical spectrum is not integrable to infinity, but in any real system the line will be cut off by band width or transition probability considerations.

Experimental spectra don't display the singularity at $\hbar\omega_o$ because it is rounded at finite temperature by the width of the Fermi function and more generally by the core hole lifetime width. When these effects are introduced by convolution, the resulting spectrum has the shape shown in Fig. 7b, which closely resembles experimental lineshapes in metals.

For a complete description the excitation of conduction band plasmons must also be included. These add broad features at higher binding energy corresponding to energy lost to a collective mode of excitation of the conduction band. One can conceptually distinguish between intrinsic plasmons which are excited during the photoemission process and extrinsic plasmons which are excited while the photoelectron moves through the metal. In addition bulk and surface plasmons can usually be seen at different energies.

The 2s spectrum of metallic Na, Fig. 8, displays these phenomena quite clearly.

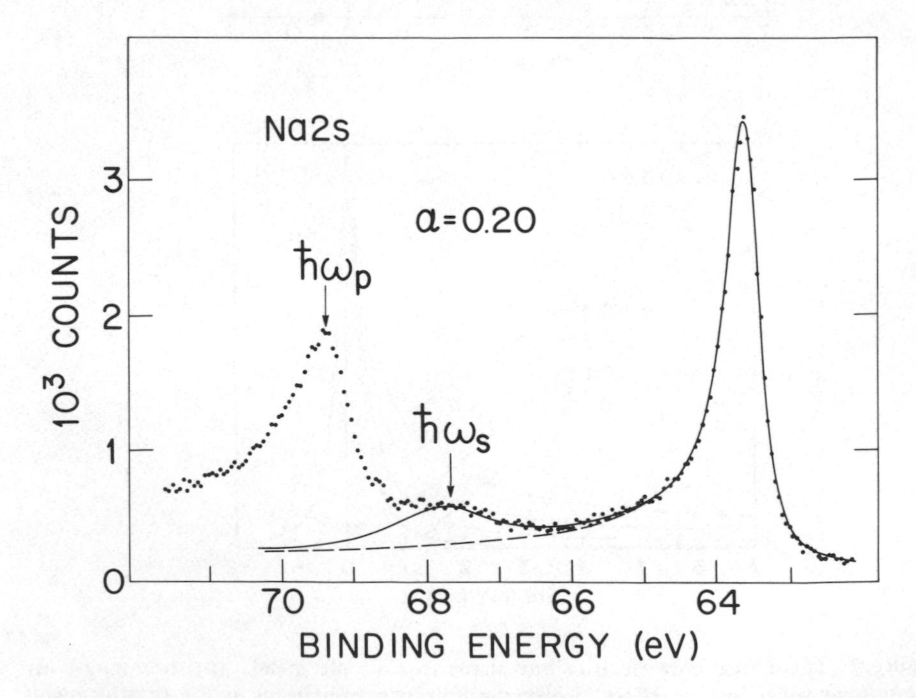

Figure 8 The 2s electron spectrum of metallic sodium. Compare the line shape with Fig. 7b. The bulk and surface plasmon losses are apparent.

The solid line through the data points is the result of a fit incorporating all the phenomena described in this section except for the bulk plasmon. It yields the singularity index α (0.20), the core hole lifetime width (0.28 eV), as well as the binding and plasmon energies. The core electron spectra of the simple metals Na, Mg, and Al have been analyzed in this way (19) and shown to be compatible with the theoretical description of the many-body phenomenon. Core hole lifetime widths have also been found (19) in good agreement with theory. In the case of Li, phonon broadening was found to be the dominant source of line width, (20) but it is a unique case predicated on the small atomic mass of Li.

The formalism used to obtain the line shape $I(\omega)$ can be extended to treat transition metals with highly structured conduction bands (21). It can also be used to gain some understanding of the line shape in semimetals (22) and semiconductors in which the singular part may vanish altogether (23).

SUMMARY

In this discussion of core electron spectroscopy we have focused on the most basic property, the binding energy. We have seen that measured binding energies contain initial state and final state components of intra-atomic and extra-atomic origin. The discussion then turned to a more detailed treatment of the outer shell excitations which accompany core electron ionization. The case of the simple metal where a closed form solution exists was examined in greater detail. Discussion of other aspects of core-electrons spectroscopy, e.g. multiplet effects and other satellites were given at an earlier NATO Advanced Study Institute (4).

Bibliography

A. Electron Spectroscopy, Theory, Techniques and Applications, Vol. 1-3, edited by C. R. Brundle and A. D. Baker (Academic Press, New York, 1977, 1978, 1980).

B. Photoemission in Solids I and II, (issued as Topics in Applied Physics Vol. 26 and 27) edited by M. Cardona and L. Ley, (Springer, Berlin, 1978 and 1979).

Notes and References

1. K. Siegbahn, et al., Nova. Acta Regiae. Soc. Sci. Ups. Ser IV. Vol. 20 (Uppsala, 1967).

2. Experimentally determined binding energies in solids are usually referred to the Fermi level. Interpretation requires correction of the vacuum level.

3. J.-N. Chazalviel, M. Campagna, G. W. Wertheim, P. H. Schmidt, and Y. Yafet, Phys. Rev. Lett. 37, 919 (1976).

4. G. K. Wertheim in "Electron and Ion Spectroscopy of Solids" edited by L. Fiermans, J. Vennik, and W. Dekeyser (Plenum, New York, 1978) p. 192.

5. A. R. Williams and N. D. Lang, Phys. Rev. Lett. 40, 954 (1978).

6. R. E. Watson, M. L. Perlman, and J. F. Herbst, Phys. Rev. B13, 2358 (1976).

7. P. H. Citrin, G. K. Wertheim, and Y. Baer, Phys. Rev. Lett. 41, 1425 (1978)

8. T. M. Duc, C. Guillot, Y. Lassailly, J. Lecante, Y. Jugnet and J. C. Vedrine, Phys. Rev. Lett. 43, 789 (1979).

9. J. F. van der Veen, F. J. Himpsel and D. E. Eastman, Phys. Rev. Lett. 44, 189 (1980).

10. G. K. Wertheim, R. L. Cohen, G. Crecelius, K. W. West, and J. H. Wernick, Phys. Rev. B20, 860 (1979).

11. G. K. Wertheim, P. M. Th. M. van Attekum, Solid State Comm. 33, 1127 (1980).

12. M. A. Butler, G. K. Wertheim, D. L. Rousseau and S. Hüfner, Chem. Phys. Lett. 13, 473 (1972).

13. G. K. Wertheim, C. W. Bates, Jr., and D. N. E. Buchanan, Solid State Commun. 30, 473 (1979).

14. S. Svensson, N. Martinson, E. Basilier, P. A. Malmqvist, U. Gelius and K. Siegbahn, J. Electron Spectrosc. 9, 51 (1976).

15. P. H. Citrin and T. D. Thomas, J. Chem. Phys. 57, 4446 (1972).

16. J. J. Hopfield, Comm. Solid State Phys. 2, 40 (1969).

17. P. Nozieres and C. T. DeDominicis, Phys. Rev. 178, 1097 (1969).

18. For more details see G. K. Wertheim and P. H. Citrin in B. vol. I p. 197.

19. P. H. Citrin, G. K. Wertheim, and Y. Baer, Phys. Rev. B. 16, 4256 (1977).

20. Y. Baer, P. H. Citrin, and G. K. Wertheim, Phys. Rev. Lett. 37, 49 (1976).

21. G. K. Wertheim, and L. R. Walker, J. Phys. F. 6, 2297 (1976).

22. P. M. Th. M. van Attekum and G. K. Wertheim, Phys. Rev. Lett 43, 1896 (1979).

23. G. K. Wertheim and D. N. E. Buchanan, Phys. Rev. B. 16, 2613 (1977).

ANGLE-RESOLVED PHOTOEMISSION FROM ADSORBATE COVERED SURFACES

N.V. Richardson

Donnan Laboratories, University of Liverpool, U.K.

Abstract
This article constitutes a review of angle-resolved photoemission from adsorbate covered surfaces. The use of photoemission as a diffraction technique for studying surface crystallography is discussed. The selection rules which apply in photoemission are introduced and their importance for interpretation of spectra from molecular adsorbates is considered. Spectra which demonstrate a dominant adatom lateral interaction are contrasted with others in which adatom substrate interactions dominate.

1. SURFACE CRYSTALLOGRAPHY AND PHOTOELECTRON DIFFRACTION

1.1. Introduction

We have seen, in Professor King's lecture, how UV-PE and X-ray PE spectroscopy of surfaces are able even in angle-integrated form, to aid the identification of surface species, to follow the progress of gas-solid interactions and to begin a characterisation of the adsorbate-substrate bond. Angle-resolved photoemission measurements are yet more powerful in unravelling the details of interactions at the solid-vacuum interface. An early expectation for angle-resolved photoemission from adsorbate covered surfaces was that adsorbate site and adsorbate-substrate bond lengths might be determined. Gadzuk [1], suggested that measurement of the intensity of an adsorbate induced band, as a function of the azimuthal angle, ϕ, would reveal a lobed structure readily related to the charge distribution in the initial state and hence the site geometry. Gadzuk assumed that single plane-wave final states were adequate in a description of the photoemission process,[2]. The measured distribution is, there-

P. Day (ed.), Emission and Scattering Techniques, 75–102.

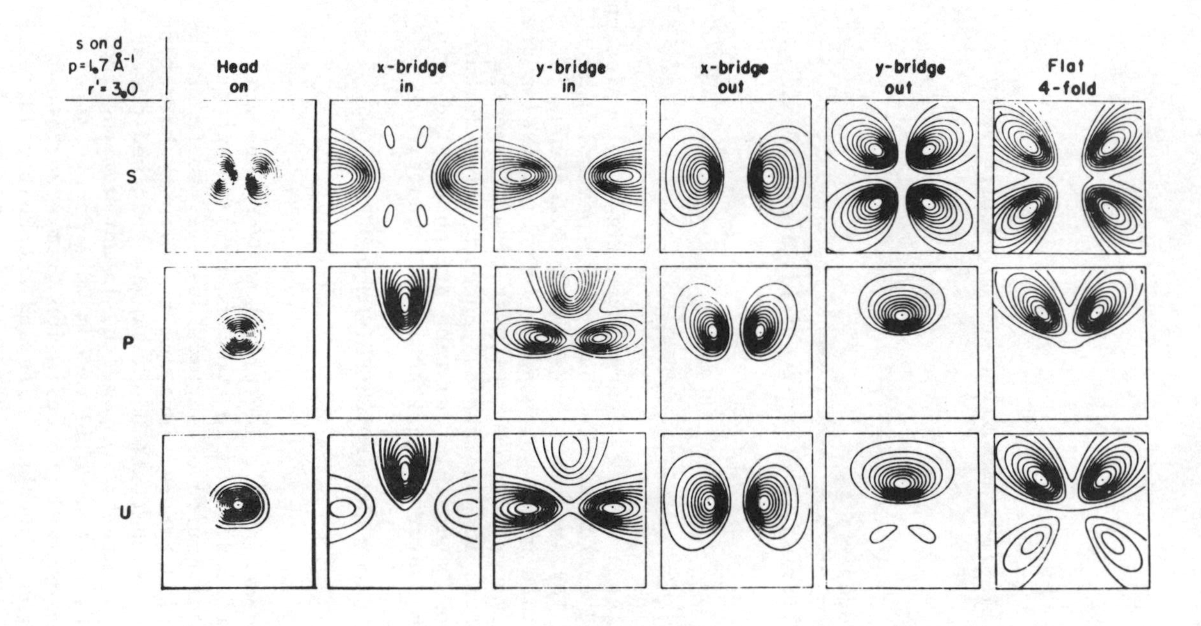

Figure 1. Curves of constant emission intensity for an s atom on the (001) face of a d-band substrate. Photons are incident at 45° in the yz-plane and either s-polarised (S), p-polarised (P) or unpolarised (U). The adsorption geometry is noted above where in(out) refer to bridge bonding sites in (out of) the plane of the substrate.

fore, simply the initial state distribution convoluted with a term $\vec{A}.\vec{k}$ relating the emission direction, via the momentum vector, k, of the emitted electron to the electric vector, $\vec{A}$, of the exciting radiation. Fig. 1 compares the distributions to be expected from an s-orbital adatom bonded to the (001) face of a d-band metal. The specific case is hydrogen on tungsten. Liebsch, [3], also with the aim of determining adsorbate site, showed that final interference effects could give rise to similar azimuthal patterns. The difference in the two mechanisms is schematically shown in Fig. 2.

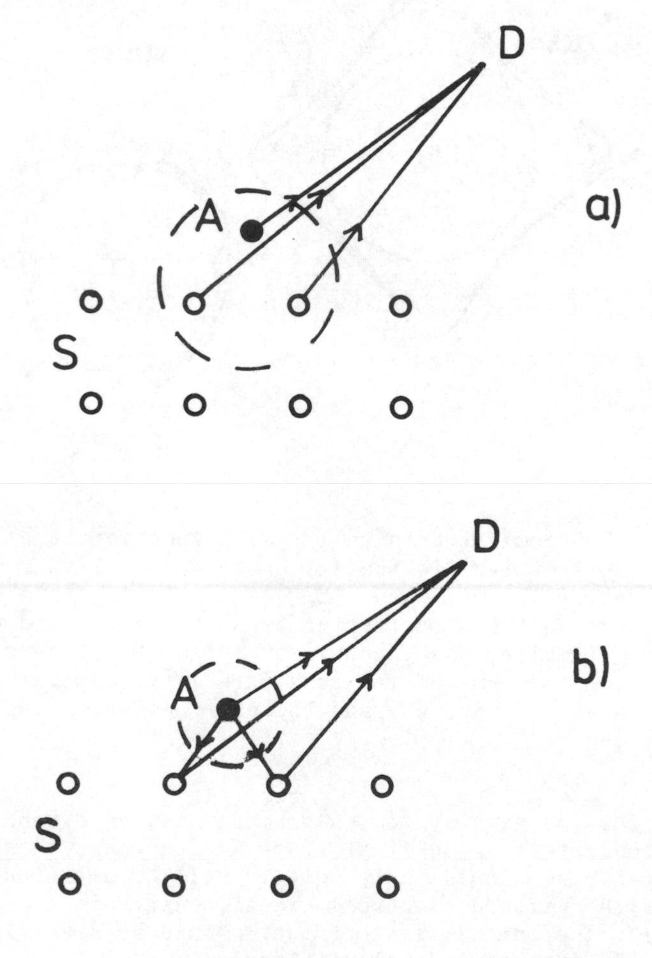

Figure 2. A schematic representation of photoemission illustrating in (a) initial state effects and in (b) final state interference effects. The adatom is labelled A, substrate atoms, S, and the detector D.

In the former case, emission is coherent from the different centres. In the latter, electrons from a single centre scatter from the substrate and interference patterns are generated, even from an initially spherical charge distribution, which reflect the adsorbate site. Fig. 3 shows the intensity variation

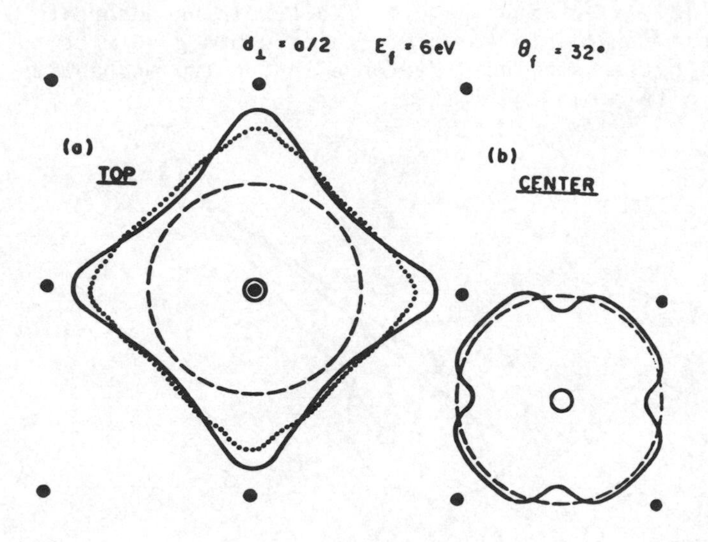

Figure 3. Photoemission intensity as a function of azimuthal angle for an s-orbital in the (a) on top and (b) four-fold hollow sites of a (001) face. Single scattering is shown by solid curves, multiple scattering by dotted curve and no scattering by the circular, dashed curves. The vertical spacing, d_1, of the adatom in terms of the substrate lattice spacing a, the final state energy, E_f, and the polar emission angle, θ_f, are given above.

expected for the same system, s-orbital atom on d-band substrate, but now concentrating on final state scattering events. The latter mechanism should apply equally well to emission from core levels as the emitted electrons kinetic energy is the primary parameter. The initial state effect should be significant only for strongly interacting valence levels.

1.2. Experimental Azimuthal Dependence

Disappointingly, little experimental evidence was obtained, at this time, for such azimuthal dependencies even though strongly lobed structures from metal chalcogenides, such as TaS_2 [4,5], provided much of the impetus for all angle-resolved photoemission studies. More recently, with the wider availability of synchrotron radiation, enabling photon energies to the tuned to maximise these effects, such "floral patterns" have been observed in emission from the core-levels of adsorbates [6-9]. With the aid of LEED-type multiple scattering programs, the results have been interpreted and shown to require a proper treatment of both initial and final state effects. Happily, site geometries are indeed reflected.

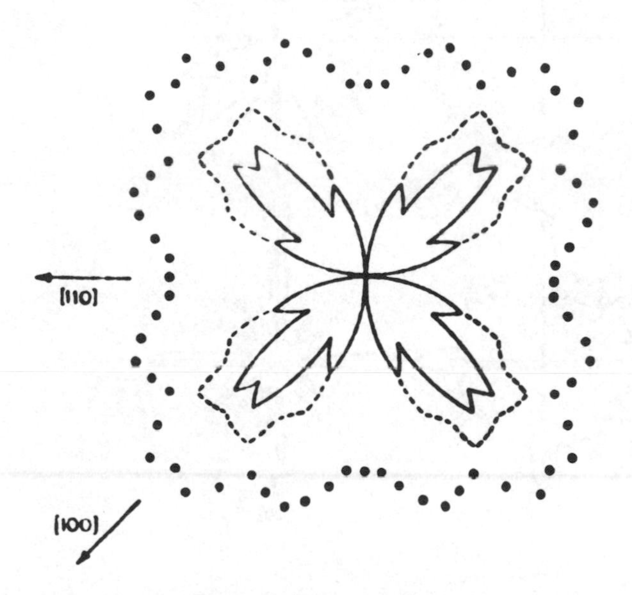

Figure 4. Comparison between theory and experiment for the azimuthal dependence of emission from the Na 2p levels of c(2x2) Na-Ni(001) at $h\nu = 80$ eV and $\theta = 30^{o}$. The full circles are calculated intensities. The full curve is the same data with minimum value subtracted and the dashed curve is the corresponding mirror-symmetrized data.

Fig. 4 shows the emission from Na-Ni(001) obtained with 80 eV p-polarised photons incident at 45° to the surface normal, [6]. Emission, at 30° to the normal, is from the sodium 2p orbitals corresponding to a final electron kinetic energy of 46 eV. The adsorption site is concluded to be the four-fold hollow site. Cs-W(001) [7], and Te-Ni(001) [8], have been similarly studied.

At higher emitted electron kinetic energies, the final state scattering is expected to be dominated by single scattering events, see Fig. 3, making interpretation somewhat simple. Fig. 5 shows the azimuthal variation of intensity obtained from O-Cu(001) with Al K_α radiation [10], and so the final state energies are now in excess of 1000 eV. Again a four-fold hollow site occupancy is concluded.

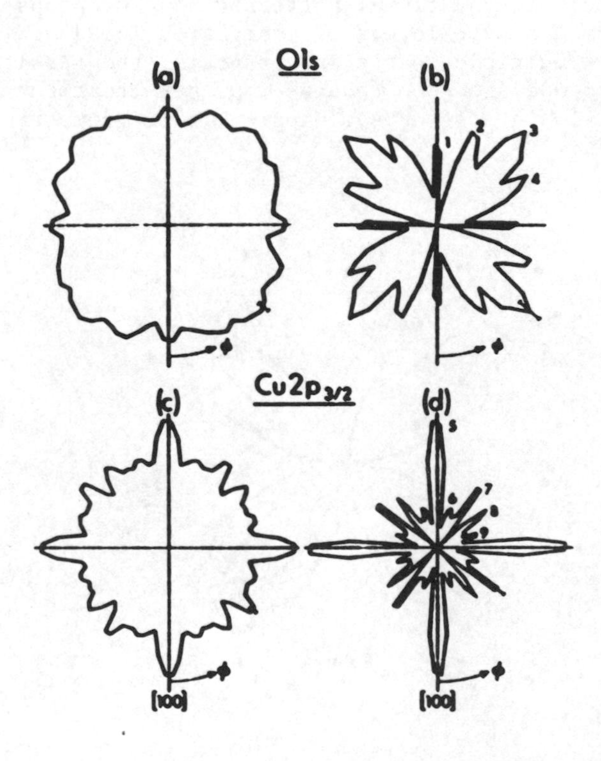

Figure 5. Azimuthal X-ray photoemission intensities for the O_{1s} and $Cu_{2p\ 3/2}$ levels of a Cu(001) surface exposed to 1200 L of O_2. The photon energy was 1486.6 eV and the polar emission angle 10°. The data has been four-fold averaged and in (b) and (d) the minimum value subtracted.

Very recently, analysis of experimental data has proceeded further to the determination of adsorbate-substrate bond lengths [11]. The azimuthal dependence of iodine 4d emission from $(\sqrt{3}x\sqrt{3})R30^o$ I-Ag(III) indicates that those three-fold hollow sites directly above a vacancy are occupied and that the I-Ag layer spacing is 2.4 Å ± 0.1 Å in good agreement with earlier LEED [12], and

SEXAFS [13], results. Calculations indicated that quite different patterns would result from on-top sites, the other three-fold sites or for different inter-layer spacings. Measurements were made over the photon energy range 90-110 eV (emitted electron energy 35-55 eV) and the patterns were found to be sensitive to this parameter.

The adsorption of ammonia on Ir(III) at 190 K has received a similar study [14]. This time however a valence level, likely to be involved in the bonding, was chosen. Strong modulation of emission from the 1e (N-H bonding) orbital with azimuthal angle, shown in Fig. 6, was taken to be indicative of adsorption in a three-fold hollow site above a vacancy and of N-H bonds projecting between neighbouring iridium atoms. This is the clearest example to date of such effects from molecular adsorbates and once again a pronounced photon energy dependence is apparent.

1.3. Photoionization Cross-section Effects

The photon energy dependence, in fact, provides an alternative means of determining adsorbate-substrate spacing from photoemission data. Structure in normal emission intensity versus photon energy curves, for adsorbate derived levels, may arise through final state scattering by the substrate atoms, [15-17]. A tuneable photon source is required for such experiments.

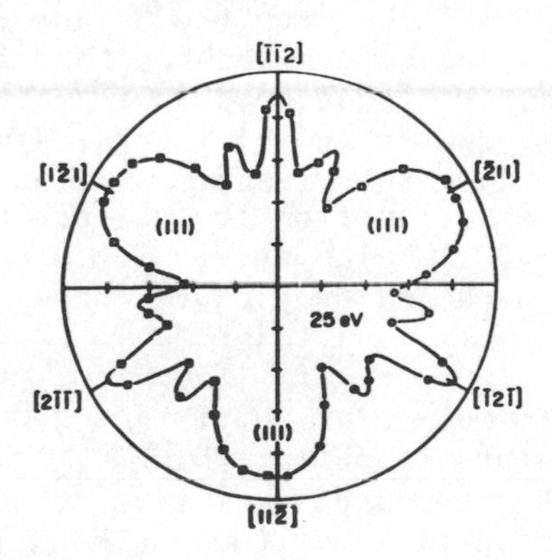

Figure 6. Azimuthal plot of photoemission from the 1e orbitals of NH_3 adsorbed on Ir(III). Crystallographic directions and the projections of the (111) vectors are shown.

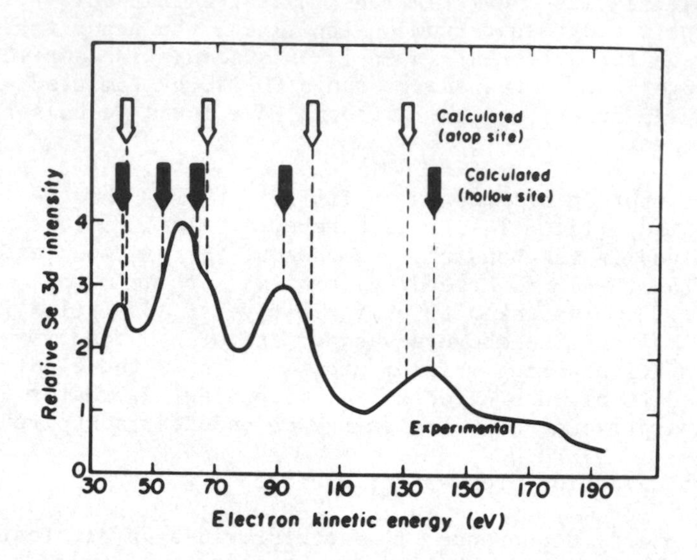

Figure 7. Plot of the relative Se 3d intensity versus electron kinetic energy for the p(2x2) Se-Ni(001) system and the calculated peak energies for the atop and hollow sites.

Fig. 7 shows some experimental data for Se 3d emission obtained by Kevan et al, [18], for p(2x2) Se-Ni(001) and the calculated curve assuming a spacing of 1.55 Å between the Se and uppermost Ni layers. Li and Tong, [19], have demonstrated that for adatom orbitals which contain nodes, atomic effects dominate the curve, whereas for emission from 1s, 2p, 3d and 4f initial states substrate scattering events are dominant. Figure 8 illustrates this effect by comparison of the emission from the valence p orbitals of chalcogens adsorbed on Ni(001). The curves for S and Se are dominated by a large atomic peak and only oxygen shows a strong substrate scattering effect.

A single application to molecular adsorption systems has been described in the literature to date, [20]. C_{1s} emission intensity as a function of $h\nu$ should be sensitive to the Ni-C separation but relatively insensitive to the C-O bond length in the c(2x2) CO-Ni(001) system. Subsequent investigation of the behaviour of the O_{1s} emission should allow the C-O spacing to be determined. The method is therefore more selective than LEED and may gain a wider appeal with the increasing availability of synchrotron sources.

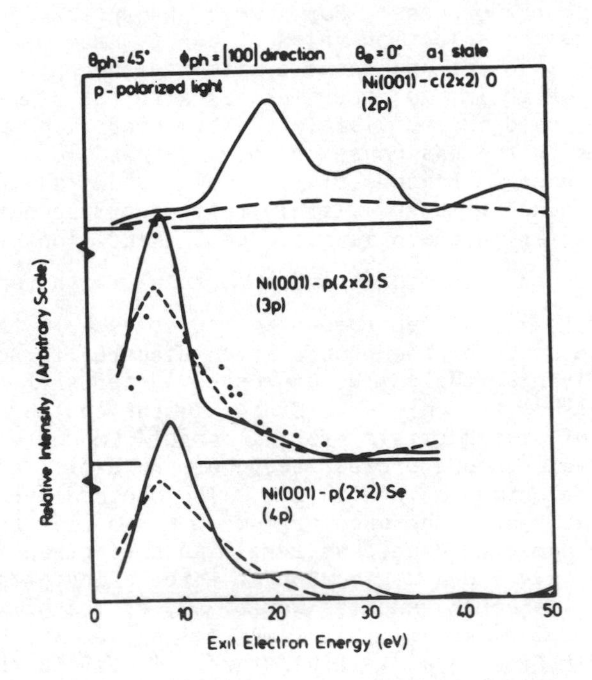

Figure 8. Emission spectra from the valence levels of p(2x2) chalcogen-Ni(001) structures as a function of final state energy. Solid curves are emission from ordered overlayers, broken curves are the atomic emission and dots are the experimental data.

2. SELECTION RULES IN PHOTOEMISSION

2.1. Introduction

As an alternative to photoelectron diffraction studies, one can, by judicious use of the selection rules governing photoemission, gain information about adsorption site and about the preferred orientation of molecular species at surfaces. An understanding of these symmetry-based selection rules makes it possible to interpret adsorbate photoemission spectra in greater detail and thereby lays the foundation for a discussion of the electronic structure of adsorbate covered surfaces. The ideas have been covered in several recent papers and review articles [21-26], but we should digress a little to examine these selection rules here.

Photons possess a single unit of angular momentum and their destruction by absorption requires that the absorbing system takes up this momentum in order that the total angular momentum be conserved during the process. For atomic absorption spectroscopy this leads to the selection rules $\Delta\ell = \pm 1$, $\Delta m = 0, \pm 1$ such that a sodium 3s electron may be excited to the 3p, 4p, 5p etc orbitals but not to 3d, 4d etc. This is also the basis of the symmetry rules used in the visible and UV absorption spectroscopy of molecules in the gas-phase or molecules and ions in the liquid and solid phases. Benzene, for example, shows a single strong absorption band in the UV-visible region corresponding to the $^1A_{1g} \rightarrow {}^1E_{1u}$ transition which is alllowed. Transitions to the $^1B_{1u}$ and $^1B_{2u}$ states are forbidden and appear only with very much reduced intensity, [27]. In photoelectron spectroscopy, the final one-electron state of the photoemitted electron is not a bound state but may nevertheless be expressed in terms of the symmetry properties of the emitter. Since, in the continuum, a non-zero density of states exists, for all angular momentum states, at all final state energies, there are no exclusions on emission from any electronic level, [28]. The photoemitted electron may always carry away the angular momentum required to hold the total angular momentum change at zero. An s electron would be emitted as a p-wave and a d electron as interfering p and f waves. From a C_{2v} molecule, such as water, a_1, b_1 and b_2 final states would result from an a_1 orbital whereas a_2, b_2 and b_1 states would result from an a_2 initial state. Though there are no exclusions, the selection rules together with the appropriate densities of states do determine the intensity of photoemission bands.

There is an extra dimension to photoelectron spectroscopy, compared with absorption spectroscopies, in the direction of electron emission and this is also controlled by the selection rules. The random orientation of atoms and molecules in the gas-phase averages out much of the intrinsic angular dependence reducing the anisotropy of emission to description by a single parameter, usually denoted β. This parameter carries information on the final states which are populated and how the different outgoing waves interfere.

For atoms and molecules adsorbed on surfaces, we can expect much more dramatic effects if only because their orientation is fixed relative to the photon beam direction and certain spatial integrations are no longer required. It is worth noting at this point that $\Delta m = 0(\pm 1)$ transitions are by photons polarised parallel (perpendicular) to the z-axis of the system. The surface normal provides the obvious choice for this axis and x and y are usually chosen to coincide with principal directions of the substrate in the surface plane or such that the photon beam defines the xz-plane.

2.2. Photoemission Intensity

The intensity of a photoemission band depends, from the golden rule, [30-34], on the square of a matrix element, M_{if}, given by

$$M_{if} = \vec{A}.<f/\vec{\mu}/i> \qquad (1)$$

where /f> and /i> are, for our purposes the one electron wave functions appropriate to the final and initial states respectively. $\vec{\mu}$ is the dipole moment operator and $\vec{A}$ is the electric vector describing the photon field at the surface. Group theoretical methods may be applied to this problem. For the matrix element to be non-zero, the totally symmetric representation must be encompassed by the direct product

$$\Gamma_f \times \Gamma_\mu \times \Gamma_i$$

where Γ_r denotes the irreducible representation for which r provides a suitable basis set. μ transforms as the translations x, y and z, and, with care, point groups can be used for the analysis. The choice of point group is left until the discussion of specific examples. It is sufficient, at this stage, to say that it depends on the relative importance of

(1) Intra-adsorbate interactions (molecular systems).
(2) Adsorbate-adsorbate and adsorbate-substrate interactions in the ground state of the system.
(3) Scattering events in the final state.

Remembering that a single co-ordinate frame can be easily used to express the wave-functions and the electric vector of the exciting radiation, equation (1) can be expanded to

$$M_{if} = A_x<f/\mu_x/i> + A_y<f/\mu_y/i> + A_z<f/\mu_z/i> \qquad (2)$$

With z as the surface normal and xz as the incidence plane, it means that for s-polarised radiation (from a synchrotron, [35], or polariser, [36]) A_x and A_z are zero and equation (2) becomes simply

$$M^s_{if} = A_y<f/\mu_y/i> \qquad (3)$$

For p-polarised light, A_y is zero and M_{if} is given by

$$M^p_{if} = A_x<f/\mu_x/i> + A_z<f/\mu_z/i> \qquad (4)$$

Unpolarised photons should be considered a superposition of s- and p-polarised fields such that A_y is always decoupled from A_x and A_z and, hence, after squaring the matrix element to obtain intensities, terms A_xA_y and A_yA_z do not appear

$$I^u \alpha / M_{if}/^2 = /A_y^s \langle f/\mu_y/i\rangle/^2 + /A_x^p \langle f/\mu_x/i\rangle + A_z^p \langle f/\mu_z/i\rangle/^2 \qquad (5)$$

Note that an interference term between A_x^p and A_z^p will occur for both p-polarised and unpolarised photons. This has been used as the basis of a method for determining band symmetries with unpolarised radiation, [37-39], and must be considered in the choice of experimental geometry for exact determinations of adsorbate orientation [40]. An independent choice of photon incidence angle allows the relative magnitudes of A_x^p and A_z^p to be made and, of course, at $\alpha = 0$ (normal incidence) A_z^p is zero. The photon field at the surface arises as a sum of incident and reflected beams with due account to be taken of the surface reflectivity and the phase changes which occur on reflection, [41, 42]. Fig. 9 shows the variation of the components of the field strength squared as a function of photon incidence angle for 21.2 eV (HeI energy) photons incident on nickel, [41].

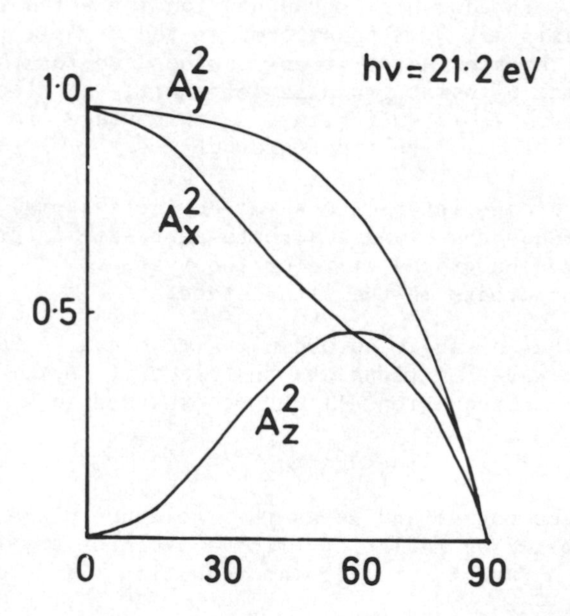

Figure 9. Macroscopic values of $\langle A_x\rangle^2$, $\langle A_y\rangle^2$ and $\langle A_z\rangle^2$ versus photon incidence angle for 21.2 eV light incident on nickel.

2.3. Influence of the Final State

In situations of high symmetry, the final states /f> which can be linked to /i> through the dipole operator, μ, may be very limited but it is the additional constraints placed by a detector of limited angular acceptance, enabling the final state to be more closely studied, which give the most profound effects. The above equations apply to band intensities when all emitted electrons are collected. In an angle-resolved experiment, intensity is further modulated by the angular behaviour of the final state, with allowance where necessary for interference effects between coherently emitted waves. A simple example might make the situation clearer. Normally incident plane-polarised (say x-polarised) light would excite electrons in the p_z orbital of an adatom into a d_{xz} final state ($\Delta\ell = +1$, $\Delta m = \pm 1$). The measured photoelectron intensity would therefore be expected to vary as $\cos^2\theta \sin^2\phi \cos^2\phi$ since this is the angular variation of charge density in a d_{xz} wave. Note that no emission would be expected along the surface normal, in the yz plane or along the surface. Using the converse of such arguments, it is possible, by measurement of the intensities of adsorbate derived bands as a function of photon polarisation, incidence angle and electron emission angles, to deduce the symmetry properties of the final state and, hence, by use of the selection rules, those of the initial state. In group theoretical terms, the important conclusions for angle-resolved photoemission are that

(1) at normal emission only final states belonging to the totally symmetric representation can contribute as for all others the surface normal lies on a nodal plane or is a nodal axis.

(2) for emission in a mirror plane of the system only symmetric (even) final states are observed. The mirror plane is a nodal plane for antisymmetric (odd) final states.

2.4. The Oriented Molecule Approximation

Having seen how symmetry effects have contributed to the understanding of photoemission from clean surfaces, we should now apply the ideas to photoemission from adsorbate covered surfaces. In this first selection, we deal with those systems, largely molecular, for which the adsorbate-adsorbate interactions are relatively weak as are the site-determining, adsorbate-substrate lateral interactions. To a first approximation the effect of the surface can be considered only to orientate the molecule and perturb the species only through an interaction in the z-direction (surface normal). The effective point group may then be that of the gas-phase molecule, as seems to be the case for carbon monoxide adsorbed on a Ni(001) surface.

Fig. 10 shows some spectra for this system under different conditions of photon incidence angle, polarisation and electron collection angle, [40]. Professor King, in his first lecture, discussed the relationship between the adsorbed CO spectra and that of the gas-phase species. We recall that the feature near

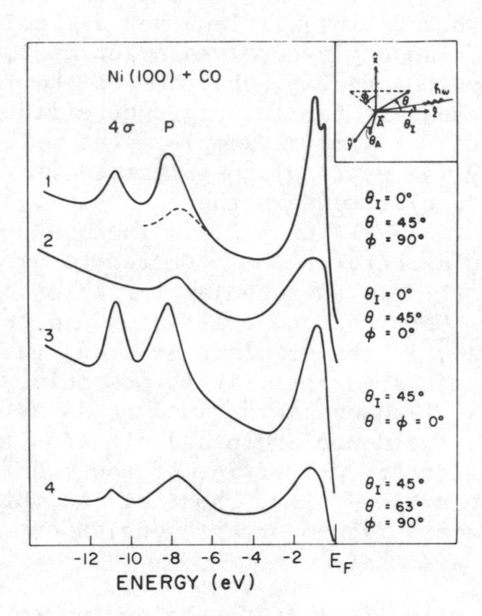

Figure 10. Photoemission spectra obtained from CO-Ni(001) at different experimental geometries.

8 eV is produced by an overlap of 1π and 5σ ionisations [43], and that at 11 eV corresponds to the 4σ ionisation. The full description of the 8 eV band, i.e. the exact ordering of 5σ and 1π ionisations had to await angle-resolved measurements [44-47], and indeed an early study of photoionisation cross-section as a function of photon energy [48], suggested an ordering of ionisation energies $5\alpha < 1\pi$ which later proved incorrect. Spectrum 2 of Fig. 10 shows a complete absence of the 11 eV feature and this can be rationalised only if the molecule is normal to the surface. Excitation of the 4σ level, symmetric with respect to <u>any</u> emission plane (say xz plane), by photons polarised perpendicular to this plane and, therefore, relying on the antisymmetric component of the dipole operator μ_y, results in an antisymmetric and therefore undetectable, final state. Since this argument applies equally well to the 5σ orbital, the observed adsorbate feature of Fig. 10.2 must correspond to the 1π level. This doubly degenerate level has symmetric and antisymmetric components and the latter may couple via μ_y to symmetric final states.

$$I_{1\pi} \alpha / A^{s}_{y} \langle f^{+} / \mu^{-}_{y} / 1\pi^{-}_{y} \rangle /^{2} \qquad (6)$$

It is not possible to select experimental conditions such that emission from the 1π level is completely suppressed but observation at normal emission (Fig. 10.3), with p-polarised light incident at 45°, produces an enhancement of the emission from levels of σ symmetry. In this arrangement only the totally symmetric, σ-type final states are observable and these are coupled to 4σ and 5σ initial states by the σ symmetry component of the dipole operator, μ_z. The intensity, therefore, depends on the normal component of the electric vector which is a maximum for $\alpha = 45^o$.

$$I_{4\sigma}^{\theta = 0^o} \quad \alpha/A_z^p \langle f^\sigma/\mu_z^\sigma/4\sigma\rangle/^2 \tag{7}$$

The 1π level may still appear because it can couple to final states of σ symmetry through the parallel component of the dipole operator, i.e. the A_x^p component of the light.

$$I_{1\pi}^{\theta = 0^o} \quad \alpha/A_x^p \langle f^\sigma/\mu_x^\pi/1\pi_x\rangle/^2 \tag{8}$$

Spectra 1 and 4 of Fig. 10 represent intermediate cases where both σ and π-type initial states may appear strongly. The conclusions are that CO is bonded with its molecular axis normal $1\pi < 5\sigma << 4\sigma$. Similar conclusions have been reached for CO/Ni-(100), [45], and CO/Pd(100), [44].

Measurement of the 4σ and 5σ photoionisation cross-sections for gas-phase CO molecules, as a function of photon energy, reveals a resonance when the final state energy is some 20 eV above the vacuum level, [49]. It has been interpreted as arising from a quasi-bound state of $\ell = 3$, $m = 0$ partial wave character (σ-symmetry) which is coupled to the σ-type initial states by photons polarised parallel to the molecular axis, [50]. Fig. 11 shows the intensity of the 4σ level of CO/Ni(001) as a function of photon energy under different experimental geometries [47]. The resonance is observed only if there is a component of the electric vector normal to the surface <u>and</u> when observation is close to normal. This strongly indicates a CO molecule normal to the surface. A similar surface resonance for molecular nitrogen has been similarly exploited by Horn et al [51], in a study of N_2/Ni(100) and they reach the same conclusion of a molecular axis normal to the surface.

A somewhat more involved example of the use and validity of the oriented molecule approximation is offered by benzene adsorption. Early angle-integrated measurements by Demuth and Eastman [52], correlated the HeI spectrum of C_6H_6/Ni(polycryst) at room temperature with that of the gas-phase benzene molecule. They showed that the molecule is undissociated and, from the differential shift of the uppermost π level (e_{1g}) relative to other levels, probably lying with the molecular plane parallel to the substrate surface. Angle-resolved studies of benzene on

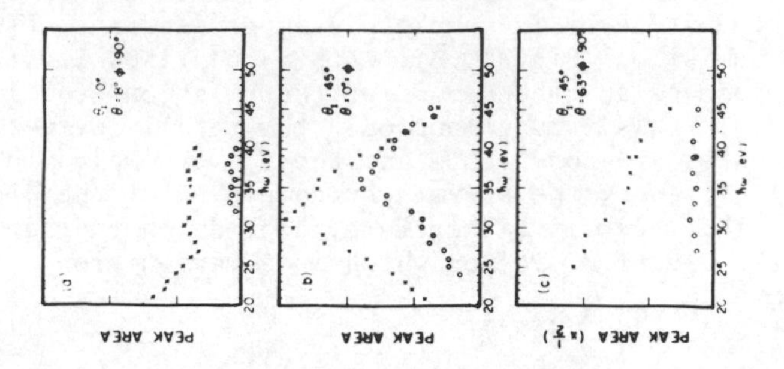

Figure 11. Peak intensities for 4σ level (O) and combined 5σ and 1π levels (X) as a function of photon energy for different experimental geometries. $\emptyset = 90^o$ corresponds to emission in the incidence plane.

Pd(001), Pd(111) and Ni(111) revealed marked, essentially substrate independent, changes in the band intensity as the photon incident angle or polar collection angle was varied, [53]. An isolated benzene molecule belongs to the D_{6h} point group but retention of such a high symmetry is not possible upon adsorption. C_{6v} would be the effective symmetry if the substrate interaction were dominated by that along the surface normal and if adsorbate-adsorbate interactions are small. C_{3v} symmetry could exist upon the (111) surface and C_{2v} on a (001) surface; all if the molecules were parallel to the surface. For any other orientation the highest symmetry possible is C_{2v}. Observation of a band in the direction normal to the C_6H_6 plane, i.e. along the surface normal for a π-bonded species, requires the final state to have a_1 symmetry (C_{6v} notation). Since the dipole operator transforms as a_1 and e_1, we expect only a_1 and e_1 initial states to appear; the former being at a maximum when photons are incident so as to produce a large A^p_z component and the latter when the electric vector is parallel to the surface as for normal incidence or s-polarisation, [54]. In the spectra (Fig. 12 is an example) taken at normal emission, irrespective of substrate, the e_2, b_1 and b_2 bands are indeed absent. There is strong confirmation that the molecule is flat on the surface and that the lateral structure of the substrate and surrounding adsorbate molecules produce only a minor perturbation, [54,55,26]. Further, the polarisation dependence [55], or photon incidence angle dependence (Fig. 12), [54], shows that the a_1 band, corresponding to the gas-phase a_{2u} π-level, appears on the shoulder of the deeper lying e_1 (e_{1u}) ionisation rather than being almost degenerate with the e_2 (e_{2g}) ionisation as in the gas-phase. It seems that both π-type orbitals (e_{1g} and a_{2u}) are involved in bonding to the substrate. Off-normal, the severe restriction on the final state is reduced and all initial states are observed. The relative inten-

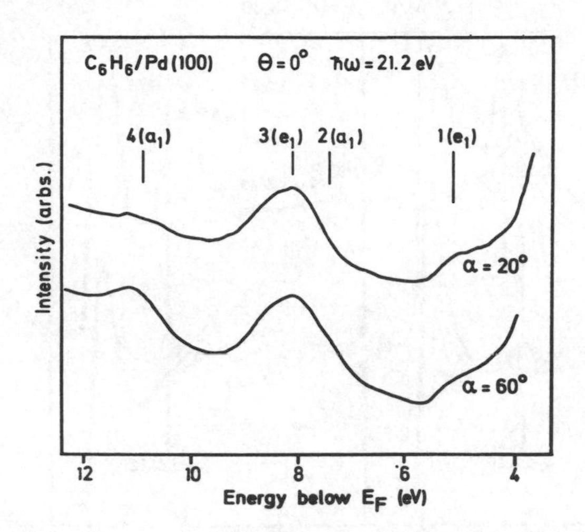

Figure 12. Valence level photoelectron spectra for C_6H_6-Pd(001) at normal emission and 21.2 eV photons incident at 20° and 60°.

sity changes have been rationalised in terms of the likely partial-wave components in the final state [54].

With this confirmation of the usefulness of the approach, other systems have been studied in a similar manner. Ethylene adsorbed on a Pd(001) surface, [26,55], gives the rather surprising result that the molecule is not flat on the surface but that the carbon double bond is somewhat inclined. This conclusion is reached because the a_2 band appears at normal emission which is incompatible with a C_{2v} point group. Pyridine adsorbed on Cu(110) has been shown to bond through the N lone-pair rather than through the π-electrons of the ring, [56], and the methoxy species (CH_3O) which appears on warming a methanol covered Ni(111) surface, [57], or Cu(110) surface, [58], to 140 K, has the O-C bond normal to the surface.

2.5. Adsorbate Band Formation

Not all adsorbate covered surfaces fail to exhibit significant effects arising from lateral interactions and site geometry. Atomic adsorbates provide the clearest examples of a more extensive interaction. There are now many examples in the literature of ordered adlayers producing surface energy bands and dispersion up to <u>ca</u> 1.5 eV. Group VI elements adsorb on the (001) face of f.c.c. metals with occupancy of four-fold hollow sites and the formation of p(2x2) and/or c(2x2) overlayers [23,59]. The sur-

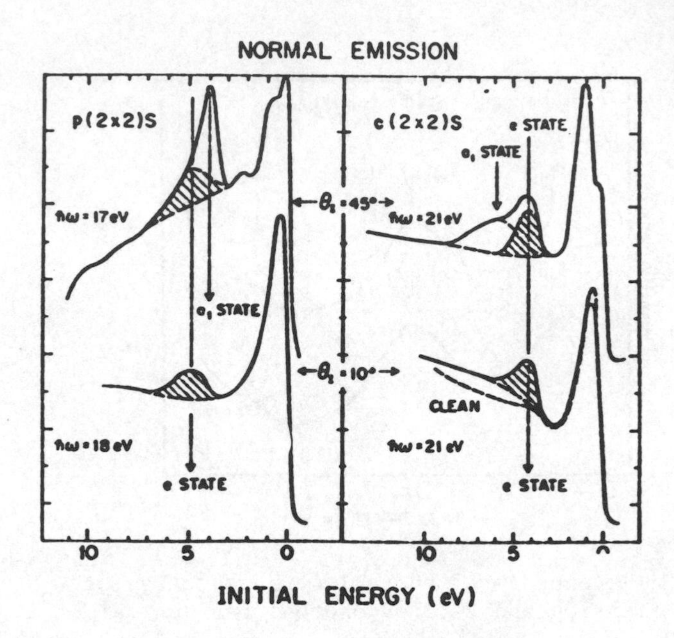

Figure 13. Normal emission spectra for p(2x2) and c(2x2) S-Ni-(001) obtained with p-polarised light at two angles of incidence.

the p_y derived band and is readily identified by the use of s-polarised radiation. Observation of a band in the mirror plane requires an even parity final state. These are only coupled to odd initial states by the odd parity component of the operator, μ_y, and so the intensity depends on the strength of the s-polarised light, assuming emission and incidence planes to be coplanar, in a similar manner to that described by equation (6).

The p_x and p_z derived bands are of even symmetry, requiring the use of p-polarised radiation for observation in the mirror plane. Note that the degeneracy of p_x, p_y, found at normal emission, is now lost. On examination of the square surface Brillouin Zone, one sees that at random points along $\overline{\Gamma X}$ and $\overline{\Gamma M}$ the point group is only C_s which contains no degenerate representations. Only at $\bar{\Gamma}$ and $\bar{M}$ is the symmetry high enough (C_{4v}) for these bands to be degenerate. The symmetry appropriate to $\bar{X}$ is C_{2v} so p_x and p_z are distinct whilst elsewhere p_x and p_z character in bands is mixed together. Fig. 15 is a comparison of the calculated and experimentally derived band structures. The experiments confirm the importance of the interactions between the Ni sp band and the p_x and p_z derived bands. The odd parity, p_y, band is unperturbed on crossing the even sp band but the others are affected in energy and band width.

face unit cell, or surface molecule, XM_4, has C_{4v} symmetry which is sufficient to lift the degeneracy of the chalcogen valence p orbitals. The p_z orbital pointing toward the surface has a_1 symmetry and the p_x, p_y orbitals remain degenerate, having e symmetry and lie parallel to the surface. At normal emission, observable final states have a_1 symmetry so immediately we see that these coupled to the p_z initial state by the normal component of the dipole operator, μ_z, and therefore depend on the strength of A_z^p

$$I_{pz} \alpha /A_z^p <f^a 1/\mu_z^a 1/p_z^a 1>/^2 \qquad (9)$$

Conversely, the e symmetry, p_x, p_y initial states couple to a final states via the x,y components of the operator. Their intensity is given by

$$I_{px,py} \alpha /A_x^p <f^a 1/\mu_x^e/p_x^e>/^2 + /A_y^s <f^a 1/\mu_y^e/p_y^e>/^2 \qquad (10)$$

The two components are readily identified as illustrated by Fig. 13 which shows the normal emission spectra, for p(2x2) and c(2x2) overlayers of S-Ni(001), at two different incident angles of p-polarised light, [60]. In this case, the second term of equation (10) is zero because A_y^s is zero. From Figs. 13b and 13d, obtained with $\alpha = 10^o$, the e states may be placed at 4.9 eV and 4.4 eV for the p(2x2) and c(2x2) overlayers respectively. At $\alpha = 45^o$, A_z^p component is maximised and the a_1 components are identified, (Figs. 13a and 13c), at 4.2 eV and 5.9 eV respectively. That the p_x, p_y ionisation energy is greater than the p_z in the p(2x2) structure, indicates that the bonding to the substrate is more important than lateral interactions and that it is the p_x, p_y orbitals which make the greater contribution. This in turn confirms the occupancy of four-fold hollow sites (on-top sites would certainly utilise the p_z orbital) and suggests the sulphur atom takes a position deep in the hollow in agreement with LEED results which give S-Ni layer spacing as 1.30 Å and therefore a diagonal Ni-S-Ni angle of 107°, [59]. The interaction between neighbouring sulphur atoms acts to stabilise the p_z level whilst destabilising the p_x, p_y orbitals and the reverse ordering is indeed found in the closer packed c(2x2) overlayer.

Fig. 14 presents a summary of the off-normal photoemission data in the form of a two-dimensional band structure plot. The spectral features show considerable energy shifts (dispersion) as a consequence of adsorbate band formation, confirming the greater importance of lateral interactions. In contrast the p(2x2) overlayer shows no such dispersion. Bands, in Fig. 14, are identified as even or odd symmetry with respect to the emission plane chosen to correspond to either the (110) or (100) azimuth of the substrate. If this is the xz-plane, then the single odd state is

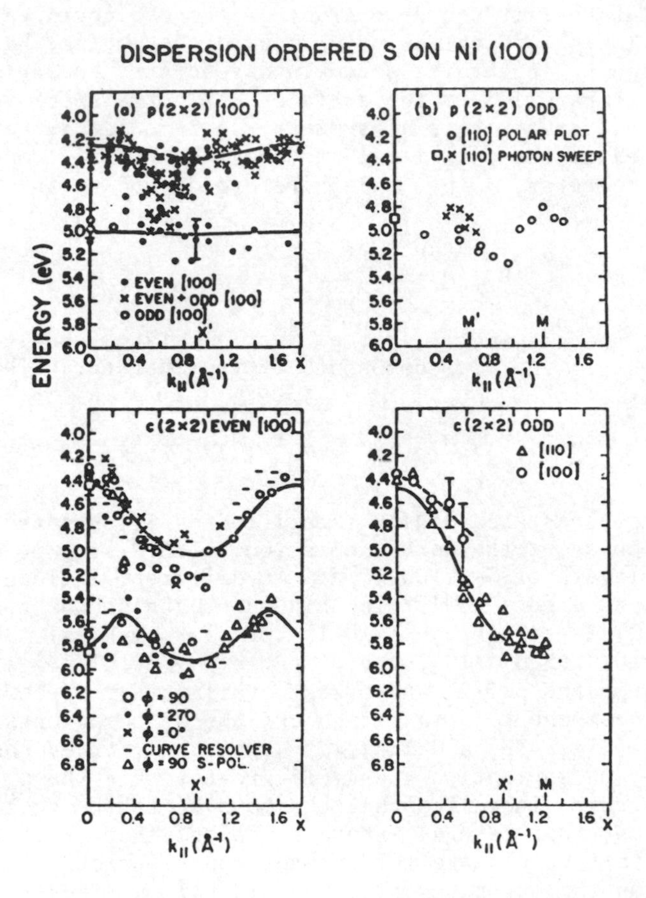

Figure 14. Measured peak position versus $k_{\parallel}$ for c(2x2) S-Ni(001)

Such an analysis can be extended to the case of ordered overlayers which lack a simple registry with the substrate and for which there is, therefore, no common site geometry. For S, Se or Te adsorbed on Al(111), [61], this arises because the substrate lattice is too small. Hexagonal structures are formed in which the adsorbate axes are aligned with those of the substrate but registry is not achieved. Nevertheless, the photoemission spectra can be analysed in terms of a two-dimensional band structure with Brillouin Zone dimensions determined solely by the adsorbate [61]. The experimentally observable influence of the substrate is to mix the even p_x and p_z states at their crossing point (Fig. 16). In an isolated hexagonal layer p_x and p_z have different symmetries with respect to the mirror plane coincident with the layer. This mirror plane is destroyed by the substrate

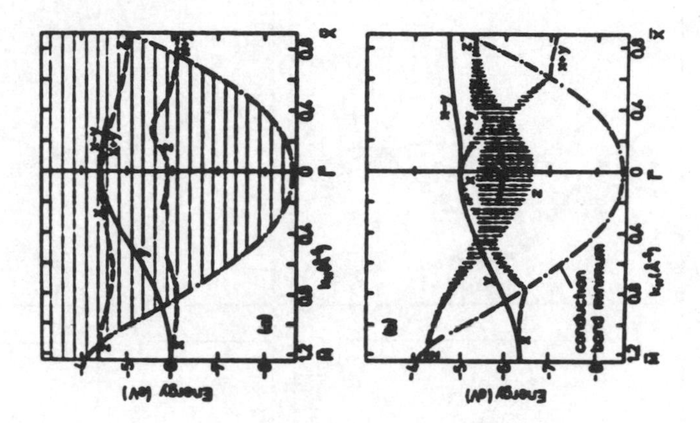

Figure 15. Comparison of experimental (a) and theoretical (b) dispersion of sulphur 3p levels for c(2x2) S-Ni(001). Solid lines are even parity states and dashed lines are odd parity states. The shaded area in (a) shows the projection of the nickel sp band.

interaction and a band gap opens up which is ca 1 eV for Se-Al-(111) but rather smaller (0.5 eV) for S-Al(111) and Te-Al(111).

A related example is provided by the physisorption of xenon on Pd(001) at 80 K, [62]. A close-packed, hexagonal layer, with two rotational domains, is formed by the xenon atoms on the square substrate lattice. Though spin-orbit coupling provides a complication in this system, analysis of the normal emission spectrum and the observation of band dispersion clearly indicates that lateral interactions are significant and that they override the interaction with the substrate in determining the direction of band splitting.

There are examples of chemisorption systems in which the substrate interactions are more significant than for those situations discussed above. So great, in fact, that it becomes very difficult to identify the band (or bands) or predominantly adsorbate character. The adsorption of hydrogen on W(001) provides a convenient example in which several features in the spectra can be induced by the presence of hydrogen. Fig. 17, shows the experimental band structure plot for the saturation coverage of hydrogen on a W(001) surface, [63]. Some polarisation dependent studies of this system by Anderson et al [64], using the by now familiar mirror-plane symmetry arguments, revealed that the hydrogen induced feature which disperses upward in energy from $\bar{\Gamma}$ to $\bar{M}$ has odd parity with respect to the emission plane (Fig. 18). At first sight, it may seem unreasonable for the hydrogen 1s

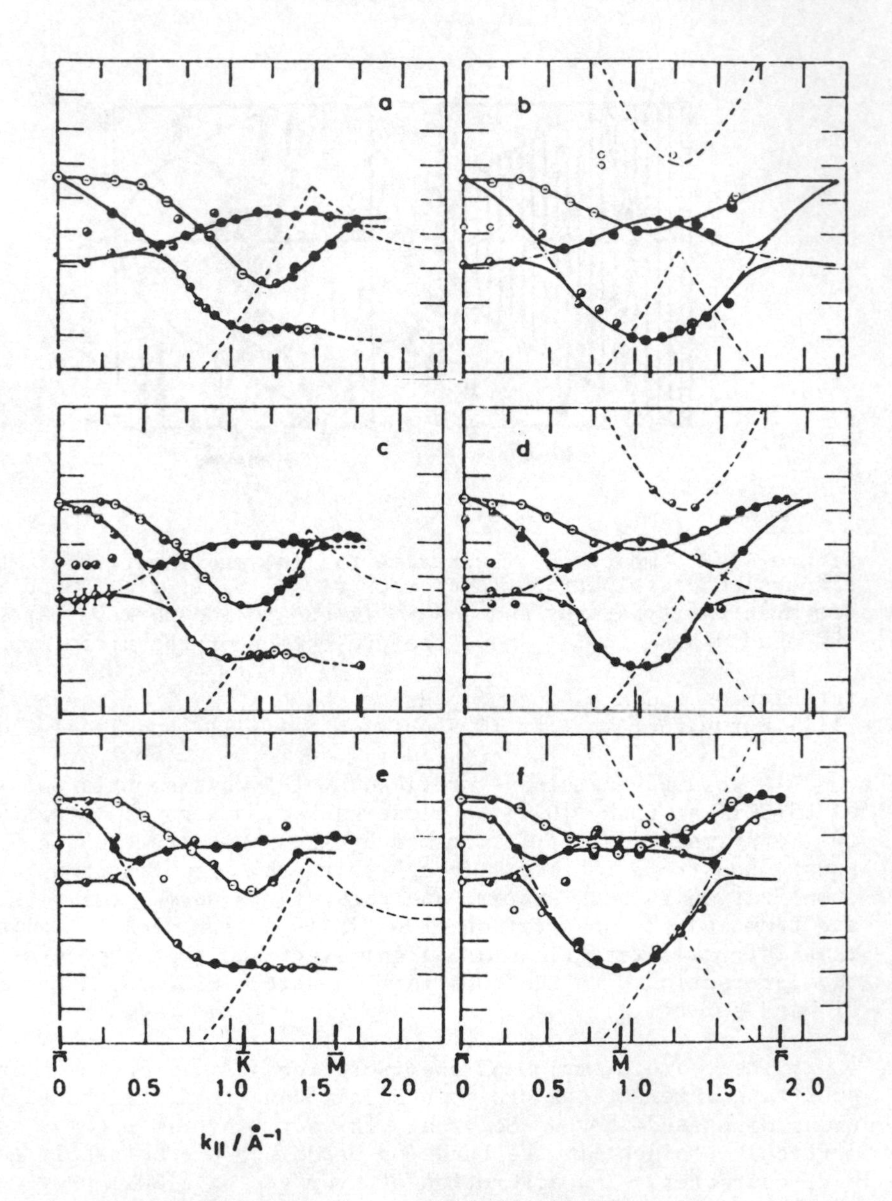

Figure 16. The experimentally determined two-dimensional band structures for S(a,b), Se(c,d) and Te(e,f) layers on Al(111). The lower lying broken lines indicate the bottom of the aluminium sp band. The bar within a circle denotes a band intense for p-polarised light.

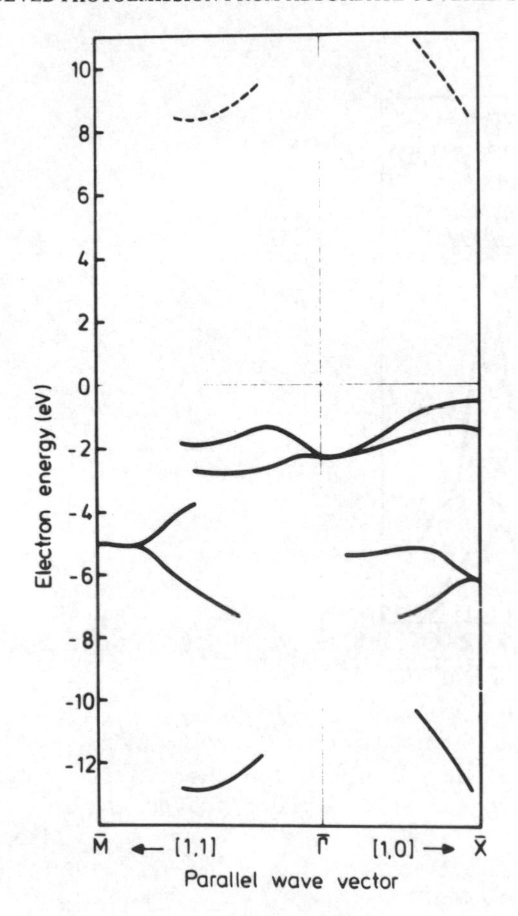

Figure 17. Experimentally determined band structure for (1x1)H-W(001) obtained from the HeI measurements along the principal direction.

orbital to contribute to a negative parity band. However, the saturation coverage structure has all the W-W bridge sites occupied, [65,67], i.e. there are two hydrogen atoms per tungsten atom or per surface unit cell. The out-of-phase combination of these gives rise to the odd parity band along $\bar{\Gamma}\bar{M}$, [68]. That this anti-bonding interaction (within the hydrogen layer) gives the deepest lying hydrogen induced band is proof that in this case adsorbate-substrate interactions are dominant. The surface bonding is envisaged as being between the tungsten $d_{x^2-y^2}$ orbital and the out-of-phase combination of hydrogen 1s orbitals as shown in Fig. 19 [64-68].

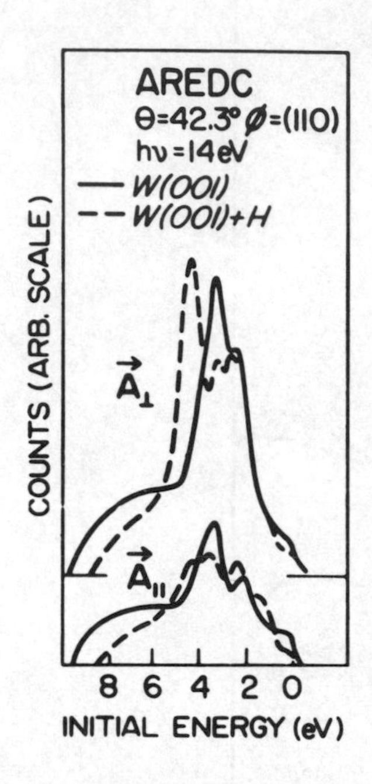

Figure 18. Angle-resolved photoemission spectra for tungsten and tungsten with hydrogen for different photon polarisations.

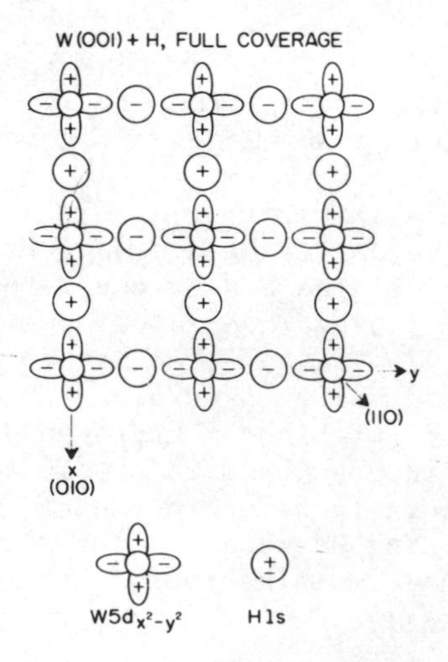

Figure 19. Proposed model of the chemisorption (1x1)H-W(001)

REFERENCES

1. J.W. Gadzuk, Phys. Rev., B10, 5030 (1974).

2. Single plane wave final states have since been shown to be totally inadequate for a description of photoemission intensities, see for example, B.W. Holland, Surf. Sci., 68, 490 (1977).

3. A. Liebsch, Phys. Rev. Letters, 32, 1203 (1974).

4. M. Traum, N.V. Smith and F.J. Disalvo, Phys. Rev. Letters, 32, 1241 (1974).

5. N.V. Smith and M.M. Traum, Phys. Rev., B11, 2087 (1975).

6. D.P. Woodruff, D. Norman, B.W. Holland, N.V. Smith, H.H. Farrell and M.M. Traum, Phys. Rev. Letters, 41, 1130 (1978).

7. N.V. Smith, P.K. Larsen and S. Chiang, Phys. Rev., B16, 2699 (1977).

8. N.V. Smith, H.H. Farrell, M.M. Traum, D.P. Woodruff, D. Norman, M.S. Woolfson and B.W. Holland, Phys.Rev., B21, 3119 (1980).

9. P.K. Larsen, N.V. Smith, M. Schluter, H.H. Farrell, K.M. Ho and M.L. Cohen, Phys. Rev., B17, 2612 (1978).

10. S. Kono, C.S. Fadley, N.F.T. Hall and Z. Hussain, Phys.Rev. Letters, 41, 117 (1978).

11. W.M. Kang, C.H. Li and S.Y. Tong, to be published (1980).

12. M. Forstmann, W. Berndt and P. Büttner, Phys. Rev. Letters, 30, 17 (1973).

13. P.H. Citrin, P.E. Eisenberger and R.C. Hewitt, Phys. Rev. Letters, 41, 309 (1978).

14. R.J. Purtell, R.P. Merrill, C.W. Seaburg and T.N. Rhodin, Phys. Rev. Letters, 44, 1279 (1980).

15. C.H. Li and S.Y. Tong, Phys. Rev., B19, 1769 (1979).

16. S.Y. Tong and N. Stoner, J. Phys. C11, 3511 (1978).

17. S.Y. Tong and M.A. Van Hove, Solid State Commun., 19, 543 (1976).

18. S.D. Kevan, D.H. Rosenblatt, D. Denley, B.-C.Lu and D.A. Shirley, Phys. Rev. Letters, 41, 1565 (1978).

19. C.H. Li and S.Y. Tong, Phys. Rev. Letters, 42, 901 (1979).

20. C.H. Li and S.Y. Tong, Phys. Rev. Letters, 43, 526 (1979).

21. "Photoemission and the electronic properties of surfaces", eds.B. Feuerbacher, B. Fitton and R.F.Willis, Wiley, (1978).

22. Topics in Applied Physics, 26 and 27, eds. L. Ley and M. Cardona, Springer-Verlag, Berlin (1979).

23. K. Jacobi and C.v. Muschwitz, Solid State Commun., 26, 477 (1978).

24. M. Scheffler, K. Kambe and F. Forstmann, Solid State Commun., 23, 789 (1977).

25. N.V. Richardson and A.M. Bradshaw, to appear in "Electron Spectroscopy: Theory. Techniques and Applications" (eds. A.D. Baker and C.R. Brundle), 4 (Academic Press,London (1980)

26. M. Scheffler and A.M. Bradshaw, "The Chemistry and Physics of Solid Surfaces and Heterogeneous Catalysis", eds. D.A. King and D.P. Woodruff, Elsevier, to be published.

27. J.K. Burdett, in "Spectroscopy", (eds. B.P. Straughan and S. Walker), 2, 120 (1976), Chapman and Hall, London.

28. J.H.D. Eland, "Photoelectron Spectroscopy", Butterworths, London (1974).

29. B. Brehm, M. Menzinger and C. Zonn, Can. J. Chem., 48, 3193 (1970).

30. G.D. Mahan, Phys. Rev. Letters, 24, 1068 (1970).

31. I. Adawi, Phys. Rev., 134, A788 (1964).

32. G.D. Mahan, Phys. Rev., B2, 4334 (1970).

33. P.J. Feibelman and D.E. Eastman, Phys. Rev., B10, 4932 (1974)

34. N.W. Schaich and N. Ashcroft, Phys. Rev., B3, 2452 (1971).

35. C.B. Kunz in Ref. 22, 27, Chapter 6.

36. K. Jacobi, P. Geng and W. Ranke, J. Phys. E11, 982 (1978).

37. N.V. Richardson and J.K. Sass, J. Phys. F8, L99 (1978).

38. N.V. Richardson, J.K. Sass, D.R. Lloyd and C.M. Quinn, Surf. Sci., 80, 165 (1980).

39. R. Nishitani, M. Aono, T. Tanaka, S. Kawai, H. Iwasaki, C. Oshima and S. Nakamura, Surface Sci., 95, 341 (1980).

40. E.W. Plummer, in Proc. 7th Intern. Congr. and 3rd Intern. Conf. on Solid Surfaces (Vienna, 1977), 647.

41. S.P. Weeks and E.W. Plummer, Solid State Commun., 21, 695 (1977).

42. J.A. Stratton, "Electromagnetic Theory", McGraw-Hill, New York (1941).

43. D.R. Lloyd, Disc. Faraday Soc., 58, 136 (1974).

44. D.R. Lloyd, C.M. Quinn and N.V. Richardson, Solid State Commun., 20, 409 (1976).

45. P.M. Williams, P. Butcher, S. Woods and K. Jacobi, Phys. Rev., B14, 3215 (1976).

46. R.J. Smith, J. Anderson and G.J. Lapeyre, Phys. Rev. Letters, 37, 1081 (1976).

47. C. Allyn, T. Gustafsson and E.W. Plummer, Chem. Phys.Letters, 47, 127 (1977).

48. T. Gustafsson, E.W. Plummer, D.E. Eastman and J.L. Freeauf, Solid State Commun., 17, 391 (1975).

49. E.W. Plummer, T. Gustafsson, W. Gudat and D.E. Eastman, Phys. Rev., A15, 2339 (1977).

50. J.W. Davenport, Phys. Rev. Letters, 36, 945 (1976).

51. K. Horn, W. Eberhard and E.W. Plummer, to be published.

52. J.E. Demuth and D.E. Eastman, Phys. Rev. Letters, 32, 1123 (1974).

53. D.R. Lloyd, C.M. Quinn and N.V. Richardson, Solid State Commun., 23, 141 (1977).

54. G.L. Nyberg and N.V. Richardson, Surf Sci., 85, 335 (1979).

55. P. Hofmann, K. Horn and A.M. Bradshaw, to be published.

56. B.J. Bandy, D.R. Lloyd and N.V. Richardson, Surf. Sci., 89. 344 (1979).

57. A.M. Bradshaw, unpublished results.

58. P. Hofmann, C. Muriani, K. Horn and A.M. Bradshaw, "Proc. of 3rd European Conf. on Surf. Science", eds. D.A. Degras and M. Costa, 541 (1980).

59. J.E. Demuth, D.W. Jepsen and P.M. Marcus, Phys. Rev. Letters, 31, 540 (1973).

60. E.W. Plummer, B. Tonner, N. Holzwarth and A. Liebsch, Phys. Rev., B21, 4306 (1980).

61. K. Jacobi, C.v. Muschwitz and K. Kambe, Surf. Sci., 93, 310 (1980).

62. M. Scheffler, K. Horn, A.M. Bradshaw and K. Kambe, Surc. Sci. Sci., 80, 69 (1979).

63. B. Feuerbacher and R.F. Willis, Phys. Rev. Letters, 36, 1339 (1976).

64. J. Anderson, G.J. Lapeyre and R.J. Smith, Phys. Rev., B17, 2436 (1978).

65. W. Ho, R.F. Willis and E.W. Plummer, Phys. Rev. Letters, 40, 1463 (1978).

66. P.J. Estrup and J. Anderson, J. Chem. Phys., 45, 2254 (1966).

67. A. Adnot and J.-D. Carette, Phys. Rev. Letters, 39, 209 (1977).

68. N.V. Smith and L.F. Matheiss, Phys. Rev. Letters, 37, 1494, (1976).

X-RAY EMISSION SPECTROSCOPY

G. Wiech

Sektion Physik der Universität München, 8 München 22, FRG

ABSTRACT

This paper reviews the applicability of x-ray emission spectroscopy as a tool for studying the electronic structure of molecules and solids. The wavelengths of interest cover the range from about 1 Å (12 000 eV) up to 600 Å (20 eV). After a general introduction the basic principles of the spectrometers used in x-ray spectroscopy are described. In a theoretical section the relationship between the observed spectra and the electronic structure are then treated in the framework of the one-electron model. Examples of line emission spectra and band emission spectra are discussed and compared with theoretical predictions.

The last section reports recent developments in the field of anisotropic x-ray emission of single crystals. The investigation of this effect provides additional information about electronic structure and permits the identification and separation of π and σ orbitals or bands. This will be demonstrated by several examples.

I. INTRODUCTION

The aim of this paper is to present an overview of the ways in which x-ray emission spectroscopy can be used for the study of the electronic structure of molecules and solids. Since this paper is the only one in the ASI meeting which deals with x-ray spectroscopy it will be necessary to restrict ourselves to a limited number of topics to avoid covering too large an area.

In principle x-ray spectra can be explained adequately only in terms of a theory which includes many-body effects. Features

P. Day (ed.), Emission and Scattering Techniques, 103–151.

caused by many-body effects (screening effects in metals, plasmons, electron-phonon interaction etc.) show up practically in all spectra and they can sometimes be quite pronounced (tailing of emission band spectra due to Auger broadening, high energy satellites). On the other hand in many spectra these effects are comparatively small and give rise only to weak extra structure. In all these cases the main structural features can be interpreted quite well in terms of the one-electron approximation which assumes the validity of Koopman's theorem.

In section II we summarize some general aspects of x-ray emission spectroscopy. More details will be given in the following sections.

In section III we briefly treat the dispersing instruments. For high-resolution studies a number of specially designed spectrometers have been built which cover the whole energy range of interest. Here only the basic principles will be considered.

Section IV is devoted to a short review of the one-electron theory of x-ray transitions and of the relation between the electronic structure of matter and x-ray spectra. These relations serve as a reference for the later sections. - For a detailed discussion of many-body effects, the interested reader is referred to literature (1-5).

In section V illustrative examples of x-ray line spectra and valence band spectra will be discussed. We shall see to what extent the spectra are influenced by changes of the chemical bonding; we also shall see that x-ray emission spectra and x-ray photoelectron spectra yield quite complementary information in the case of second period elements.

The information provided by x-ray emission spectra can be improved considerably by studying the anisotropic emission of the characteristic radiation of single crystals. The aim of section VI is to survey the way in which anisotropic emission has extended our ability to investigate the electronic structure of molecules and solids, and to illustrate this with a selection of recent experimental work.

II. QUICK COURSE THROUGH X-RAY SPECTROSCOPY

1. Electronic structure and allowed x-ray transitions

Suppose we have a third period element, for instance semiconducting silicon which we shall meet repeatedly in the following. A schematic view of the most relevant features of the electronic structure of silicon is shown in Fig. 1. In Fig. 1a we see the band structure

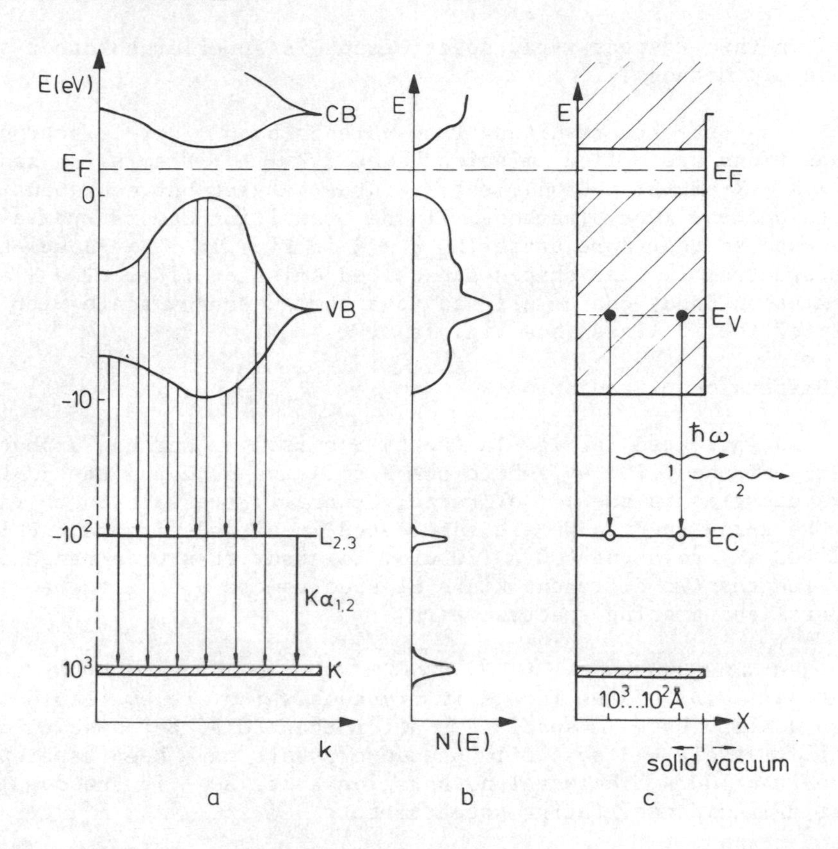

FIG. 1. Electronic structure of solid matter (schematically); a: band structure and transitions permitted by dipole selection rule, E_F Fermi energy, CB conduction band, VB valence band; b: density of states N(E); c: effective thickness of target; the probability of being absorbed is higher for photon 1 than for photon 2.

E(k): the unoccupied conduction bands CB and the occupied valence bands VB are k-dependent and separated by an energy gap, while the more tightly bound core levels $L_{2,3}$(2p) and K(1s) can be considered as being k-independent. Fig. 1b shows the corresponding density-of-states curves N(E).

X-ray transitions are governed by strong selection rules. The electrons of the $L_{2,3}$ and the K states have p-like and s-like symmetry, respectively. As a consequence of the so-called dipole selection rule electronic transitions from the valence band to a vacancy in the $L_{2,3}$ level are allowed only for s- and/ore d-like electrons, and transitions to the K level are allowed only for p-like electrons. The transition $L_{2,3} \rightarrow K$ is also dipole allowed. - Thus by measuring the different spectra we probe valence electrons of different symmetry. In this respect x-ray spectroscopy is superior to other spectroscopic methods.

X-ray spectra resulting from valence band → core electronic transitions are called emission bands (VB → L : L-emission band; VB → K : K-emission band), their widths varying between about 2 eV (lithium) and 20 eV (carbon). If the transition occurs from a core state to another core state ($L_{2,3}$ → K in Fig. 1a : Kα -doublet) the spectrum is line shaped and called emission line.- The K and L emission bands can be aligned to a common energy scale with the help of the Kα lines (see Fig. 1a).

2. Dispersion of radiation

As indicated in Fig. 1a the $L_{2,3}$ emission band has a photon energy of about 10^2 eV, while the K emission band and the Kα lines have energies in the 10^3 eV range. Generally the wavelength range of the x-ray spectra we are interested in extends from about 1 Å (12 000 eV) to about 600 Å (20 eV). To study this wide range of wavelengths two different kinds of spectrometers have to be used: crystal and grating spectrometers.

For wavelengths $\lambda < 20$ Å crystals are used to disperse the radiation . Most of the inorganic crystals which are available in high quality have 2d spacings smaller than 20 Å, but special organic crystals and soap film pseudo-crystals have been used up to more than 100 Å. The wavelength region above 20 Å is the domain of grazing-incidence grating spectrometers.

In x-ray spectroscopy it has become customary to distinguish between hard ($\lambda < 1$ Å), soft (1 Å $< \lambda <$ 20 Å) and ultra-soft ($\lambda > 20$ Å) radiation.

3. Modes of excitation

For excitation of x-ray spectra an atom has to be ionized in a core shell either by electron bombardement or by photons. These two modes of excitation are called primary and secondary excitation, respectively.

If primary excitation is used it often happens that chemical compounds decompose under the influence of the electron impact. Secondary excitation has proved to be much less destructive. In the region of soft x-rays it is usual nowadays to use secondary excitation. In the region of ultra-soft x-radiation, however, electron excitation is used predominantly because in conventional x-ray tubes the intensity of long wavelength radiation available for excitation is low. In recent years the synchrotron radiation from the storage ring DORIS at Hamburg has been used successfully for excitation (6,7) and also the radiation of a rotating anode

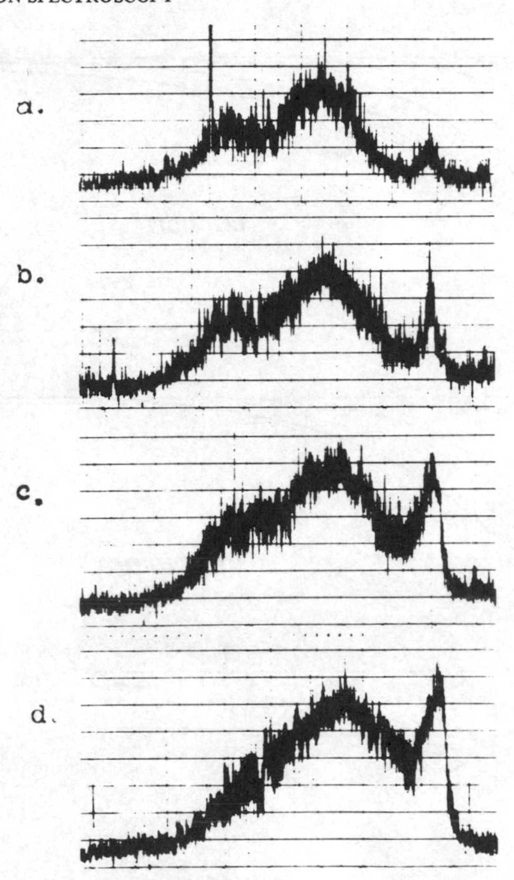

FIG. 2. Time-dependent variation of the shape of the Al $L_{2,3}$-emission band of Al_2O_3 due to decomposition; primary excitation, 3 kV, 2.2 mA; from top to bottom: 1.5, 3.5, 9.5 and 13.5 hours after start of measurement (9).

high power x-ray tube (8).

For illustration we consider the decomposition of Al_2O_3 when primary excitation was used (9). The x-ray tube was operated at 3 kV, 2.2 mA; the area of the focal spot was about 15 mm^2 and the pressure in the tube was some 10^{-8} torr. Fig. 2 shows the Al L-emission band of aluminium oxide and its variation with time. The spectrum at the top is only slightly affected by decomposition. The last spectrum, however, is very similar to that of pure aluminium. Similar changes, which sometimes can occur in a very short time, have been observed in numerous cases (10). Though secondary excitation provides much better results it is no panacea; in particular if one wants to study organic compounds one has to take precautions (7,11).

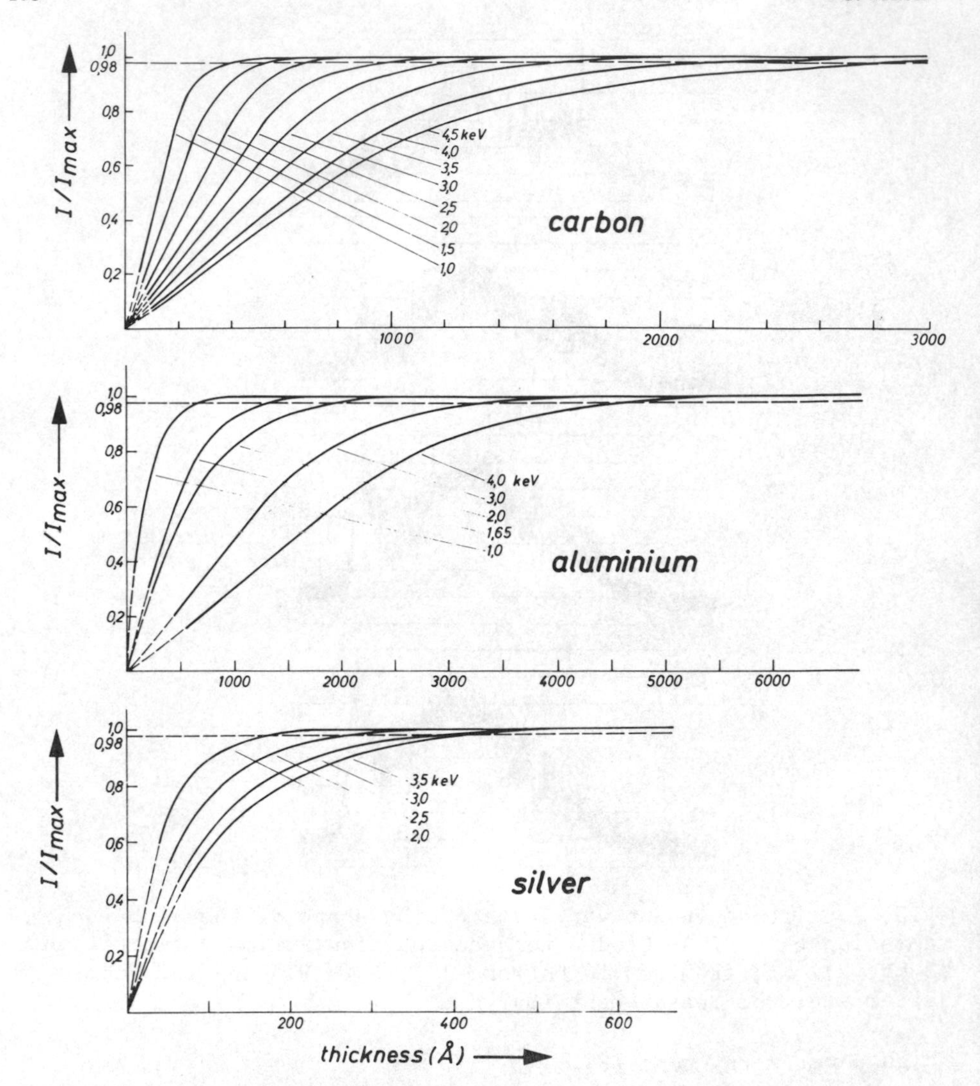

FIG. 3. Peak intensity of the K emission band of carbon, the $L_{2,3}$ emission band of aluminium and the Mζ line of silver excited by electron bomardment of evaporated targets as a function of target thickness (12).

4. Thickness of layers contributing to x-ray emission

Finally we may ask: what depth of a sample does x-ray spectroscopy probe when primary excitation is used? To answer this question the intensities of the carbon K band (44 Å), the aluminium L band (171 Å) and the silver Mζ line (40 Å) were measured as a function of the thickness of evaporated targets (12). In Fig. 3 the intensity (normalized to the intensity of an infinite thick layer) is plotted versus target thickness. Voltages between 1 kV and 4,5 kV were used. For carbon and aluminium the target thickness which contributes to x-ray emission varies from about 200 Å up to some 10^3 Å according to electron energy, and even for the medium Z element silver the effective thickness of the target is representative of the bulk properties.

The limiting factor is the penetration depth of the electrons, while x-ray absorption does not play the dominant role. In the case of soft x-rays, for which the ionization energies are in the keV range, even thicker layers do contribute to emission, and the effective target thickness is increased still more if secondary excitation is used.

5. Self-absorption

On its way to the surface of the sample the radiation is partially absorbed. This effect is called self-absorption (see Fig. 1c). Especially in the vicinity of the high energy edges of emission bands of metals, where the absorption coefficient changes discontinuously, the spectra can be modified considerably by self-absorption. Before quantitative conclusions can be drawn from the spectra they have to be corrected for self-absorption effects (12). Self-absorption effects and other interfering effects - such as satellite structure - can be reduced to a minimum by near-threshold measurements (13).

III. INSTRUMENTATION

In x-ray spectroscopy plane crystal spectrometers, bent crystal spectrometers and grating spectrometers are used.

1. Plane crystal spectrometers

In all crystal spectrometers the x-rays fo wavelength λ are dispersed according to the Bragg law

$$n\lambda = 2 d_n \cdot \sin \varphi_n$$

where n is the order of diffraction, d_n the distance of atomic planes for the nth order and φ_n is the angle between the incident ray and the diffracting plane of the crystal.

In a plane one-crystal spectrometer the general mounting is closely related to the early Bragg spectrometer (Fig. 4). A narrow beam of radiation passes through slits S_1 and S_2 and hits the surface of the crystal. Another slit S_3 is put in front of the detector. A spectrum is measured continuously or by the step-scanning technique by rotating the crystal through an angle φ and simul-

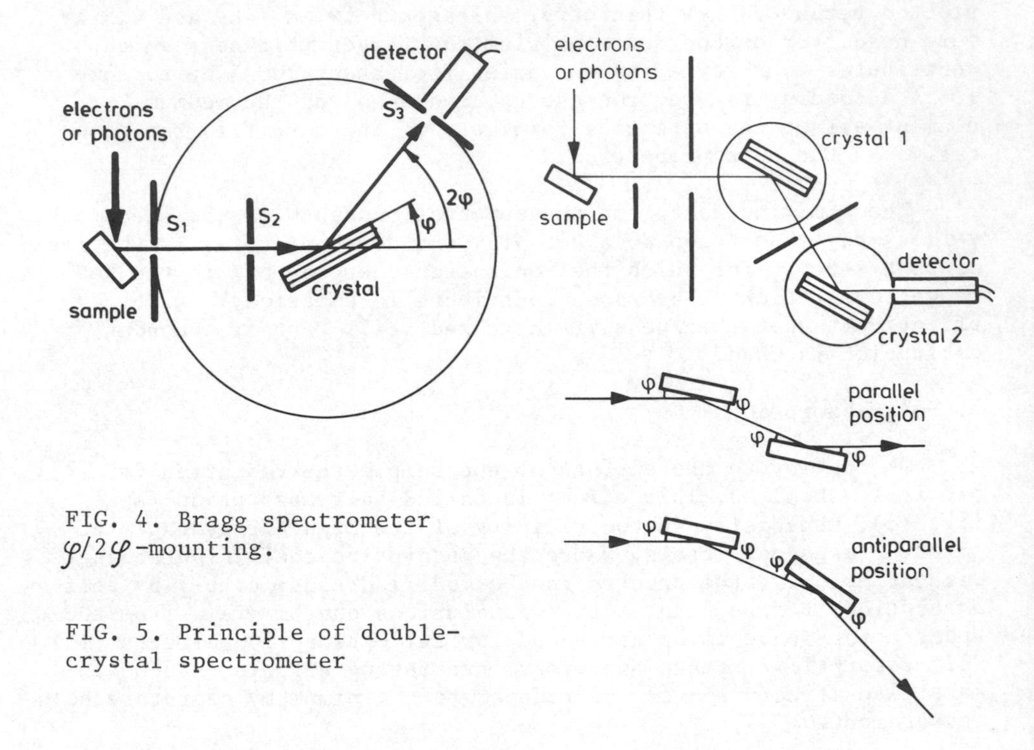

FIG. 4. Bragg spectrometer $\varphi/2\varphi$-mounting.

FIG. 5. Principle of double-crystal spectrometer

taneously rotating the detector through an angle 2φ.

Commercial instruments mostly are equipped with Soller collimator slits instead of the slits S_1, S_2 and S_3. In this way the intensity is improved considerably, because more extended samples can be used. The gain of intensity is compensated, however, by a reduction of resolution.

For high-precision measurements double-crystal spectrometers are used. The principles are easily understood from Fig. 5. A double-crystal spectrometer consists of two consecutive one-crystal spectrometers. In the so-called antiparallel position the dispersion will be twice that of a single-crystal instrument. If the two crystals are put in the parallel position there will be no dispersion. The outgoing beam is parallel to the incident beam. In this case the measured line results from a convolution of the diffraction patterns of the two crystals, called the rocking curve of the crystal pair. The halfwidth of this line tells us the smallest detail in an x-ray spectrum which can be resolved. This information about the window function of the instrument is the main advantage of the double-crystals spectrometer over other types of spectrometers.

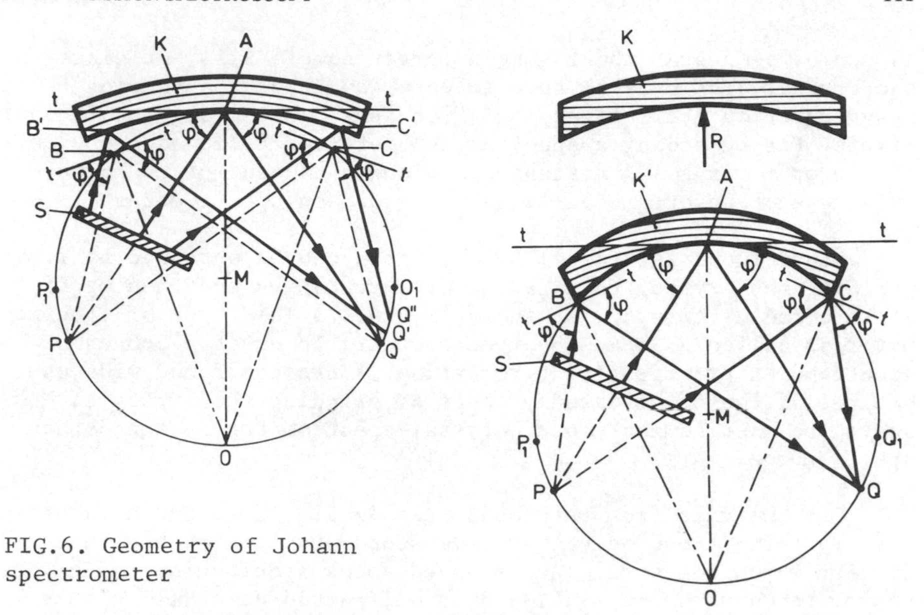

FIG.6. Geometry of Johann spectrometer

FIG. 7. Geometry of Johansson spectrometer; K ground crystal before bending, K' crystal bent to radius R/2

2. Bent crystal spectrometers

To increase the luminousity of crystal spectrometers the crystals are bent to a cylindrical curvature. Two types of focussing by surface reflection are illustrated in Figs. 6 and 7.

As shown in Fig. 6 a crystal slab with the atomic planes parallel to the surface is bent to a radius of curvature of $\overline{AO} = 2\ \overline{AM} = R$. If the radiation of wavelength λ_o emitted from a source point P on the circle of radius R/2 were diffracted by atomic planes t-t (glancing angle φ) at points A, B and C (and other points lying between B and C on the circle) the divergent beams would be focussed to the image point Q (dashed lines). The same is true for other wavelengths.For instance Q_1 is the image of P_1, the wavelength being $\lambda_o - \Delta\lambda$. The circle with radius R/2 therefore is called focussing circle or Rowland circle. Diffraction, however, cannot occur at B and C. It can occur at the same angle φ at points B' and C', the incident and the diffracted beams (full lines) then running parallel with the beams diffracted at B and C. As illustrated in Fig. 6 these beams come from different source points, and are focussed to different image points Q, Q' and Q", i.e. the focussing is not exact.- It may be mentioned that exact focussing is obtained for one wavelength at least, if the crystal is bent ot a logarithmic spiral (14).

Actually the sample is not at points P on the focussing circle.

As can be seen from the figure a larger sample S is put close to the crystal. The crystal then selects wavelengths according to the Bragg equation (full lines) and "focusses" them along the focussing circle. For recording a spectrum the detector is moved along this circle or - for small distances - along a tangent of the circle. This type of spectrometer is called a Johann spectrometer.

Exact focussing for all wavelengths can be obtained by first grinding the crystal to a radius of curvature R and then bending it to a radius $\overline{AM} = R/2$ as shown in Fig. 7. This type of spectrometer is called a Johansson spectrometer. Though the Johansson type spectrometer provides exact focussing it has not found wide use, because of the complicated process of grinding the crystals. Nowadays the most commonly used crystal spectrometer is the Johann spectrometer.

Let the rays from the focus spot in Fig. 6 be extended behind the crystal. It can be easily understood from the figure that radiation which comes from an extended focus spot behind the crystal, passes through the crystal, and is diffracted at atomic planes lying perpendicular to the crystal surface, will be focussed also in Q, Q' and Q". A spectrometer using this kind of focussing is called a Cauchois spectrometer. - Cauchois transmission spectrometers provide good focussing for small glancing angles φ (small λ), while the focussing defects become smaller with increasing φ for the Johann spectrometers.

In crystal spectrometers (flow-) proportional counters and scintillation counters serve as detectors.

3. Grating spectrometer

In the region of ultrasoft x-rays ($\lambda > 20$ Å) grazing incidence concave grating spectrometers are used. The essential elements of a grating spectrometer are illustrated in Fig. 8. The dispersion of radiation of wavelength λ is based on the grating equation

$$n\lambda = \sigma(\cos\varphi - \cos\psi_n)$$

where n is the order of diffraction, σ is the grating constant, and φ and ψ_n are the angles of grazing incidence and diffraction, respectively.

Entrance slit, grating and exit slit are set up in the so-called Rowland mounting, i.e. they must lie on a circle with radius $\overline{GM} = \overline{GO}/2 = R/2$, R being the radius of curvature of the spherical surface of the grating. The radiation emitted by the sample passes through the entrance slit, is then dispersed by the grating and each wavelength is focussed on the Rowland circle. The grazing incidence mounting ($\varphi < 5^{\circ}$) is applied for two reasons, (i) to increase the angular dispersion by reducing the effective grating constant, and

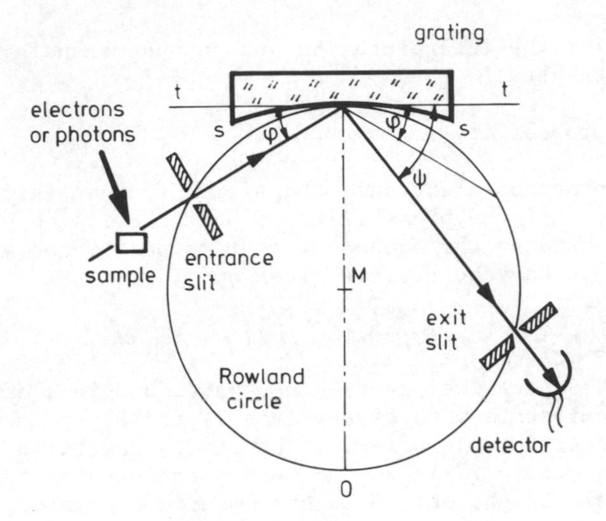

FIG. 8. Schematic diagram of the grazing-incidence concave grating spectrometer

(ii) to improve the reflectivity of the grating by total reflection of the radiation. So-called blazed gratings are used, either replicas of mechanically ruled master gratings or holographically produced gratings. The number of lines per mm varies between several hundreds and more than 3000. The detectors most commonly used nowadays are the flow proportional counter and the open secondary electron multiplier.

The pressure in the sample chamber (x-ray tube) should be better than 10^{-7}-10^{-8} torr to avoid or to reduce target contamination. As for instance carbon is almost omnipresent in the form of hydrocarbons; these are cracked by the electron or photon impact, and the cracking products deposited on the target give rise to carbon K emission (44 Å). Due to its short wavelength and because it appears in several spectral orders, the carbon K emission band often interferes with the spectra under investigation.

It must be emphasized that in this section only the essential principles of the spectrometers have been described. For more details, in particular for the description of instruments developed for special purposes the interested reader is referred to useful reviews (15-19) and to the original papers cited there.

IV. THEORETICAL CONSIDERATIONS

In this section we collect some theoretical relations which

are essential for the interpretation and an understanding of x-ray spectroscopic results.

1. Transition probability

For a spontaneous electronic transition from an initial state i to a final state f the probability of photon emission dP_{fi} with energy $\hbar\omega$, measured as the number of photons per second and solid angle $d\Omega$ based on Davydov (20) is given by

$$dP_{fi} = \frac{\omega e^2}{2\pi\hbar m_e^2 c^3} |\langle \psi_f | e^{-i\vec{q}\vec{r}} (\vec{u}\vec{p}) | \psi_i \rangle|^2 d\Omega \tag{1}$$

where ψ_i and ψ_f denote the one-electron wave functions of the initial and final state with eigenvalues E_i and E_f, respectively, $\vec{u}$ is a unit vector for the polarization of the radiation field (direction of electric field vector, and perpendicular to the wave vector $\vec{q}$ of the photon), $\vec{p}$ is the momentum operator, the other symbols having the usual meaning.

The wave functions of discrete states differ appreciably from zero only within the dimensions of atoms or molecules. The integration in (1) therefore needs to be taken only over a region $r \lesssim a_o$, where a_o represents the dimensions of the orbitals: this is typically of the order of 1 Å for valence orbitals and less for core orbitals.

In the ultrasoft x-ray range the wavelength of the radiation considerably exceeds the dimensions of an atom or molecule. For a wavelength of 100 Å we have $qa_o = 2\pi a_o/\lambda \approx 6.10^{-2}$. Therefore the exponential term in equation (1) can be expanded as a power series

$$\exp(-i\vec{q}\vec{r}) = 1 + (-i\vec{q}\vec{r}) + (-i\vec{q}\vec{r})^2/2 + \cdots \tag{2}$$

and set equal to unity, since the terms other than the first term are small. Equation (1) then is reduced to

$$dP_{fi} = \frac{\omega e^2}{2\pi\hbar m_e^2 c^3} |\langle \psi_f | \vec{p} | \psi_i \rangle|^2 d\Omega . \tag{3}$$

This approximation is called the dipole approximation.

Let λ be 10 Å in the soft x-ray range. In this case $qa_o \approx 6.10^{-1}$, and the second term in equation (2) also contributes to the transition probability P_{fi}. This contribution is called the quadrupole radiation (which is governed by other selection rules than dipole radiation).

As can be seen it follows from equation (2) that, as the wavelength decreases the contribution of the quadrupole radiation is increased compared to the dipole radiation. Quadrupole radiation predominates if $\lambda \lesssim a_o$, i.e. if the wavelength of the emitted radiation becomes smaller than the dimensions of the atom. Then $qa_o \gg 1$, and in equation (2) the first term can be neglected.

The expression $\langle \psi_f | \vec{p} | \psi_i \rangle$ is called the matrix element of the momentum operator $\vec{p}$. The matrix element of the momentum operator can be expesssed by the matrix element of the position operator $\vec{r}$:

$$\langle \psi_f | \vec{p} | \psi_i \rangle = -i m_e \omega \langle \psi_f | \vec{r} | \psi_i \rangle \qquad (4)$$

with $\omega = (E_i - E_f)/\hbar$, E_i and E_f being the energies of the two states i and f, respectively.

By inserting (4) in (3) we obtain for the transition probability

$$dP_{fi} = \frac{\omega^3}{2\pi\hbar c^3} e^2 |\langle \psi_f | \vec{r} | \psi_i \rangle \vec{u}|^2 d\Omega = \frac{\omega^3}{2\pi\hbar c^3} (\vec{M}\vec{u}) d\Omega \qquad (5)$$

The vector

$$\vec{M}_{fi} = e \langle \psi_f | \vec{r} | \psi_i \rangle \qquad (6)$$

is called the electric dipole moment of the transition i → f, and the emitted radiation is called electric dipole radiation.

Denoting the angle between $\vec{q}$ and $\vec{M}$ (i.e. the angle between the propagation direction of the radiation and the dipole moment) by θ, we obtain instead of (5)

$$dP_{fi} = \frac{\omega^3}{2\pi\hbar c^3} |\vec{M}_{fi}|^2 \cdot \sin^2\theta \cdot d\Omega \qquad (7)$$

Taking the integral over Ω we obtain from (7) the total number of photons $N(\omega)$ emitted by a single radiatior per unit time

$$P_{fi} = \frac{4\omega^3}{3\hbar c^3} |\vec{M}_{fi}|^2 \qquad (8)$$

To obtain the rate of emission of energy, (8) must be multiplied by $\hbar\omega$. The expression for the intensity then assumes the form

$$I_{fi} = \hbar\omega P_{fi} = \frac{4\omega^4}{3c^3} |\vec{M}_{fi}|^2 \qquad (9)$$

It should be noted that the formulae for the intensity contain a factor ω^3 or ω^4, according as the intensity is measured as the number of photons per time or as the energy per time.

2. The selection rules

As shown by equations (5) and (6) the probability for the emission of dipole radiation depends upon the direction of the emitted radiation and the magnitude of the matrix element. For systems where the electron moves in a central field the matrix element is non-zero only if the following rules are satisfied:

$$\ell_i = \ell_f \pm 1 \,, \quad \Delta\ell = \pm 1 \;; \quad \left.\begin{matrix} m_i = m_f \\ m_i = m_f \pm 1 \end{matrix}\right\} \Delta m = 0, \pm 1 \qquad (10)$$

These are the selection rules for electric dipole radiation.

If these selection rules (10) are not satisfied electric dipole transitions are not allowed. In this case, however, electric quadrupole radiation or magnetic dipole radiation may be allowed because they are governed by different selection rules. - In this paper only electric dipole transition will be considered, as they are in general the most intense ones. First we shall take into account only the l-selection rules. The m-selection rules, connected with the polarisation of the radiation, will be discussed in section VI. Polarisation effects become important if single crystal samples are studied. In the case of polycrystalline samples with isotropic distribution of crystallites, they are irrelevant.

3. Intensity of x-ray emission spectra

The intensity of an x-ray emission spectrum (number of photons) at frequency ω is obtained by summing up all transition probabilities P_{fi}, for the states i and f which have the same energy difference $E_i - E_f = \hbar\omega$

$$I(\omega) \sim \sum_i \sum_f \delta(E_i - E_f - \hbar\omega)\,\omega^3 |M_{fi}|^2 \tag{11}$$

the δ-function is responsible for the energy conservation.

In atoms and molecules a definite number of atomic and molecular orbitals are occupied by electrons of discrete energies. In solids the possible energy values are very close-spaced; thus they can be considered a quasi continuum. Instead of discrete energy levels we have energy bands, the energy E in an energy band being a function of the reduced propagation vector $\vec{k}$ of the electrons:

$$E_n = E_n(\vec{k})$$

n is the band index. The expression for the intensity then can be written in the form

$$I(\omega) \sim \omega^3 \sum_n \int_{BZ} \delta(E^i(\vec{k}) - E_n^f(\vec{k}) - \hbar\omega)\,|M_{fi}|^2\, d^3k \tag{12}$$

the integral has to be taken over the whole Brillouin zone BZ. An expression equivalent to (12) is

$$I(\omega) \sim \omega^3 \frac{V}{(2\pi)^3} \int_S \frac{dS}{|\mathrm{grad}\,(E_i(\vec{k}) - E_f(\vec{k}))|} |M_{fi}|^2_{E_i - E_f = \hbar\omega} \tag{13}$$

where dS is the elementary area of an isoenergetic surface and V the volume of the sample. The integral has to be taken over all surfaces S with energy differences dE. In x-ray spectroscopy only vertical transitions (conservation of $\vec{k}$) have to be considered.

4. The density of states

The expression

$$N(E) = \frac{V}{(2\pi)^3} \int \frac{dS}{|\text{grad}_k E(\vec{k})|} \tag{14}$$

is called the density of states. N(E) is the number of electronic states of an energy band n per unit energy in the volume V.

When electron transitions are considered, the density of states of the initial and the final states is involved, expressed by the joint density of states

$$N(E_i, E_f) = \frac{V}{(2\pi)^3} \int_S \frac{dS}{|\text{grad}_k (E_i(k) - E_f(k))|_{E_i - E_f = \hbar\omega}} \tag{15}$$

Sometimes it is useful to discuss the results in terms of a density of states of s like, p like etc. electrons, $N_s(E)$, $N_p(E)$,..., respectively. (14) then can be written as

$$N(E) = N_s(E) + N_p(E) + N_d(E) + \cdots \tag{16}$$

The terms $N_s(E)$, $N_p(E)$, ... are called the partial densities of states.

In x-ray spectroscopy the core levels can be considered as k-indenpendent, i.e. E_i in expressions (12), (13) and (15) is not a function of k. Consequently the joint density of states equals the density-of-states function (14). This means that the intensity distribution of an x-ray emission band, (12) and (13), reflects the density of states modified by a matrix element. If the matrix element is constant, then because of the selection rules x-ray spectra reflect the partial densities of states.

5. Information provided by x-ray spectra

Since in K emission spectra the core state has s symmetry, in the dipole approximation the K emission bands reflect the distribution of p-like electrons in the valence band. The core state of the L spectra has p symmetry. Therefore the L emission bands reflect the s- and/or d-linke valence electrons with respect to the atom under consideration. Thus by studying the various emission bands we get different and complementary information about the valence bands.

The matrix element is only non-zero, if the wavefunctions ψ_i and ψ_f overlap. Since the core orbitals are strongly localized, electronic transitions can occur only from orbitals of the same atom or of closely adjacent atoms (1st coordination sphere). X-ray spectra therefore probe an atom and its close environment, that is, they act as a local probe.

When compounds are studied then by investigating the K, L, ... spectra of the component atoms we get a wealth of information about the molecular orbitals and their atomic orbital composition. Suppose that in a compound AB there are bonding and non-bonding orbitals. If transitions are allowed to core states of atoms A <u>and</u> B from a bonding orbital features from this orbital are expected to show up in the spectra of both atoms at the same relative energies. This property can show whether that particlar electrons participate in bonding or not. Because of its localization a non-bonding orbital can contribute only to the x-ray spectrum of the atom to which it belongs.

V. ILLUSTRATIVE EXAMPLES

The theoretical aspects described in the preceeding section will now be illustrated by a number of examples of emission line spectra and emission band spectra.

It may be mentioned that only a selected number of examples will be presented. No effort will be made at any kind of completeness. For details we refer to the proceedings of the x-ray conferences held every two years: Paris 1970 (21), Munich 1972 (22), Helsinki 1974 (23), Washington 1976 (24), Sendai 1978 (25), and Stirling 1980 (26). These proceedings also contain a great number of recent results and references of original papers.

1. Emission line spectra

As already mentioned emission line spectra arise from electronic transitions from a core state into a hole in another more tightly bound core state. Suppose we have solid samples of a pure element and one of its compounds. A schematic energy level diagram of the valence and core states is shown in Fig. 9 (the core states of other elements of the compound being omitted). As compared to the element the orbital energies in the compound are changed due to chemical bonding. According to the figure we get $\hbar\omega_o$ for the photon energy of the emission line in the element, resulting from electronic transitions between inner shells of energies E_1 and E_2

$$\hbar\omega_o = E_1 - E_2$$

In the compound the original inner shell energies may be shifted by ΔE_1 and ΔE_2 towards higher binding energies ($\Delta E < 0, |\Delta E_1| < |\Delta E_2|$). The energy of the x-ray photon then is

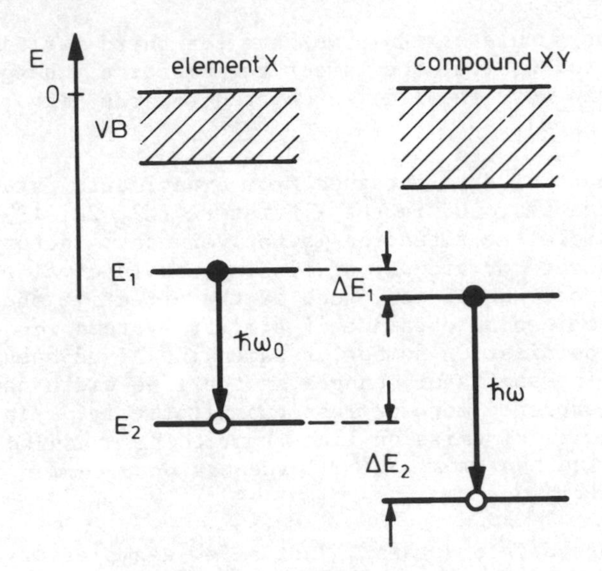

FIG. 9. Simplified energy level diagram of solid elemet X and compound XY: the chemical shift of the emission line of atom X results from the difference between the shifts ΔE_1 and ΔE_2 of the individual energy levels

$$\hbar\omega = (E_1 + \Delta E_1) - (E_2 + \Delta E_2)$$
$$= E_1 - E_2 + (\Delta E_1 - \Delta E_2)$$
$$= \hbar\omega_0 + \Delta\varepsilon$$

The photon energy has changed with chemical bonding by $\Delta\varepsilon$, which is the difference of the two differences ΔE_1 and ΔE_2. $\Delta\varepsilon$ may therefore be small, although ΔE_1 and ΔE_2 are not small, but of about the same magnitude. Since, however, the peak position of an x-ray line can be measured with high accuracy, even samll changes in photon energy are accessible to measurement.

The spectra depend upon the "chemical state" of the emitting atom. The chemical state comprises (i) the valency of the atom, (ii) the kind of neighbouring atoms and their electronegativities, and (iii) the coordination number. From all these quantities results an effective charge Q on the atom. If this effective charge is changed the lines are shifted and/or their width is changed. The predominant effects are the line shifts, the so-called chemical shifts.

In x-ray spectrsocopy the line shifts are determined with reference to the line positions for neutral atoms in the solid element. The line shift $\Delta\varepsilon = \hbar\omega - \hbar\omega_0$ is a function of the change in the atomic charge. As a rule of thumb we have for third period elements: $\Delta\varepsilon$ proportional to +Q, i.e. the larger the positive charge on the emitting atom, the more the line is shifted towards higher energies (shorter wavelengths).

A large amount of data obtained from experimental studies on emission lines can be found in the literature (27, 28, 19). Nevertheless on the whole the situation is not very satisfactory because there is no straight forward way of finding absolute values of physical or chemical quantities, such as the effective charge. Reliable information can be obtained if similar systems were compared, e.g. with same coordination number or same kind of adjacent atoms. In these cases line shifts or changes in the line width can be analyzed and interpreted more or less quantitatively. - Independent of that the energies of emission lines have to be measured in order to be able to align the emission band spectra on a common energy scale (see Fig. 1.)

We shall not go into details; just a few examples may give a feeling of how large the observed changes are.

1.1 Sulphur. Fig. 10 shows the averaged results for the $K\alpha_1$ and $K\alpha_2$ lines of sulphur in various groups of compounds. With increasing positive valency the lines are shifted towards smaller wavelengths (higher energies). The only group with negative valency are the sulphides, and therefore - because of the negative charge on the sulphur atom - the lines are shifted towards longer wavelengths. As can be seen there are small differences in wavelengths within the different groups of sulphates or sulphides, depending upon the bonding properties in the particular case. The energy difference between the fluorosulphates and the metal sulphides is about 1.3 eV, which is approximately the spacing between the $K\alpha_1$ and the $K\alpha_2$ line.

1.2 Aluminium. As a second example Fig. 11 shows the wavelengths and widths of metallic aluminium and a number of aluminium compounds.

As for the sulphur compounds the chemical shifts are very large in those compounds where the aluminium atom is bound to oxygen atoms, and the shifts are even larger for the fluor compounds. This is because oxygen and fluorine have the largest values of electronegativity (3.5 and 4.0, respectively). The whole range of line shifts comprises about 4 XE which is equivalent to 0.7 eV. - Within the third period elements the magnitude of chemical shift decreases with atomic number. The widths of the Al $K\alpha$-doublet are about 5 XE (0.89 eV), the width being smallest for the doublet of metallic aluminium (0.81 eV) and largest for AlP (0.96 eV).

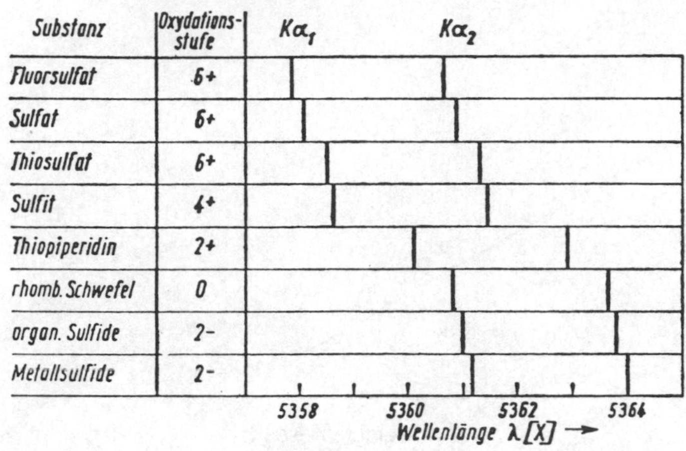

FIG. 10. Mean wavelengths of S $K\alpha_{1,2}$ for the different oxydation states of sulphur; after (19), based on (29)

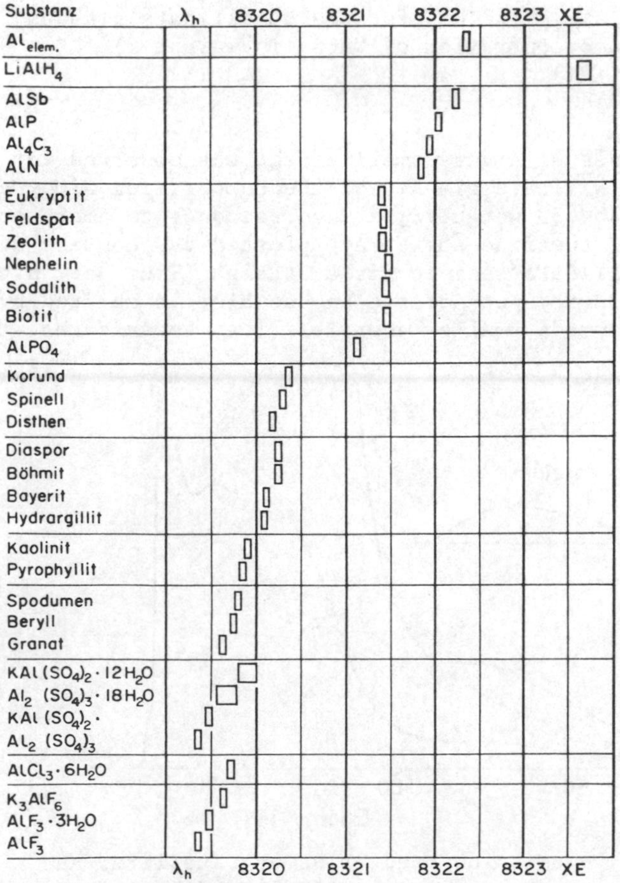

Substanz λh 8320 8321 8322 8323 XE

$Al_{elem.}$
$LiAlH_4$
AlSb
AlP
Al_4C_3
AlN
Eukryptit
Feldspat
Zeolith
Nephelin
Sodalith
Biotit
$AlPO_4$
Korund
Spinell
Disthen
Diaspor
Böhmit
Bayerit
Hydrargillit
Kaolinit
Pyrophyllit
Spodumen
Beryll
Granat
$KAl(SO_4)_2 \cdot 12H_2O$
$Al_2(SO_4)_3 \cdot 18H_2O$
$KAl(SO_4)_2 \cdot$
$Al_2(SO_4)_3$
$AlCl_3 \cdot 6H_2O$
K_3AlF_6
$AlF_3 \cdot 3H_2O$
AlF_3

λh 8320 8321 8322 8323 XE

FIG. 11. Wavelengths of maximum of Al $K\alpha_{1,2}$-doublet of different aluminium compounds (3o)

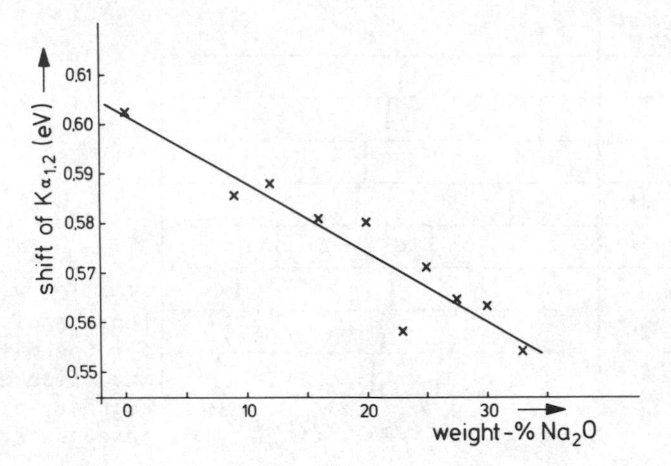

FIG. 12. Shift of Si $K\alpha_{1,2}$ (in eV) for sodium silicate glasses as a function of the concentration of Na_2O (in weight-%), ΔE = E(glass)-E(element), (31)

1.3 Silicate glasses. A very small effect was observed for the Si $K\alpha$-line shifts in silicate glasses of the composition SiO_2xNa_2O, x varying between 9 and 33 weight %; x=0 corresponds to amorphous SiO_2 (31). In Fig. 12 the line shifts are plotted vs. content of Na_2O. The experimental data seem to fit a straight line very nicely. Between x=0% and x=33% the peak of the $K\alpha$ - doublet is shifted by only about 0.05 eV towards smaller energies (i.e. towards the $K\alpha$-

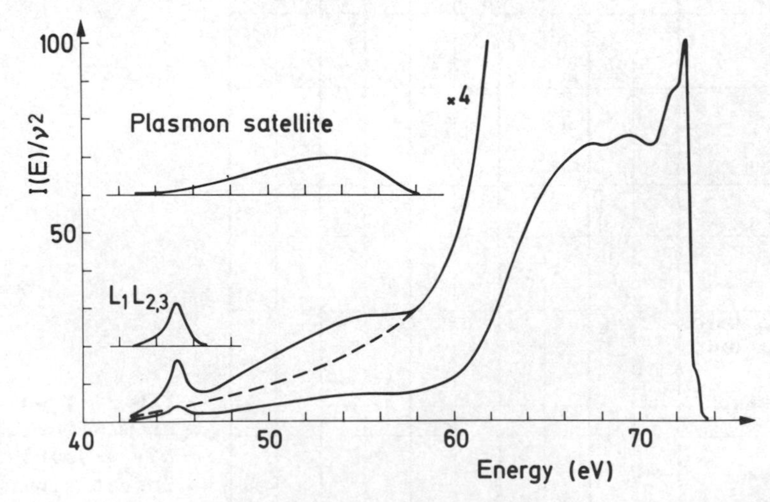

FIG. 13. Aluminium $L_{2,3}$-emission band, plasmon satellite, and $L_1L_{2,3}$-line. Low energy region also shown enlarged by a factor 4 (33)

doublet position of pure silicon). These small effects are supposed to be near the limit of what can be observed with x-ray spectrometers without taking a great deal of care. Much larger effects were observed for the Si Kβ-emission bands.–The purpose of this investigation was to study the influence of an expected increase of non-bridging Si-O-Na bonds compared to the bridging bonds -O-Si-O-Si- as the content of Na_2O in the glassed is increased (31,32).

2. Emission band spectra

2.1 Aluminium. As an example of a typical metal in Fig. 13 is shown the $L_{2,3}$ emission band of aluminium in an energy range of about 30 eV. The figure shows not only the main band, but also a number of features which are to be associated with many-body phenomena.

The main $L_{2,3}$ emission band extends from photon energies of about 61 eV to 73 eV. Due to the fact that aluminium can be regarded as a nearly free-electron like metal ($E^{1/2}$-behavior of density-of-states curve) the lower half of the spectrum does not exhibit any structural features; these show up only in the upper half of the spectrum. After a spike-shaped peak there is a two-step cut-off in intensity due to the Fermi edges of the L_3 band (high edge) and the L_2 band (small edge), these being different in energy by about 0.4 eV due to the spin-orbit splitting of the $L_{2,3}$ levels. At the bottom of the main band one observes some tailing caused by Auger processes, a plasmon satellite and the weak $L_1L_{2,3}$ line. The spike at 72,5 eV has been the subject of much discussion in the literature (34,35). The spike is thought to be the consequence of screening effects, although it is also present in the density-of-states curve (Fig. 14). At still higher energies, 76 eV to 89 eV, a satellite is observed which is caused by an electron transition in a multiply ionized atom, its measured intensity being less than 1% of the peak intensity of the main band.

In Fig. 14 the Kβ- and the $L_{2,3}$-emission band of aluminium are compared with the calculated density of states. The Al Kβ-band does not show any fine structure. The high-energy cut-off is not so abrupt as in the Al $L_{2,3}$-band due, to a broader width of the K level, and to the resolution of the spectrometer. As can be seen there is a close similarity between the structural features of the Al $L_{2,3}$-emission band and the density of states. It seems to be justified therefore to compare experimental data with theoretical results based on the one-electron model in other cases as well.

2.2 Silicon. Solid silicon has the diamond structure. Because there are four valence electrons per atom and two atoms per unit cell, the band structure consists of four energy bands. In the case of silicon we have the opportunity for a direct comparison of measured with calculated spectra. Fig. 15a shows a plot of the total

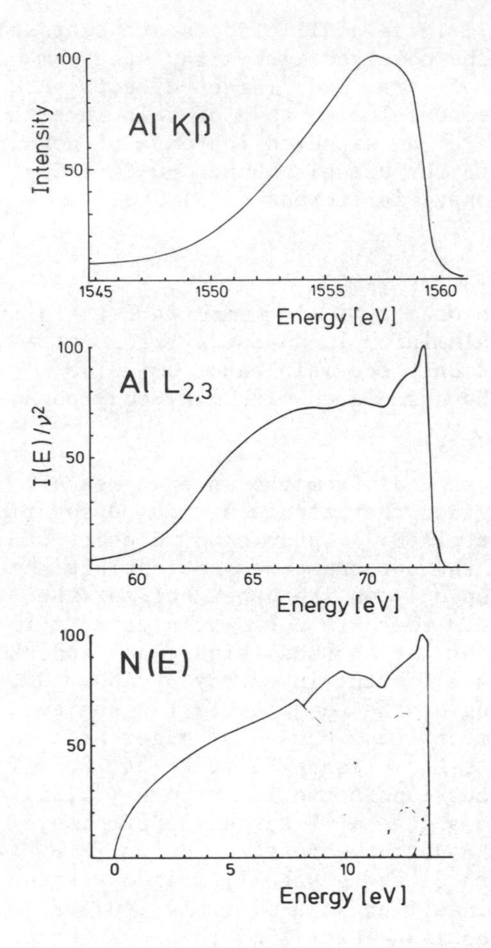

FIG. 14. Al Kβ-emission band (36), Al $L_{2,3}$-emission band (33), and density of states (37) of metallic aluminium

density of states and the density-of-states curves ν_1, ν_2, ν_3 and ν_4 of each energy band, as calculated by Klima (39). In Figs. b and d are shown the calculated Si $L_{2,3}$- and Si Kβ-emission bands, respectively (40). The large differences between the intensity distributions of the spectra and the density of states are primarily a consequence of the selection rules. Roughly speaking the K spectrum represents $N_p(E)$ and the L spectrum $N_{s+d}(E)$. As we can see the lowest energy bands are mainly s-like (ν_1 and ν_2). At the top of the valence band (ν_3 and ν_4) the p-like electrons predominate. The sharp structure in the density of states is smeared out in the spectra, because life-time broadening due to the Auger effect has been taken into account; this is also responsible for the tailing off at the low energy end of the spectra.

The experimental data were obtained by Wiech and Zöpf (38) (Fig. c) and by Läuger (Fig. d) (36). As can be seen the experimen-

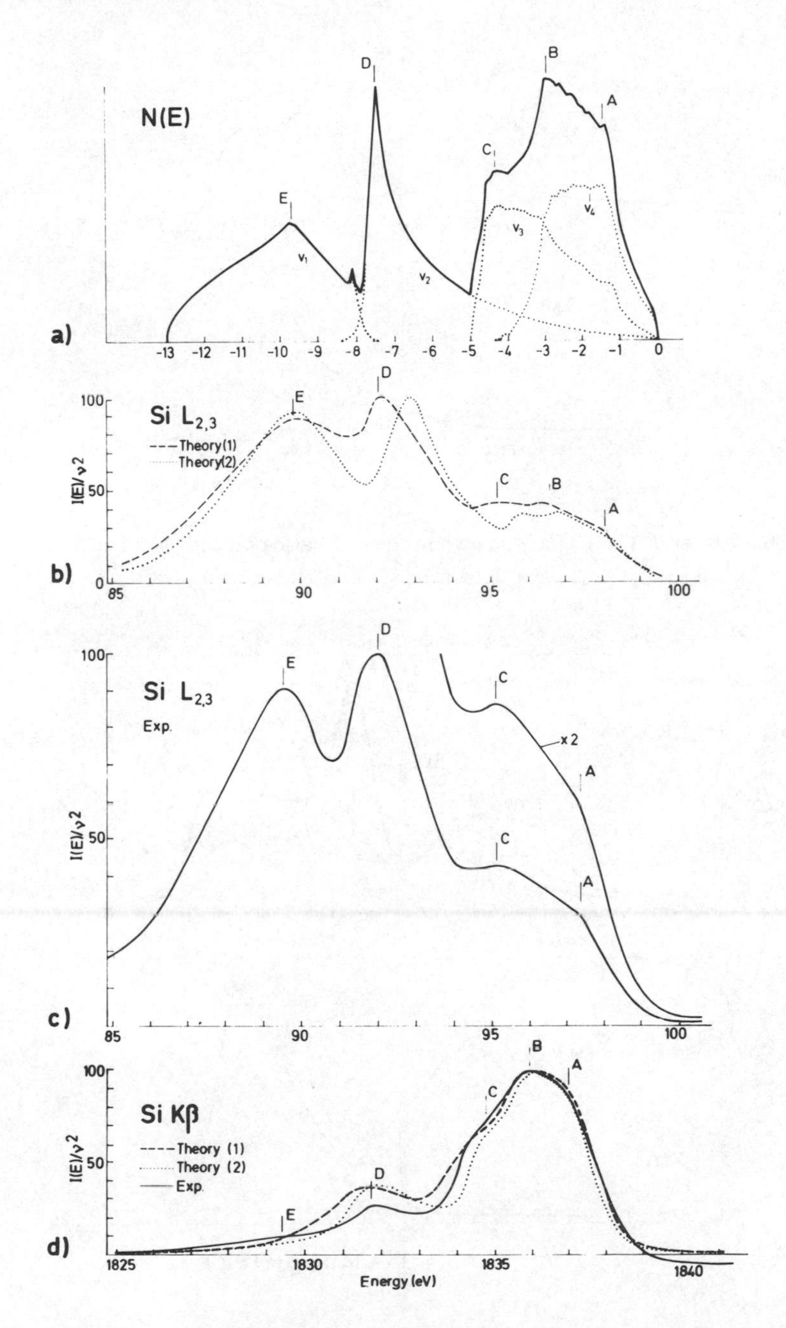

FIG. 15. Valence band structure of silicon (38). a) Density of states N(E) (39), b) Calculated Si $L_{2,3}$-emission bands (4o), c) Experimental Si $L_{2,3}$-emission band, d) Calculated Si Kβ-emission bands (dashed and dotted lines) (4o), and experimental Si Kβ-emission band (full line) (36,3o)

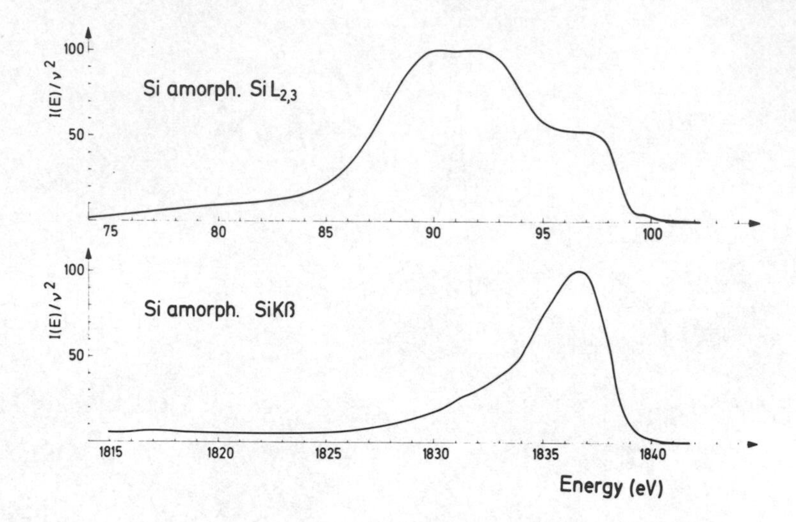

FIG. 16. Kβ and $L_{2,3}$ emission bands of amorphous silicon (41)

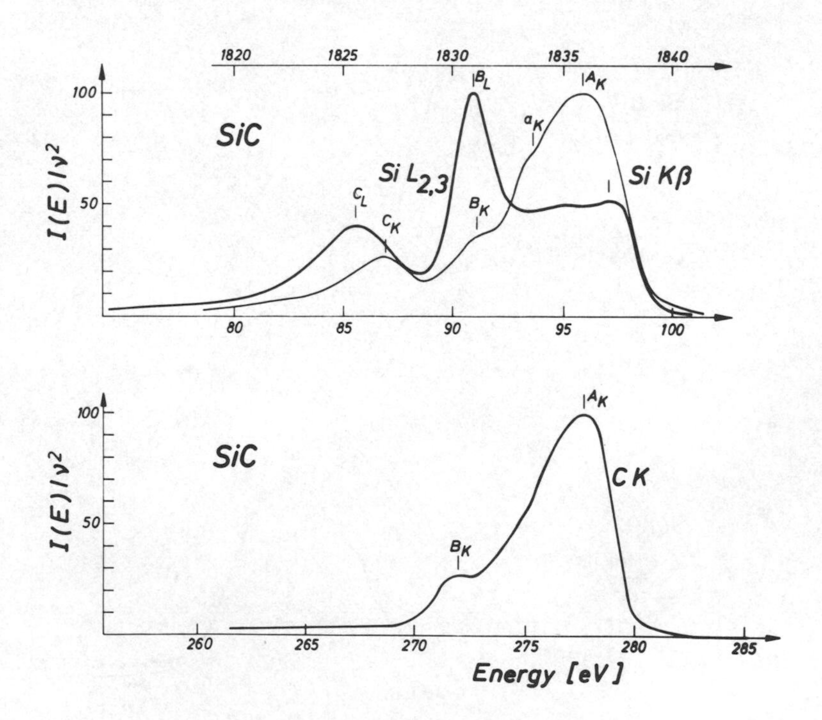

FIG. 17. Si Kβ-, Si $L_{2,3}$- and C K-emission bands of silicon carbide SiC (42, 43)

tal and the theoretical results, in particular those denoted as theory I, agree fairly well.

The Si K- and Si L-emission bands of amorphous silicon are shown in Fig. 16. The overall shapes of the spectra agree with those measured for crystalline silicon although the structural details are smeared out. This is generally observed for amorphous samples.

Because of the similarity of the spectra of both phases the band structure and the density of states must have the same gross features for crystalline and amorphous silicon. The breakdown of long-range order has not changed the electronic structure drastically. Since it is assumed that in amorphous silicon only short-range spacial order exists with predominantly tetrahedrally arranged interatomic bonds, the electronic structure in the crystalline phase results essentially from the particular bonding of adjacent atoms, and not so much from their periodic arrangement. On the other hand the similarity of the emission bands of amorphous and crystalline silicon strongly supports the assumption that the strong tetrahedral bonding orbitals in the crystalline phase are preserved in the amorphous state.

2.3 Silicon carbide SiC. SiC has a zincblende type structure; each silicon atom is surrounded tetrahedrally by four carbon atoms and vice versa. The Si K- and Si L-spectra are therefore expected to be related to the corresponding spectra of pure silicon. Due to the lifting of degeneracy at point X in the Brillouin zone of the zincblende structure the valence band is split into two subbands separated by a gap. The lower subband consists of band ν_1, the upper one of bands ν_2, ν_3 and ν_4 (see for comparison Fig. 15). The Si Kβ-, Si $L_{2,3}$- and the C K-emission bands of SiC are shown in Fig. 17. Indeed the upper parts of the silicon spectra resemble the respective spectra of pure silicon, indicating a simlilar distribution of s- and p-electrons in both substances. The peaks E_K and E_L which are connected with the low lying band ν_1 are shifted towards lower energies, and the minima between features E and C reflect the gap. That the intensity does not fall off to zero is due to the Auger tail of the upper subband. The shape of the C K-band is similar to the Si Kβ-band, with the exception of feature E, the intensity of which is zero.

Hemstreet and Fong (44) have calculated the density of states for the three highest valence bands of SiC and have compared it with a superposition of the silicon K and L emission bands of Fig. 17. The result is shown in Fig. 18. There is good agreement between the calculated and the measured features down to -6 eV. At lower energies the two curves diverge: the peak energies differ by more than 2 eV, indicating that in the theory the third energy band (from the top of valence band) extends too far to low energies.

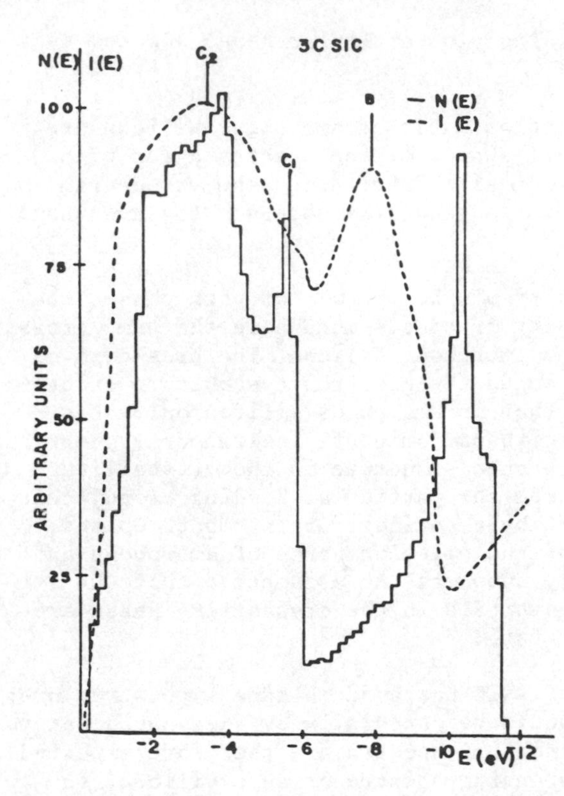

FIG. 18. Theoretical density-of-states curve for the three highest valence bands of 3C SiC (full line) (44). The dashed line represents a superposition of the Si K- and Si $L_{2,3}$-emission bands of Fig. 17 (42)

2.4 Diamond, graphite and glassy carbon. Diamond and silicon exhibit the same crystal structure. Since their bonding properties agree qualitatively, their x-ray spectra are also expected to be similar. Since, however, the L electrons of carbon are the valence electrons, carbon has only one x-ray spectrum: the K emission band. We can only check our supposition with respect to the K spectra therefore. As Fig. 19 shows, there is indeed close agreement between the features of the C K-spectrum of diamond and the Si Kβ-spectrum of silicon (Fig. 15). The figure also shows that the features of the density-of-states curve of diamond all are present in the C K-spectrum. From this one can conclude that, if diamond had a L spectrum (which represents the s-like valence states) it would have the same shape as the Si L-emission band.

In this particular case x-ray photoelectron spectroscopy (XPS) is a nearly ideal complementary spectroscopic technique. In XPS the photoionization cross-section σ differs greatly for valence states of different orbital character and wave function extent. For the low-Z elements of the second period the ionization cross-section of s-like electrons is considerably larger than for p-like electrons; $\sigma(2s)/\sigma(2p)$ is of the order of 20. The XP-spectrum of diamond arises predominantly from the s-like electrons (while the C K-spectrum represents the p-like electrons) and is therefore expected to have a

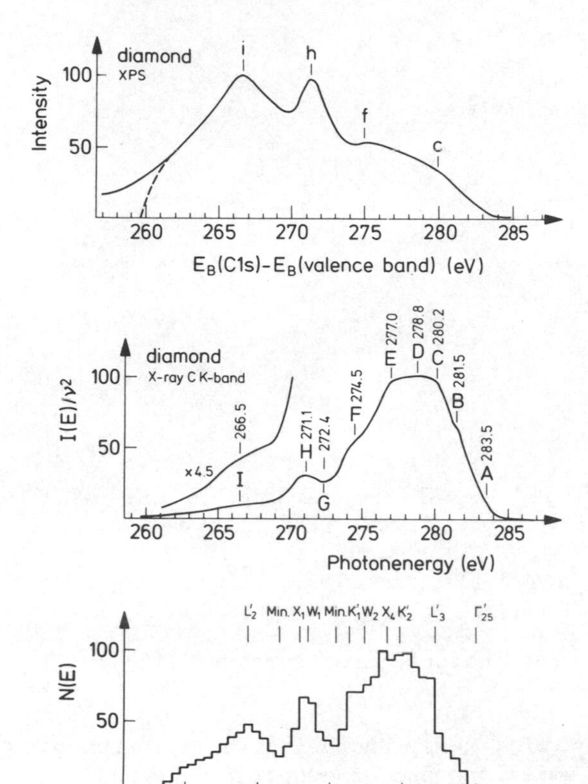

FIG. 19. Diamond (45); x-ray photoelectron spectrum (top) (46), x-ray K emission band (47), and density of states (bottom) (48)

shape similar to that of the Si L-emission band. This turns out to be true as Fig. 19 shows.

Thus from second period elements and their compounds information about the s-like electrons can be gained from XPS-studies, while x-ray spectra yield information about the p-like electrons!

In the same way can we obtain information about the electronic structure of graphite and glassy carbon. In Figs. 2o and 21 are shown the x-ray photoelectron spectra (46) and the C K-bands (45) of graphite and glassy carbon, respectively. At energies higher than about 275 eV the intensities are higher in the x-ray spectra and the features are more pronounced than in the XP-spectra, while for lower energies the reverse is true.

The electronic structure of graphite will be discussed in more detail in section VI. The spectra of glassy/evaporated carbon resemble those of graphite; all details, however, are smeared out as has been observed already for amorphous silicon (Fig. 16).

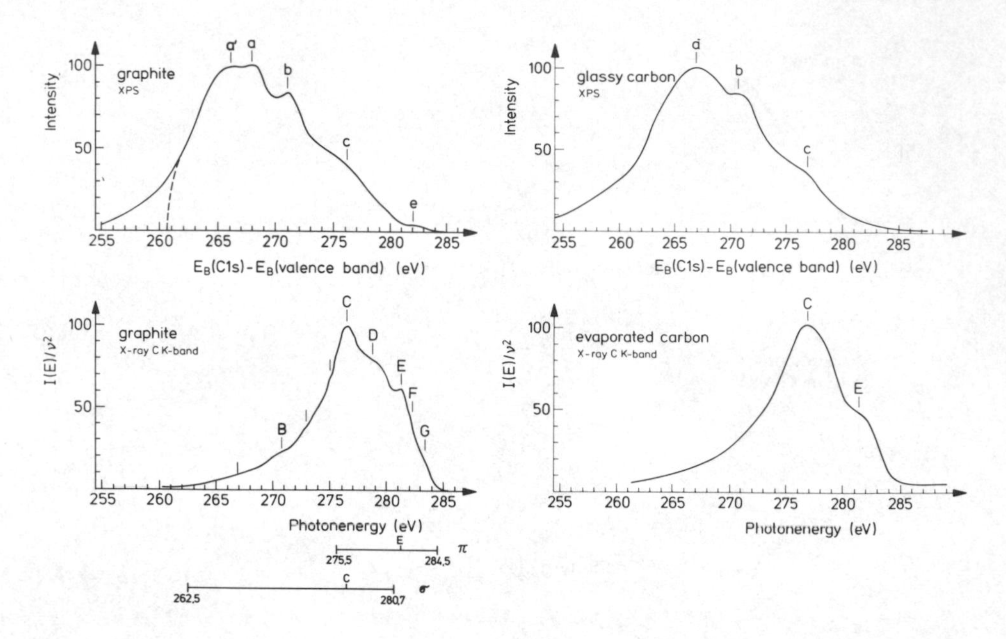

FIG. 20. Graphite (45); x-ray photoelectron spectrum (46) and K emission band (49)

FIG. 21. Amorphous carbon (45); x-ray photoelectron spectrum of glassy carbon (46) and K emission band of evaporated carbon

2.5 The sulphate ion SO_4^{2-}. As a last example let us consider the sulphate ion SO_4^{2-}. This example illustrates how the electronic structure can be analysed from a set of complementary experimental data. In Fig. 22 are shown the O K-, S Kβ- and S L-emission bands together with x-ray and ultraviolet photoelectron spectra (XPS, UPS) measured by different authors. The measurements were carried out using samples of different alkali metal sulphates. From several investigations it appears that the nature of the cation is of minor influence on the x-ray spectra of the oxyanion, and they can therefore be considered as being representative of the particular oxyanion.

The spectra in the figure were aligned to a common energy scale using the binding energies of the core levels and the energy of the S $K\alpha_1$-line. At the top are added the denotations of the molecular orbitals of the tetrahedral SO_4^{2-}ion and a list of the populations of atomic components of the molecular orbitals calculated by Johansen (55).

The shapes of the x-ray spectra differ conspicuously because they probe different symmetries of the molecular orbital electrons.

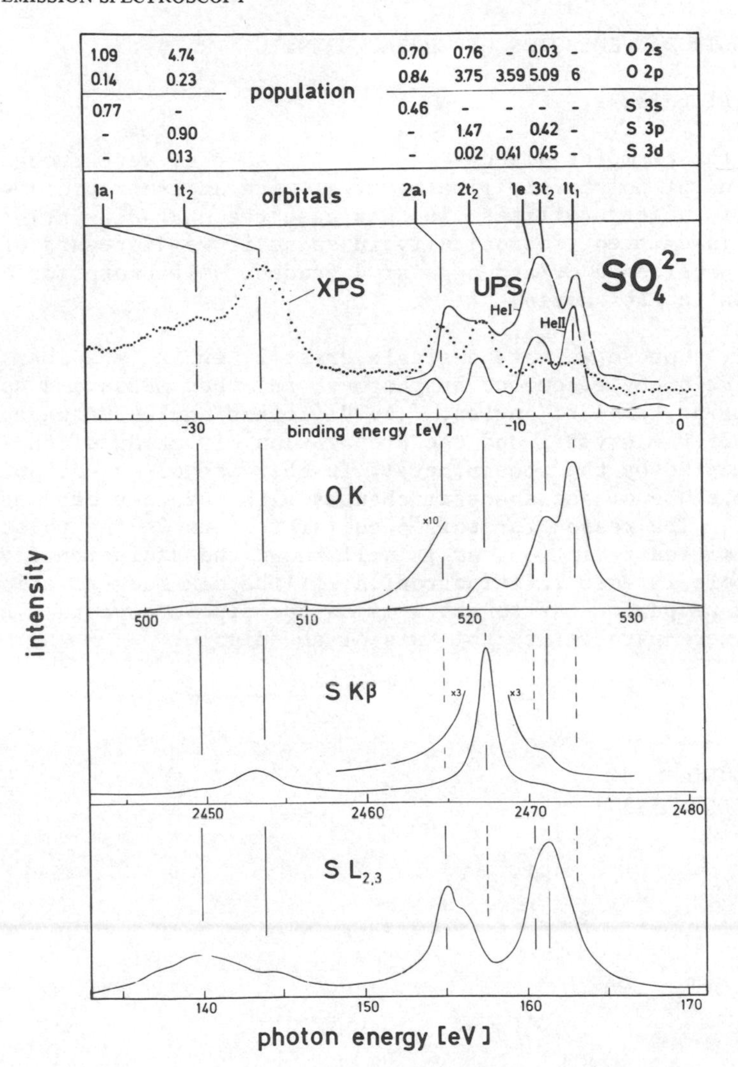

FIG. 22. Electronic structure of the sulphate ion SO_4^{2-} (50). O K-emission spectrum, S Kβ-emission spectrum (51), S $L_{2,3}$-emission spectrum (52), x-ray photoelectron spectrum (53), ultraviolet photoelectron spectra (54), and (top) calculated electronic structure (55).

Thus the lone pair O 2p electrons of orbital $1t_1$ only contribute to the O K-spectrum, and the high S 2p population of orbital $2t_2$ shows up in the S Kβ-spectrum. In the photoelectron spectra high s populations are reflected by the features of the XP spectrum, while the UP spectrum (He II) resembles the O K-spectrum, since UPS predominantly shows the p components, and also because the population of O 2p concentrated in the upper orbitals is much larger than that of S 3p.

VI. ANISOTROPIC EMISSION OF CHARACTERISTIC X-RAYS

1. General remarks

In the examples discussed sofar the samples were thought as consisting of polycrystalline material with an isotropic distribution of the crystallites. In this case the characteristic x-radiation is emitted isotropically in space if we disregard effects such as wavelength or/and angular dependent selfabsorption of the radiation in the sample.

Now, suppose we have a single crystal, excite its characteristic radiation by electrons or photons, and then measure a spectrum at different take-off angles, i.e. different angles between the surface of the crystal and the propagation direction of the radiation accepted by the spectrometer. In the case of anisotropic emission the shape of the spectrum changes with the take-off angle (Fig. 23). The reason for this peculiar feature is the polarisation of the emitted radiation. As is well known the radiation of an electric dipole is emitted anisotropically: the maximum radiation is observed perpendicular to the axis of the dipole, and the intensity is zero parallel to the axis of the dipole.

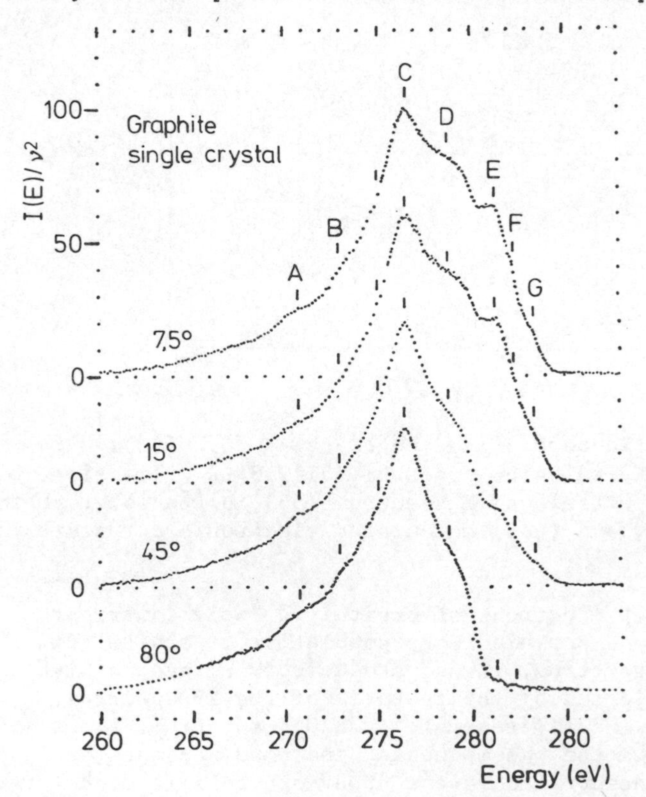

FIG. 23. K emission band of graphite single crystals measured at different take-off angles (49)

Anisotropic emission of x-radiation is a very general phenomenon. It is to be expected for all molecules and solids the symmetry of which is higher than triclinic and lower than cubic. In the non-triclinic and non-cubic systems the components of the dipole moment vector belong to at least two different irreducible representations. Those components which belong to different irreducible representations are perpendicular to each other, and therefore give rise to radiation of different polarizations. We now shall treat some theoretical aspects of allowed transitions in molecules and the polarization of the radiation accompanying these transitions.

2. Theoretical considerations

In section IV we discussed the l selection rules for dipole radiation. We now have to focus our interest on the m selection rules, which describe the polarization of the radiation. We shall argue in the language of group theory, from which one can work out the appropriate selection rules (57-59). Let the eigenfunctions of the initial and the final state be ψ_i and ψ_f, respectively (both non-degenerate).

The intensity I of an x-ray transition between states i and f is proportional to the square of the matrix element (transition moment) M_{fi}:

$$I \sim |\vec{M}_{fi}|^2$$

with
$$\vec{M}_{fi} = \langle \psi_f | \vec{p} | \psi_i \rangle = \int \psi_f \vec{p} \psi_i \, d\tau \tag{1}$$

$\vec{p}$ is the vector of the electric dipole moment with the components

$$\begin{aligned} p_x &= \sum_j e_j x_j = e x \\ p_y &= \sum_j e_j y_j = e y \\ p_z &= \sum_j e_j z_j = e z \end{aligned} \tag{2}$$

e_j represents the charge of particle j, and x_j, y_j, z_j are its coordinates. On inserting (2) into (1) we obtain equations

$$\begin{aligned} M_{fi}^x &= e \int \psi_f^* \, x \, \psi_i \, d\tau \\ M_{fi}^y &= e \int \psi_f^* \, y \, \psi_i \, d\tau \end{aligned} \tag{3}$$

$$M_{fi}^{z} = e\int \psi_f^{*} z \, \psi_i \, d\tau \tag{3}$$

The matrix element is identically zero if the integrand is an odd function. The matrix element is non-zero only if the integrand remains unchanged under all symmetry operations of the symmetry group to which the molecule or solid belongs. In the language of group theory this means: The matrix element is non-zero if the direct product of the irreducible representations Γ to which ψ_f, ψ_i, x, y, z belong is the totally symmetric representation (or contains the totally symmetric representation in case of degeneracy). Thus we have

$$\begin{aligned} &\text{(a)} \quad \Gamma(\psi_f) \times \Gamma(x) \times \Gamma(\psi_i) = \Gamma(A), \text{ or } \supset \Gamma(A) \\ &\text{(b)} \quad \Gamma(\psi_f) \times \Gamma(y) \times \Gamma(\psi_i) = \Gamma(A), \text{ or } \supset \Gamma(A) \\ &\text{(c)} \quad \Gamma(\psi_f) \times \Gamma(z) \times \Gamma(\psi_i) = \Gamma(A), \text{ or } \supset \Gamma(A) \end{aligned} \tag{4}$$

These are the selection rules for electric dipole transitions.

If only equation (a) is satisfied the emitted radiation is polarized with the dipole (vector) oscillating along the x axis. If equations (a) and (b) are satisfied the radiation is polarized with the dipole oscillating within the xy plane. If all equations (4) are satisfied the radiation is not polarized. Equations (4) are satisfied if we have

$$\begin{aligned} &\Gamma(\psi_f) \times \Gamma(\psi_i) = \Gamma(x) \\ \text{and/or} \quad &\Gamma(\psi_f) \times \Gamma(\psi_i) = \Gamma(y) \\ \text{and/or} \quad &\Gamma(\psi_f) \times \Gamma(\psi_i) = \Gamma(z) \end{aligned} \tag{5}$$

The symmetry species of the components of the dipole moment we can take from the character tables.

In addition we have to find out what atomic orbitals the atoms of a molecule may use to form σ- and π-molecular orbitals. By applying all symmetry operations of the point group to which the molecule belongs to the atomic orbitals we get the characters for the reducible representation generated. After reduction the character table tells us the atomic orbitals falling into the particular symmetry classes. On inserting these results into (5) we get the allowed and non-allowed dipole transitions. If a transition is allowed the intensity may be high or low. To find out how large the intensity is one has to calculate the integral in equation (1).

For illustration let us consider a planar molecule (or ion) AB_3 belonging to point group D_{3h} (Fig. 24)

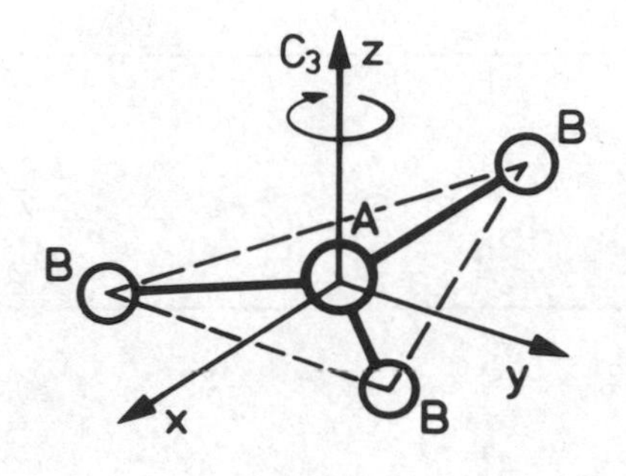

FIG. 24. Geometry of planar molecule AB_3 belonging to point group D_{3h}

The allowed transitions and the orientations of the dipoles are summarized in Table 1. In the first two colums are listed the symmetries of the core orbitals of the central atom A and the ligand atoms B, respectively, and in the third column their symmetry species. The remaining columns contain the symmetry species of the valence molecular orbitals and the polarization of specific valence → core state transitions.

When we ask for the K-emission bands, we can find from the table that the s-like K state of atom A belongs to the irreducible representation a_1'. To an orbital of a_1' symmetry transitions are allowed from e' and a_2'', the resultant radiation being xy and z polarized, respectively. The s core states of atoms B belong to the two representations a_1' and e'. The following electronic transitions are allowed: $e' \to a_1'$ (xy polarized), $a_2'' \to a_1'$ (z polarized), and a_1', a_2', $e'' \to e'$ (xy polarized), $e'' \to e'$ (z polarized). - Similar statements

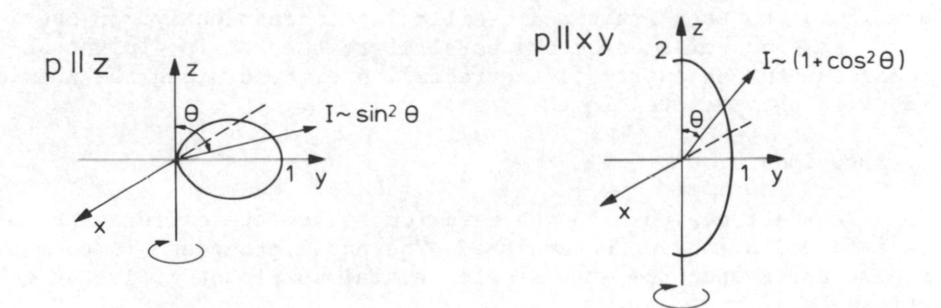

FIG.25. Intensity distribution of dipoles oscillating along the z axis (left) and parallel to the plane xy

TABLE 1. Allowed dipole transitions and orientation for the dipole moment of planar AB_3^{2-} ions belonging to point group D_{3h} Index ϱ denotes the directions A-B, and index φ means: within plane of ion and perpendicular to directions A-B

orbitals of central atom A	ligand atoms B	D_{3h}	σ a_1'	π^n a_2'	σ e'	a_1''	π a_2''	π e''
$s;d_{z^2}$	$s;p_\varrho;d_{\varphi^2-z^2},d_{\varrho^2}$	a_1'			xy		z	
	$p_\varphi;d_{\varphi\varrho}$	a_2'			xy	z		
$p_x,p_y;d_{x^2-y^2},d_{xy}$	$s;p_\varrho,p_\varphi;d_{\varphi^2-z^2}$ $d_{\varrho^2},d_{\varphi\varrho},d_{\varphi z}$	e'	xy	xy	xy			z
	$d_{\varphi z}$	a_1''		z				xy
p_z	$p_z;d_{z\varrho}$	a_2''	z					xy
d_{xz},d_{yz}	$p_z;d_{z\varrho}$	e''			z	xy	xy	xy

can be made for all the other transitions into a given core state.

z polarized means that the radiation characteristic is equivalent to that of a radiating dipole with the dipole axis parallel to the z axis, as shown in the upper part of Fig, 25. xy polarized means that the dipole is oscillating within the xy plane. The resultant radiation characteristic is shown at the bottom of Fig. 25. It can be thought as being generated by two dipoles oriented perpendicular to each other and oscillating independently. In the first case no emission occurs parallel to the z axis, in the second case the intensity of the radiation emitted along the z axis is twice that emitted in the plane xy.

3. Experimental

We shall now turn to the question of how the differently polarized radiation can be analyzed. The basic procedure is to measure a particular spectrum of a single crystal sample at different take-off angles.

Suppose we have a sample consisting of planar molecules or ions all oriented parallel to each other. Let the z axis be per-

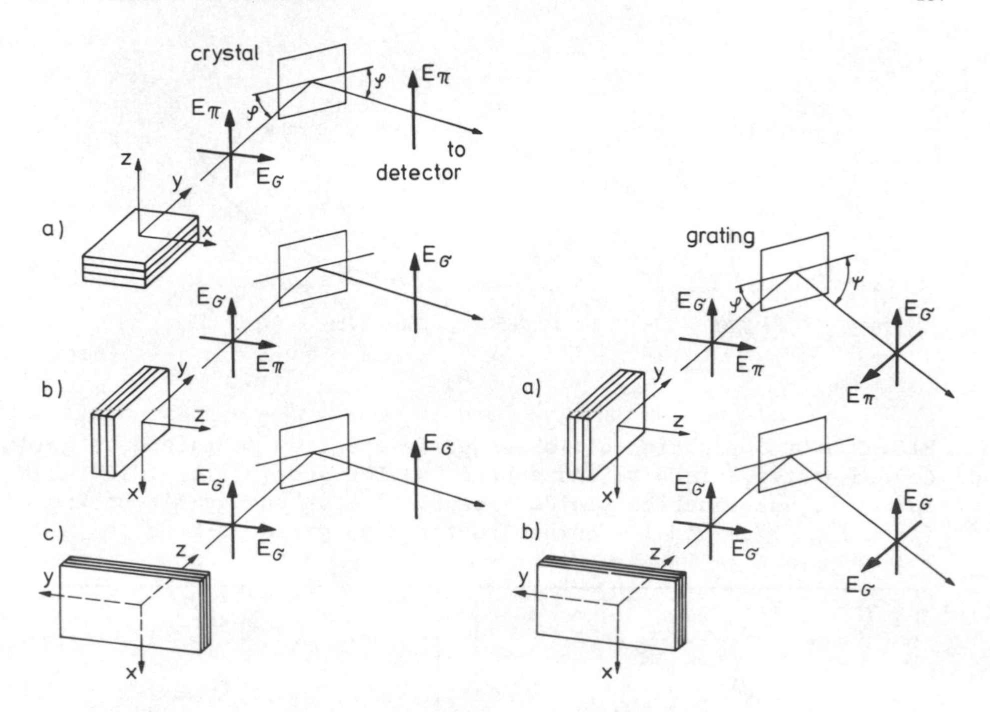

FIG. 26. Schematic view of a crystal spectrometer (left) and a grazing-incidence grating spectrometer (right) with different orientations of the single crystal samples (56)

pendicular to the planes of the molecules, and the x and y axis within these planes. The electric vector oscillating along the z axis is denoted as E_π. The resultant vector E_{xy} which oscillates within the xy plane can be considered as consisting of two identical components E_σ oscillating along the axes x and y. This is illustrated in Fig. 26.

We now have to distinguish between the application of a crystal and a grating spectrometer (Fig. 26). Let us first consider the crystal case, and assume the Bragg angle φ to be 45°. When the sample is mounted as shown in position a) E_π oscillates perpendicular to the plane of incidence, and one component of E_σ in the plane of incidence (parallel to x) while the other component of E_σ is zero along y. Since E_σ is suppressed by Bragg reflection only E_π remains. In position b) the Bragg-reflected beam contains only E_σ, and the same is true also for position c). - Thus by mounting the sample in positions a) and b) (or c)) E_π and E_σ can be measured separately. If a Bragg angle of 45° cannot be made available by choosing a proper analyzing crystal or because the spectrum extends too far in wavelengths, a separation of E_π and E_σ by measurement alone is not possible.

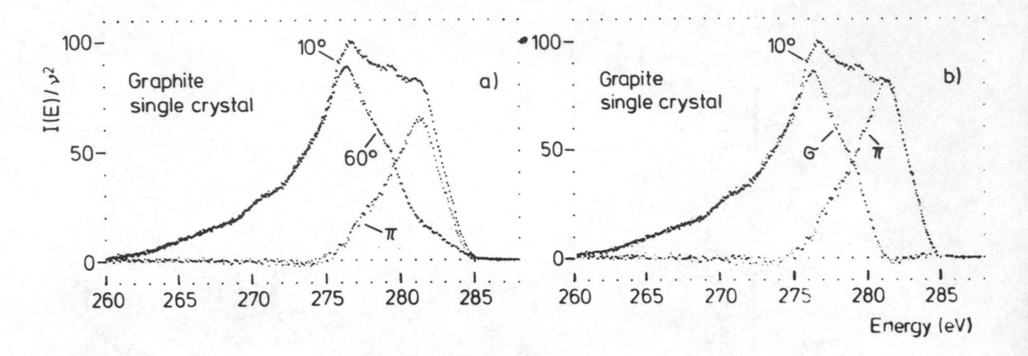

FIG. 27. Decomposition of 10° graphite spectrum (obtained by gold coated grating) into σ- and π-band. a) Low energy fit of 10° and 60° curve, subtraction yields π-band. b) High energy fit of π-band (from Fig. 27a) and 10° curve, subtraction yields σ-band (60)

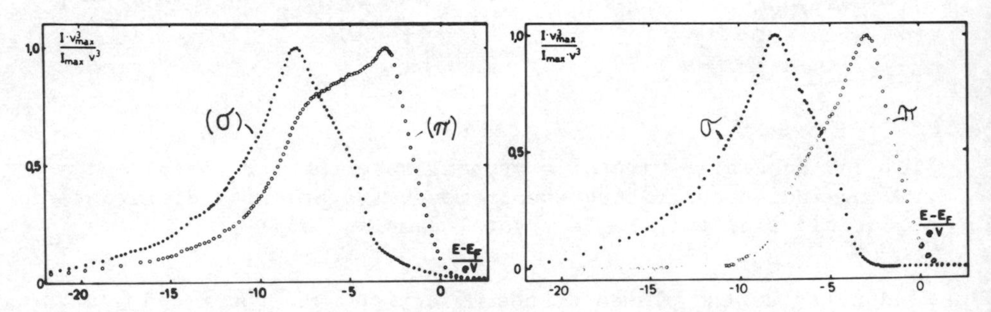

FIG. 28. Left: measured C K-spectrum of graphite, crystal orientation for curves π and σ according to a) and b), c) in Fig. 26 (left side), respectively; right: corrected π and σ band; after Kieser (61)

In the grating case no component of the radiation can be suppressed by the analyzer. Since the grating is used at grazing incidence, the radiation undergoes total reflection, and radiation with the electric vector within the plane of incidence or perpendicular to it is reflected to approximately the same degree. Let the sample be in position a) (take-off angle 0°), then E_{π} and E_{σ} are emitted by the sample and reflected by the grating. In position b) (take-off angle 90°) E_{π} is zero along the z axis, the emitted and reflected radiation consists of only the two E_{σ} components. If positions are choosen intermediate between a) and b) a varying amount of E_{π} is present in the registered radiation.

Summarizing, we can say that in a realistic case (i) a crystal spectrometer can provide information about E_{π} plus some contribution of E_σ, and about E_σ plus some contribution of E_π and (ii) a grating spectrometer can provide information about E_σ plus some contribution of E_π, and about a combination of E_π and E_σ.

4. Decomposition of spectra into their σ- and π-subbands

In practice the measured spectrum is nearly a combination of a π- and a σ-subband. The subbands can be decomposed in the following way.

For the intensity distribution observed at a take-off angle δ we can write (neglecting the wavelengths dependence of self-absorption)

$$I(\omega\quad) = f_\sigma(\delta)\cdot I_\sigma^0(\omega) + f_\pi(\delta)\cdot I_\pi^0(\omega) \tag{6}$$

$I_\sigma^0(\omega)$ and $I_\pi^0(\omega)$ are the intensity distributions of the σ and π contributions at maximum intensity, repsectively, $f_\sigma(\delta)$ and $f_\pi(\delta)$ are angular dependent intensity factors with $0 \leq (f_\sigma, f_\pi) \leq 1$, and ω is the frequency. The meaning of expression (6) is: Each spectrum measured at a given take-off angle δ is built up of two basic intensity distributions I_σ^0 and I_π^0 multiplied by appropriate intensity factors, varying between zero and unity.

To decompose spectra into their σ- and π-subbands we need (at least) two measurements carried out at take-off angles δ_1 and δ_2:

$$\begin{aligned} I(\omega,\delta_1) &= f_\sigma(\delta_1)\cdot I_\sigma^0(\omega) + f_\pi(\delta_1)\cdot I_\pi^0(\omega) \\ I(\omega,\delta_2) &= f_\sigma(\delta_2)\cdot I_\sigma^0(\omega) + f_\pi(\delta_2)\cdot I_\pi^0(\omega) \end{aligned} \tag{7}$$

We now multiply the second intensity distribution $I(\omega,\delta_2)$ by a factor α as to match it with $I(\omega,\delta_1)$ in a region where f_π is zero (σ-fit: $\alpha f_\sigma(\delta_2)=f_\sigma(\delta_1)$), and then subtract the two spectra. The result is

$$I(\omega,\delta_1) - \alpha\cdot I(\omega,\delta_2) = \left(f_\pi(\delta_1) - \alpha\cdot f_\pi(\delta_2)\right)\cdot I_\pi^0(\omega) = \Delta_\pi \tag{8}$$

i.e. the difference spectrum Δ_π is proportional to $I_\pi^0(\omega)$

$$\Delta_\pi \sim I_\pi^0(\omega) \tag{9}$$

By a similar procedure (π-fit) we obtain

$$\Delta_\sigma \sim I_\sigma^0(\omega) \tag{10}$$

From (9) and (10) we obtain (i) the shape and the width of $I_{\pi}^{o}(\omega)$ and $I_{\sigma}^{o}(\omega)$, and (ii) the amount of overlap of these two subbands. The angular dependency of a spectrum permits (iii) the identification of particular orbitals.

5. Experimental results

Table 2 gives a view of subjects appropriate to studies of anisotropic emission and includes a number of examples. In the following we shall discuss some of these examples.

TABLE 2. Subjects and examples for studies of anisotropic emission

A. Solids	
1. single crystals	graphite (49, 60, 61) MoS_2 (62); metallic Be (63), Mg and Zn (64)
2. samples with texture highly oriented crystallites	hexagonal BN (65, 66)
B. Molecules (molecular crystals)	
1. One molecule per unit cell all molecules parallel to each other	CO_3^{2-} in $CaCO_3$ (67)
2. Two molecules per unit cell molecules inclined to each other by certain angles	paradichlorobenzene (68) naphthalene (69, 70)

5.1 Graphite. As an example of the decomposition of the σ- and π-subbands let us first consider graphite. In Fig. 27a are shown the C K-emission bands of graphite measured with a grating spectrometer at take-off angles of 10° and 60°. The spectra were fitted in the low-energy region; there is fair agreement between 260 eV and about 275 eV, but then the spectra diverge. The difference curve Δ_{π} is denoted π. Departing from what has been said Δ_{σ} (denoted σ in Fig. 27b) was obtained by fitting the curve Δ_{π} to the 10° curve and then subtracting the two curves. The resultant curves allow us to determine the widths of the π-band, the σ-band (more strictly speaking the σp-band), the overlap of the two bands and their structural features.

Graphite was also measured with a crystal spectrometer by Kieser (61). His results are presented in Fig. 28. The measured spectra were decomposed by applying a fitting procedure as described

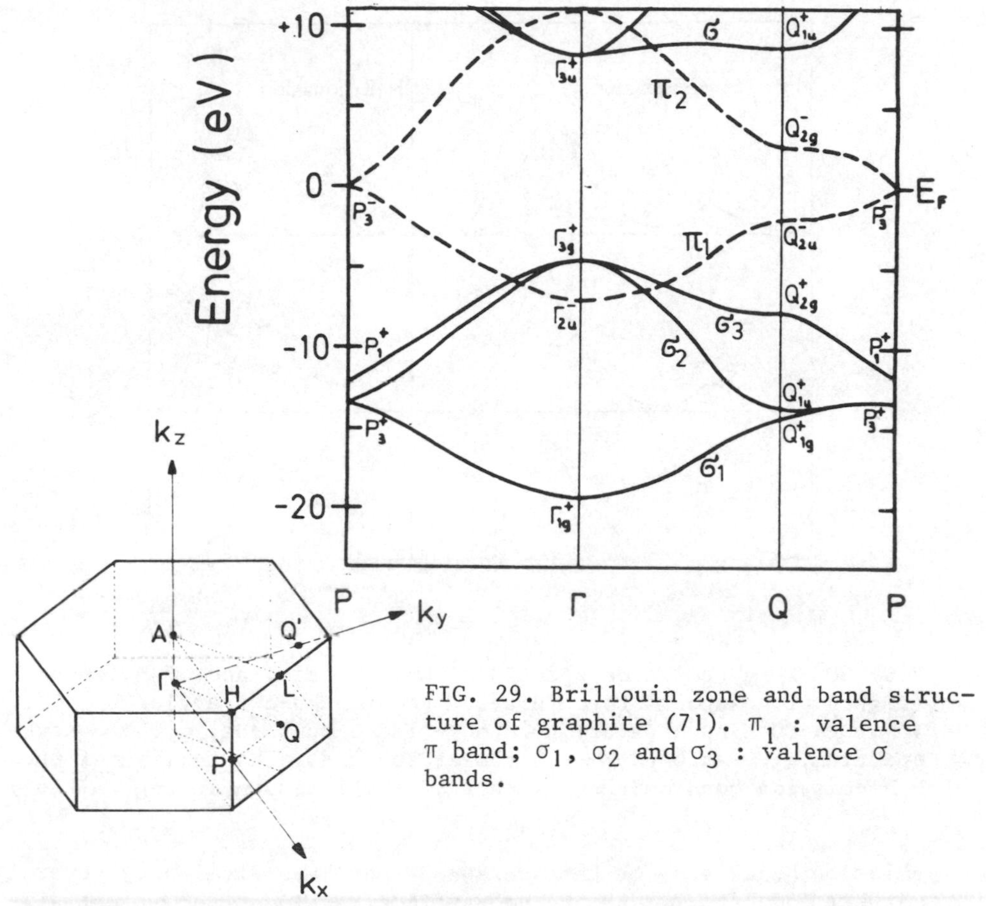

FIG. 29. Brillouin zone and band structure of graphite (71). π_1 : valence π band; σ_1, σ_2 and σ_3 : valence σ bands.

above. As can be seen the corrected σ- and π-bands agree quite well with those obtained with a grating spectrometer (Fig. 27).

Fig. 29 shows the band structure of graphite as calculated by Painter and Ellis (71). The valence energy bands are denoted as π_1 (π-band), and σ_1, σ_2 and σ_3 (σ-bands). The top part of the valence band consists of only the π_1-band, while the bottom part consists of only the σ-bands and between these two parts there is a region of overlap of π_1- and the σ-bands. The same result was obtained from the experimental data. For a quantitative discussion and comparison of theoretical and experimental results see ref. 60.

5.2 Hexagonal boron nitride BN. The crystal structure of hexagonal BN is the same as that of graphite. Its electronic structure is therefore also similar to the one shown in Fig. 29, with the exception of the band σ_1 which is separated from the other bands by a gap of several eV. BN is of particular interest because anisotropic behaviour is to be expected for both the N K- and the B K-

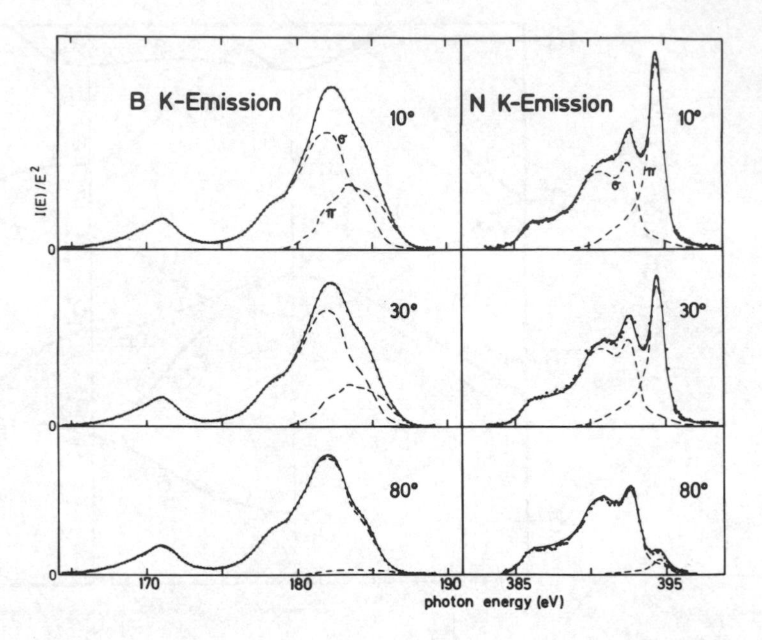

FIG. 30. X-ray emission spectra of boron (left) and of nitrogen (right) of hexagonal boron nitride BN at take-off angles of 10°, 30°, and 80° and separation into π- and σ-subbands. ..... observed spectrum, – – –subbands, ——— sum of subbands. The tailing of the N K-emission band on the low energy side is shown in Fig. 31 (66)

emission band, and for this reason BN has been studied by several authors (60, 65, 72). Since no single crystals of sufficient size are available textured polycrystalline samples had to be sued. The samples proved to be highly oriented, and the spectra exhibit intensity variations similar to those expected for single crystals.

The B K- and N K-spectra of BN measured at take-off angles 10°, 30° and 80° are shown in Fig. 30 together with their respective σ- and π-bands. In the N K-spectrum the variation of intensity is particularly pronounced. In the B K-spectrum the peak at about 171 eV is caused by the low-lying σ_1-band, and the gap extends in the region around 174 eV. A similar low-energy peak is also observed for the N K-spectrum, but it is very weak and therefore not shown in the figure (see, however, Fig. 31).

In Fig. 31 the x-ray spectra and an x-ray photoelectron spectrum (66) are aligned on a common energy scale and compared with the density-of-states curve calculated by Nakhamson and Smirnov (73). The dashed and the dotted curves represent the partical densities of states (σ- and π-bands). As the figure shows there is fair quali-

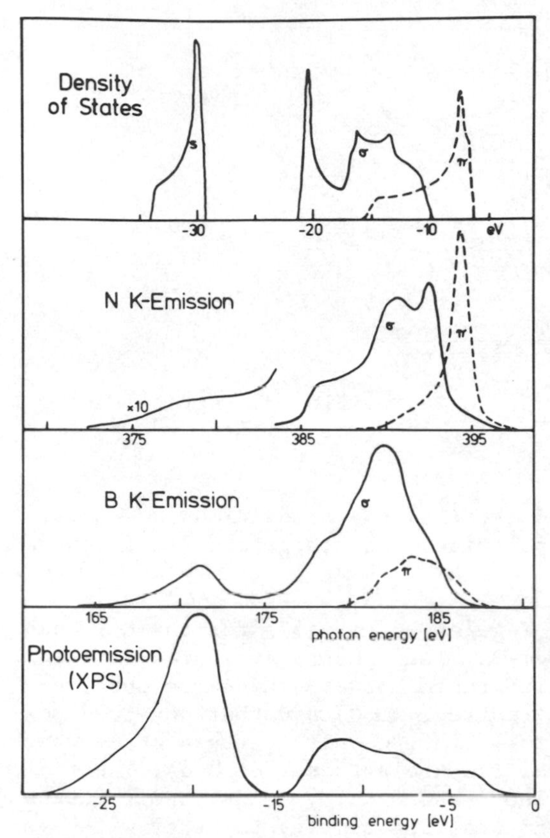

FIG. 31. Electronic structure of hexagonal boron nitride (66). π- and σ-bands of a) density of states; after (73), bands shifted. b) N K- and c) B K-emission, d) x-ray photoelectron spectrum. σ-band: full curves, π-bands: dashed curves

tative agreement between the features of the calculated and the measured curves. For a quantitative discussion of the results the reader is referred to ref. 66.

<u>5.3 The ion CO_3^{2-}.</u> The crystal structure of calcite $CaCO_3$ is shown in Fig. 32. The planar CO_3^{2-} ions (D_{3h}) are all oriented parallel to each other. Because of the crystal structure measurements are feasible at take-off angles of 0° and 90°. To a first approximation we can neglect the influence of the crystal environment on the electronic structure of the complex ion, and consider the CO_3^{2-} ion as being isolated.

The ion CO_3^{2-} has 12 occupied valence orbitals. Since some are degenerate, they have 8 distinct binding energies. From Table 1 we can learn: (i) there are two π-orbitals (a_2'' and e") the others are σ-orbitals, (ii) there is no orbital with a_1'' symmetry, (iii) transitions into the O 1s level (a_1', e') are allowed from all valence orbitals, but transitions into the C 1s level are allowed

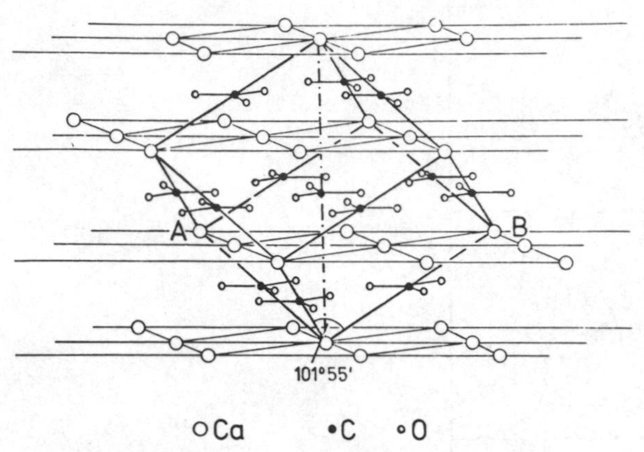

FIG. 32. Crystal structure of calcite $CaCO_3$

only from the orbitals e' and a_2'', (iv) the transitions $a_2'' \rightarrow$ C 1s, $a_2'' \rightarrow$ O 1s and $e_2'' \rightarrow$ O 1s are z polarized, and therefore are absent in the 90° spectrum.

The results obtained by Tegeler et al. (67) for the C K- and O K-spectrum are shown in Fig. 33. The spectra are normalized in the low-energy region. The difference between the curve observed parallel to the layer planes (dotted, $\pi + \sigma$) and that observed perpendicular to these planes (full line, σ only) gives a curve which represents the contribution of the π electrons (dashed). The results are shown in Fig. 34: The π and σ x-ray bands (their intensities corresponding to a polycrystalline sample) together with available photoelectron spectra (53, 74) as well as calculations of the binding energies (75, 76) and the composition of the orbitals, all spectra being aligned to a common energy scale.

By an analysis of the data we can unambiguously identify the π-orbitals $1a_2''$ and 1e", we can clear up the controversial sequence of the adjacent orbitals 3e' and $1a_2''$ (see top of Fig. 34) including a separation of these orbitals with an energetic spacing of only 0.4 eV, although the instrumental resolution is only 0.7 eV, and we can determine the binding energies of all orbitals. - For a more detailed discussion we refer to ref. 67.

5.4 Naphthalene. With naphthalene we are faced by a comparatively complicated system: The unit cell contains two planar molecules of different orientation as shown in Fig. 35, there are 24 occupied orbitals (19σ and 5π), and because it is an organic compound a number of additional experimental problems arise. Consequently the carbon K radiation is emitted anisotropically. Since, however, the angle between the planes of the two molecules is 52.6° the observed effect is smaller than if all molecules were parallel to each other.

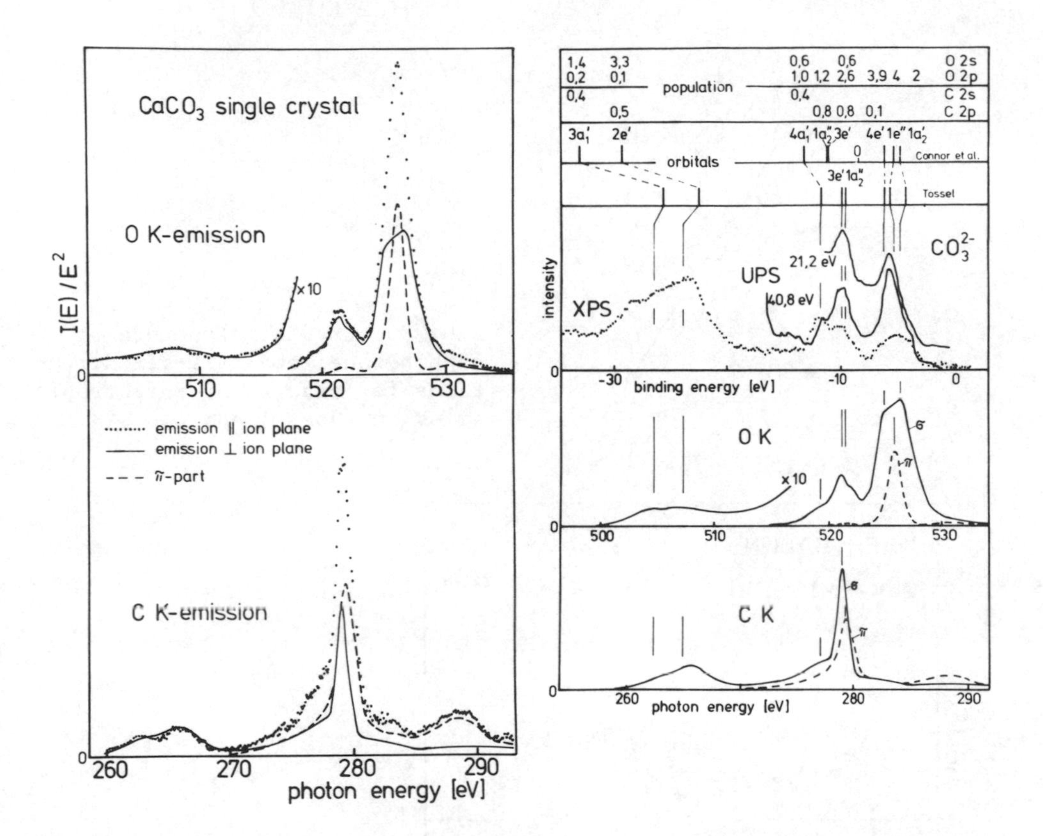

FIG. 33. O K- and C K-emission spectrum of calcite single crystals, parallel (dotted) and perpendicular (full line) to the plane of the CO_3^{2-}-ion. The dashed line shows the π-part of the emission parallel to the plane of the ion (67)

FIG. 34. Electronic structure of CO_3^{2-} (67). π- and σ-parts of the oxygen and carbon spectrum, together with XPS (53) and UPS (74) measurements, calculated positions (75, 76) and population (75) of the orbitals

The experimental results obtained by Tegeler et al. (69) for single-crystal and polycrystalline specimen are shown in Fig. 36. Radiation from the single crystals was studied along the c-axis and in the ac-plane with an angle of 66° to the c-axis, the two positions being denoted as pos. 1 and pos. 2 in Fig. 35. By the procedure described above the spectra can be separated into a π- and a σ-band. The result is shown in Fig. 37. This figure also shows the results of UPS measurements for solid (77) and gaseous (78) naphthalene. There is good agreement between the energies of corresponding features in all spectra. It is of particular interest

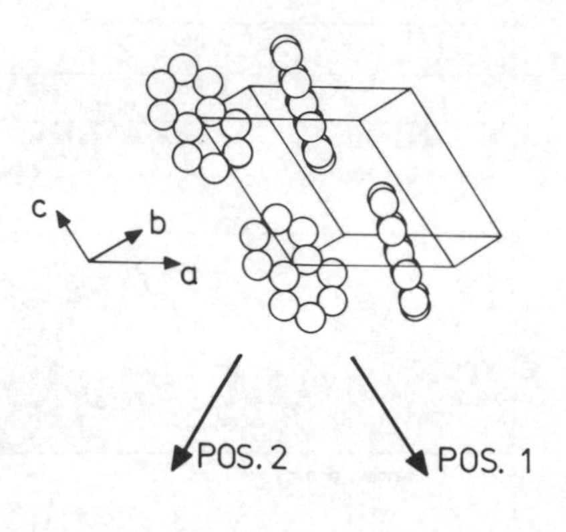

FIG. 35. Crystal structure of naphthalene, and take-off directions for studies of anisotropic emission (pos. 1 and pos. 2)

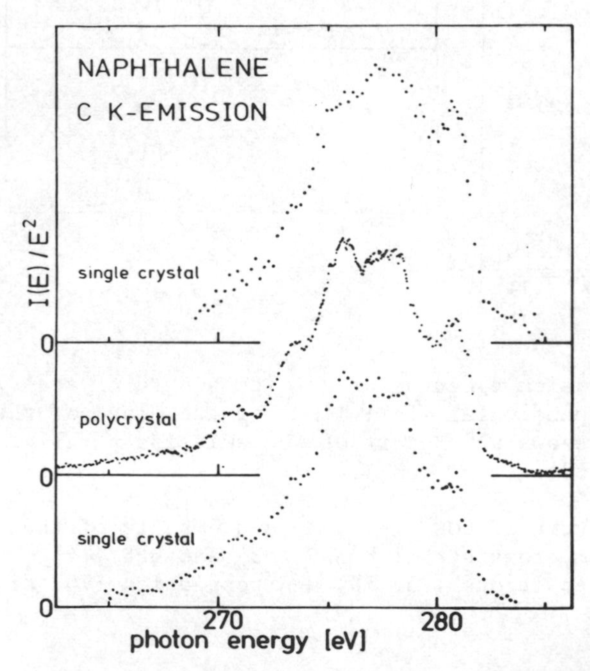

FIG. 36. C K-emission spectrum of solid naphthalene (69); top: single crystal, emission parallel to the c-axis; middle: polycrystalline naphthalene; bottom: single crystal, emission in the ac-plane with an angle of 66° to the c-axis

that the calculated π-orbitals (80) agree quite well with those obtained from an analysis of the x-ray data, and that the lowest π-orbital b_{3u} (feature H) has the comparatively high binding energy of about 15 eV.

VII. CONCLUSION

It has been the purpose of this paper to demonstrate how x-ray

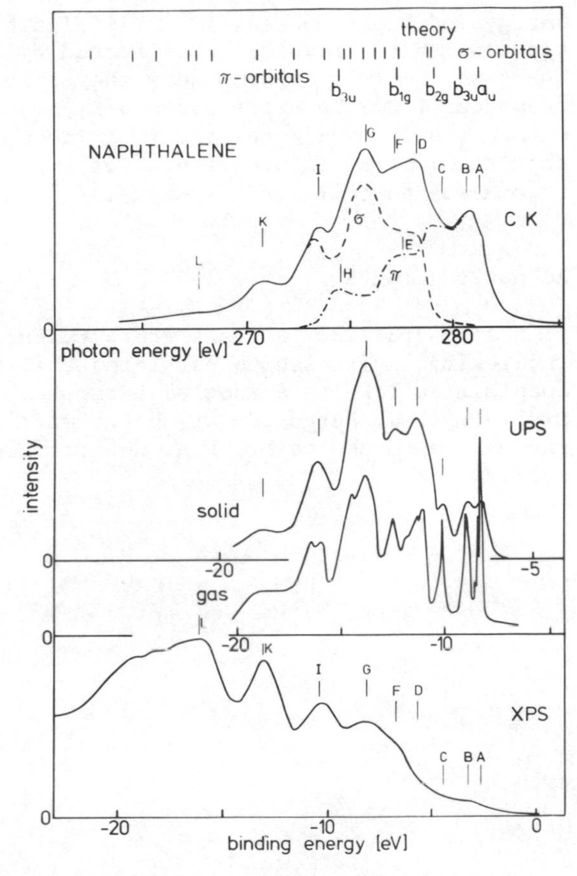

FIG. 37. Electronic structure of naphthalene (70). C K-emission spectrum of polycrystalline naphthalene and its decomposition into π- and σ-bands, together with the UP-spectra of solid (77) and gaseous (78) naphthalene, and the XP-spectrum of solid naphthalene (79); top: calculated energies for the π- and σ-orbitals (80)

emission spectroscopy can be applied to study the electronic structure and chemical bonding in molecules and solids. Although x-ray spectroscopy is a valuable tool for approaching these problems, there are also some factors which limit its applicability. The most important should be mentioned here.

First, with increasing atomic number Z the core states become broader, and their width seriously limits the resolution. While the width of the K level of third period elements is some tenths of an eV, it is about 2 eV for germanium, about 7 eV for silver and more than 50 eV for gold (81). Similar variations of the core level width with Z are found for other shells.

Spin-orbit splitting of the core levels also can reduce information. Suppose that the features in a spectrum are separated by about 2 eV; if spin-orbit splitting is of the same order of magnitude we obtain two spectra of different intensity, shifted in energy and superimposed. It is difficult to disentangle the measured spectra and the results are affected by additional uncertainties. The same is true for compounds where the spectra of the constituent atoms can (and therefore often will) overlap. That the second period elements only have K spectra is fortunately compensated by the fact that the missing information about s-like electrons can be obtained from x-ray photoelectron spectra.

Acknowledgements

I should like to thank Prof. C. Furlani and Dr. P. Day for their kind invitation to participate in this NATO symposium. I should also like to acknowledge the help of Dr. W. Zahorowski and Dipl.-Phys. W. Burghard during the writing of this paper, and to express my thanks to Dr. P.A. Cox for his criticisms on the manuscript.

REFERENCES

1. Mahan, G.D.: 1967, Phys. Rev. 163, 612
2. Nocières, P. and De Dominicis, C.T.: 1969, Phys. Rev. 178, 1o97
3. Hedin, L.: X-ray Spectroscopy, edited by Azaroff, L.V. (Mc Graw Hill) 1974, p. 226
4. Hedin, L.: 1978, J. Physique 39, C4-1o3
5. Almbladh, C.O. and Minnhagen, P.: 1978, Phys. Rev. B7, 929
6. Kosuch, N., Tegeler, E. Wiech, G., and Faessler, A.: 1978, Nucl. Instr. Meth. 152, 113
7. Tegeler, E.: Thesis 1978, Ludwig-Maximilians-Universität München
8. Gilberg, E., Hanus, M.J., and Foltz, B.: 1978, Japan. J. appl. Phys. 17-2, 1o1
9. Wiech, G.: 1967, Z. Physik 2o7, 428
1o. Andermann, G., Henke, B., Urch, D.S., and Wiech, G.: 1978, Japan. J. appl. Phys. 17-2, 428
11. Tegeler, E., Wiech, G., and Faessler, A.: 198o, J. Phys. B 13, 4771
12. Hoffmann, L., Wiech, G., and Zöpf, E.: 1969, Z. Physik 229, 131.
13. Liefeld, R.J.: in Soft X-Ray Band Spectra, edited by Fabian D.J. (Academic Press) 1968, p. 133
14. Feser, K., and Faessler, A.: 1968, Z. Physik 2o9, 1
15. Handbuch der Physik, edit. S. Flügge (Springer Verlag) 1957, Vol. XXX
16. Blochin, M.A.: Methoden der Röntgenspektralanalyse (B.G. Teubner Verlagsgesellschaft Leipzig) 1963
17. Thomsen, J.S.: in X-ray Spectroscopy, edit. Azaroff, L.V. (Mc Graw Hill) 1974, p. 26
18. Cuthill, J.R.: in X-ray Spectroscopy, edit. Azaroff, L.V. (Mc Graw Hill) 1974, p. 133
19. Meisel, A., Leonhardt, G., and Szargan, R.: Röntgenspektren und chemische Bindung (Akademische Verlagsgesellschaft, Leipzig) 1977
2o. Davydov, A.S.: Quantum Mechanics (Oxford: Pergamon) 3rd ed., 1965
21. Proc.Int.Conf. on Processus Electroniques Simples et Multiples du Domaine X et X-UV, 1971, J. Physique 32, C-4
22. Proc.Int.Conf. on X-Ray Spectra and Electronic Structure of Matter, ed. A. Faessler and G. Wiech, München 1973, 2 Vol.
23. Proc.Int.Conf. on X-Ray Processes in Matter, 1974, Physica Fennica 9-S1
24. Extend.Abstr.Int.Conf. On the Physics of X-Ray Spectra, 1976, NBS, Gaithersburg, Md. USA
25. Proc.Int.Conf. on X-Ray and XUV Spectroscopy, 1978, Japan. J. appl. Phys. 17-2
26. Proc.Int.Conf. on X-Ray Processes and Inner-Shell Ionization, 1981 (in press)
27. Faessler, A.: Röntgenspektrum und Bindungszustand, in: Landoldt-Börnstein, Bd. I/4, 6. Aufl. (Springer) 1955, p. 769
28. Meisel, A.: 1965, phys.stat.sol. 1o, 365

29. Faessler, A.: 1931, Z. Physik 72, 734; Proc. 10th Coll.Spectrosc. Internat. Washington 1963 (Spartan Books, Washington) p. 3o7
3o. Läuger, K.: 1971, J. Phys. Chem. Solids 32, 6o9
31. Alter, G. and Wiech, G.: ref. 25, p. 288
32. Brückner, R., Chun, H.-U. and Goretzki, H.: ref. 25, p. 291
33. Neddermeyer, H. and Wiech, G.: 197o, Physics Lett. A 31, 17
34. Mahan, G.D.: Vacuum Ultraviolet Radiation Physics, ed. E.E.Koch, R. Haensel and C. Kunz (Pergamon Vieweg), 1974, p. 635
35. Dow, J.D.: l.c. ref. 34, p. 649
36. Läuger, K.: Thesis 1968, Ludwig-Maximilians-Universität München
37. Ashcroft, N.W.: Optical Properties and Electronic Structure of Metals and Alloys, ed. F. Abeles (North Holland Publ. Comp.) 1966, p. 336
38. Wiech, G. and Zöpf, E.: ref. 22, p. C4-200
39. Klima, J.: private communication
4o. Klima, J.: 197o, J. Phys. C - Solid State Phys. 3, 7o
41. Wiech, G. and Zöpf, E.: Band Structure Spectroscopy of Metals and Alloys, ed. D.J. Fabian and L.M. Watson (Academic Press) 1973, 629
42. Wiech, G.: 1967, Z. Physik 2o7, 428
43. Wiech, G.: Rentgenowskie spektry i elektronnaja struktura weschtschestwa, ed. IMF AN USSR, Kiew 1969
44. Hemstreet, L.A. and Fong, C.Y.: 1971, Solid State Commun. 9, 643
45. Wiech, G.: X-Ray Photoelectron Spectroscopy, ed. Academy of Sciences of the USSR, Kiev 1977, p. 74
46. McFeely, F.R., Kowalczyk, S.P., Ley, L., Cavell, R.G., Pollak, R.A., and Shirley, D.A.: 1974, Phys. Rev. B9, 5268
47. Umeno, M. and Wiech, G.: 1973, phys.stat.sol. (b) 59, 145
48. Painter, G.S., Ellis, D.E., and Lubinsky, A.R.: 1971, Phys.Rev. B4, 361o
49. Beyreuther, Chr. and Wiech, G.: l.c. ref. 34, p. 517
5o. Kosuch, N., Wiech, G., and Faessler, A.: 198o, J. Electron Spectrosc. Relat. Phenom. 6, 1
51. Manne, R., Karras, M., and Suoninen, E.: 1972, Chem. Phys. Lett. 15, 34
52. Henke, B.L. and Taniguchi, K.: 1976, J. Appl. Phys. 47, 1o27
53. Calabrese, A. and Hayes, R.G.: 1975, J. Electron Spectrosc. Relat. Phenom 6, 1
54. Connor, J.A., Hillier, I.H., Wood, M.H., and Barber, M.: 1978, J. Chem. Soc. Faraday Trans. II 74, 1285
55. Johansen, H.: 1974, Theor. Chim. Acta 32, 273
56. Wiech, G.: ref. 26 (in press)
57. Hollas, J.M.: Die Symmetrie von Molekülen (W. de Gruyter) 1975
58. Cotton, F.A.: Chemical Applications of Group Theory (John Wiley and Sons, Inc.) 1963
59. Streitwolf, H.-W.: Gruppentheorie in der Festkörperphysik (Akademische Verlagsgesellschaft) Leipzig 1967
6o. Beyreuther, Chr. Hierl, R., and Wiech, G.: 1975, Ber. Bunsenges. phys. Chem. 79, 1o82

61. Kieser, J.: 1977, Z. Phys. B 26, 1
62. Haycock, D., Kasrai, M., and Urch, D.S.: ref. 25, p. 138
63. Brümmer, O., Dräger, G., Fomichev, W.A., and Schulakov, A.S.: ref. 22, Vol. I, p. 78
64. Dräger, G., Brümmer, O., and Bonitz, J.: 1979, phys.stat.sol. (b) 94, K 111
65. Tegeler, E., Kosuch, N., Wiech, G., and Faessler, A.: 1977, phys.stat.sol. (b) 84, 561
66. Tegeler, E., Kosuch, N., Wiech, G., and Faessler, A.: 1979, phys.stat.sol. (b) 91, 223
67. Tegeler, E., Kosuch, N., Wiech, G., and Faessler, A.: 198o, J. Electron Spectrosc. Relat. Phenom. 18, 23
68. Gilberg, E.: 1977, Chem. Phys. Letters, 51, 246
69. Tegeler, E., Kosuch, N., Wiech, G., and Faessler, A.: ref. 25, p. 97
7o. Tegeler, E., Wiech, G., and Faessler, A.: 1981, J. Phys. B (in press)
71. Painter, G.S. and Ellis, D.E.: 197o, Phys. Rev. B1, 4747
72. Borovskii, I.B., Semochkin, P.N., and Rzaeva, M.: 1975, Fiz. Metallov i Metallovedenie 4o, 537
73. Nakhmanson, M.S., and Smirnov, V.P.: 1972, Soviet Phys. - Solid State 13, 2763
74. Connor, J.A., Considine, M., and Hillier, I.H.: 1978, J. Chem. Soc., Faraday Trans. II 74, 1285
75. Connor, J.A., Hillier, I.H., Saunders, V.R., and Barber, M.: 1972, Mol. Phys. 23, 81
76. Tossel, J.A.: 1976, J. Phys. Chem. Solids 37, 1o43
77. Grobman, W.D., and Koch, E.E.: Photoemission in Solids II, ed. L. Ley and M. Cardona (Springer) 1979, p. 261
78. Brundle, C.R., Robin, M.B., and Kuebler, N.A.: 1972, J. Am. Chem. Soc. 94, 1466
79. Riga, J., Pireaux, J.J., Caudano, R., and Verbist, J.J.: 1977, Physica Scripta 16, 346
8o. Hayashi, T. and Nakajima, T.: 1975, Bull.Chem.Soc.Jpn. 48, 98o
81. Sevier, K.D.: Low Energy Electron Spectrometry (Wiley-Interscience) 1972

BREMSSTRAHLUNG ISOCHROMAT SPECTROSCOPY IN SOLIDS

Yves Baer

Laboratorium für Festkörperphysik, ETH
CH-8093 Hönggerberg, Zürich, Switzerland

The problems involved in the use of bremsstrahlung isochromat spectroscopy (BIS) as a tool for probing the unoccupied electronic states of solids are briefly reviewed and some simple applications of this technique are discussed.

INTRODUCTION

In 1915 Duane and Hunt (1) demonstrated for the first time that the continuous x-ray emission produced by a x-ray tube has a high energy or short wave length limit (SWL) corresponding precisely to the acceleration voltage applied between the electrodes of the tube. This observation which gave a fundamental information about the nature of the x-rays has been quite often used subsequently to determine the value of e/h. In the following decades the technological progress allowed considerable improvement in the reliability and accuracy of such measurements and in 1942 Ohlin (2) succeeded in resolving structure in the emission spectrum of continuous x-rays just above the SWL. The observation that the details of the structures depended on the cathode led to the conclusion that they must reflect some property of the bombarded material. Effects due to the density of states (3) and to characteristic energy losses (4) were proposed. The credit for showing the usefulness of this continuous x-ray radiation for investigating vacant states in solids must be attributed to K. Ulmer and collaborators (5-7) who have been active continuously in this field for more than 20 years. Today very few groups have taken advantage of the accumulated experience to apply this

P. Day (ed.), Emission and Scattering Techniques, 153–170.

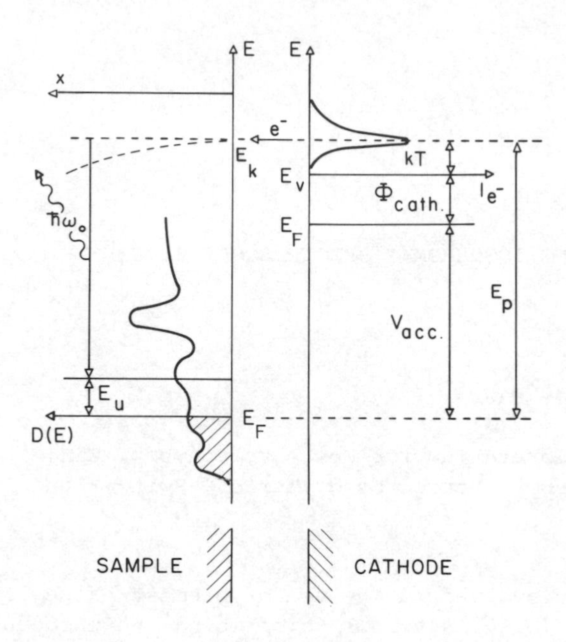

Fig. 1 Energy diagram of the system cathode-sample used to produce bremsstrahlung (22)

method to current problems in solids. In fact this radiation, which yields unique information about vacant states above the Fermi energy (E_F), can be measured with good resolution by modern techniques for any stable sample.

ELECTRON DIFFUSION IN SOLIDS

A schematic representation of the process involved in the emission of continuous radiation produced by electron bombardment of a solid is given if Fig. 1. A nearly monoenergetic electron beam emitted by a thermionic cathode at the energy ϕ + kT (ϕ:work function) above its vacuum level E_v is accelerated (V_{acc}) toward the surface of the sample. A fraction of the impinging electrons are expected to be elastically reflected by the potential step at the surface whereas the other electrons will diffuse into the solid. The interaction of high energy electrons (100 - 5000 eV) with solids is strong and gives rise to many inelastic processes leaving the solid in excited states. We are looking for a very particular process which is the direct radiative decay (Bremsstrahlung) of the incident electron into an empty state above E_F. The Bremsstrahlung intensity of a given energy $\hbar\omega_o$ can therefore be produced by any electron having a kinetic energy above the Fermi level $E_k \geq \hbar\omega_o$. At threshold, i.e. $E_k = \hbar\omega_o$, only electrons which have kept their initial energy before the bremsstrahlung process contribute to the emitted x-ray intensity of energy $\hbar\omega_o$. For higher primary energies, the slowing-down of the electrons resulting from the energy losses during the transport within the sample is symbolically represented in Fig. 1 by the dashed line falling in the x direction. On the average, the distance through which an electron has a sufficiently high energy to emit a bremsstrahlung photon increases with the primary energy above the threshold value and consequently the bremsstrahlung intensity also increases. An exact calculation of this behaviour would require precise knowledge of the probabilities of all inelastic processes in the particular material. At the present time this problem is intractable and only a qualitative discussion of the situation can be given (8). As we shall see below, the useful information on vacant states is only provided by bremsstrahlung processes of electrons which have not previously lost energy; all other processes contribute to the background. A practical and simple way (9) to separate these two contributions approximately is based on XPS core level spectra measured at kinetic energies comparable to the primary energy of the beam. The tail of inelastically scattered electrons emitted on the low energy side of any XPS line is used to estimate the number of inelastically scattered electrons which can contribute to the bremsstrahlung intensity. This procedure makes many approximations (9), in particular, many-body effects in XPS (10) and bremsstrahlung spectra are implicitely assumed to be identical and erroneously analyzed in term of energy losses. From the practical point of view, the existing bremsstrahlung results show that the background originating from the slowing-down of electrons is a reasonably smooth increasing function and is not a serious obstacle to the observation of superimposed structure far above the threshold, produced by the radiative decay of unscattered electrons (see for example Fig. 4). In the case of a very intense discrete energy loss, however, the inelastically scattered electrons can produce sizable discrete structures appearing as shifted replica of the spectrum attributable to unscattered electrons. This situation most likely arises in simple metals (11). For the purpose of a qualitative interpretation of the spectra it is usually sufficient to describe the electron transport within the solid by a single paramater, the inelastic mean free path Λ. The energy distribution of the inelastically scattered electrons is then completely ignored and Λ is a simple measure of the surface sensitivity, i.e. of the thickness below the sample surface where bremsstrahlung processes are mainly due to electrons which have kept their primary energy. The magnitude (~ 5 - 30 Å) and energy dependence of Λ are rather well known from photoemission experiments and will not be discussed here. It is still useful to notice that if the kinetic energy of the emitted electrons in photoemission and the primary energy of the beam producing bremsstrahlung are similar, both types of experiments will have precisely the same surface sensitivity. The soft x-ray absorption in solids can be neglected since it is much weaker than the electron attenuation discussed above.

THE BREMSSTRAHLUNG PROCESS

The measurements of the bremsstrahlung intensity in a narrow energy range allows us to separate these weak transitions from all non-radiative processes and from the other radiative processes producing photons of different energies. The theoretical description of the Bremsstrahlung process is usually performed for relativistic energies and is restricted to the atomic or ionic case (12-14). In the studies of the vacant states in solids, the primary energy of the incident beam does usually not exceed a few keV so that the relativistic effects can be neglected and the dipole approximation can be used. Under the assumption that the process can be described as a single-electron transition and neglecting many-body effects, the bremsstrahlung matrix element has the simple form

$$m_{f,i} \propto \int \psi_f^* \vec{r}\,\vec{\varepsilon}\,\psi_i\, d^3r$$

where $\vec{\varepsilon}$ is the photon polarization vector, ψ_i and ψ_f are the one-electron wave functions of initial and final states. For not too low kinetic energies ($E_k > 100$ eV), the initial state can be considered to be a nearly-free electron state containing all symmetries so that dipole selection rules are not important. This is no longer true for low energy initial states which may keep a more pronounced atomic character. To the best of my knowledge, no systematic calculations of bremsstrahlung matrix elements has been performed with realistic wave functions in solids. In our simple approach, it is very helpful to notice that these matrix elements are formally identical to the matrix elements for the photoexcitation. One can easily take advantage of this fact in a restricted energy range around E_F by considering in first approximation that for a given symmetry they remain identical on both sides of E_F (only absolutely exact at E_F), provided that the same photon energy is involved in bremsstrahlung and photoexcitation processes. In this sense one technique is the natural continuation of the other one and all the experimental informations accumulated on the cross-sections of the outermost states in photoemission can be used for bremsstrahlung processes near the SWL and will not be discussed here. The energy dependence of the matrix elements over a wide energy range has been shown to be important (15) and cannot be ignored. A more complicated situation will be encountered in the UV range where it becomes necessary to consider the momentum concervation which is responsible for the selection of the accessible vertical transition.

The bremsstrahlung intensity is also proportional to the product of the densities of initial and final states. For high primary energies, the density of initial states has no influence since it is nearly constant in a restricted energy range and is populated by an incident beam of constant intensity. This is no longer true for low-energy initial states; they are still markedly influenced by the periodic potential of the solid and the mechanisms for populating them by the incident electron beam must be explicitely taken into account. The energy dependence of the electron transmission coefficient at the surface and the transport within the sample are no longer negligible. In the high-energy limit, the bremsstrahlung intensity produced by the unscattered electrons at a fixed energy is proportional to the squared matrix elements and the density of states of the final states. This makes the symmetry of photoemission and bremsstrahlung processes obvious if the roles of initial and final states are exchanged. At this point it becomes necessary to define more precisely the parameters which are varied in the measurement of the bremsstrahlung intensity. There are essentially three different ways (7) to measure a spectrum: i) The primary energy of the electron beam is maintained constant and the energy distribution of the emitted bremsstrahlung is recorded. This is a measurement of the short wave length limit (SWL). ii) The primary energy of the electron beam is varied and the emitted bremsstrahlung intensity is recorded at a fixed wave lenght. This method is usually called Bremsstrahlung Isochromat Spectroscopy (BIS). iii) The bremsstrahlung intensity is measured as a function of a simultaneous and equal variation of the beam energy and of the analyzed x-ray energy. The interest of this constant initial state method (7) will not be discussed here. In the high-energy limit, the initial state has been shown previously to be unimportant so that the two first methods (SWL and BIS) can be expected to be equivalent in a restricted energy range. This equivalence is demonstrated experimentally by the similarity of the spectra of W (Fig. 2) obtained by the two different methods (16). This result, however, is no longer valid if the binding energy of some core level of

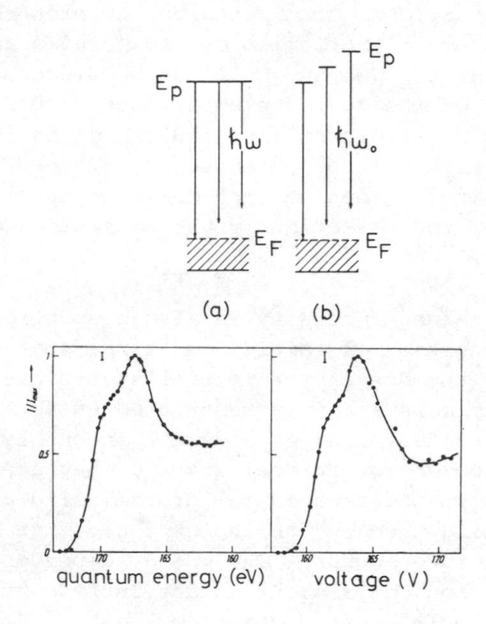

Fig. 2 Continuous x-ray emission of W measured (a) at constant electron energy (SWL) and (b) at constant photon energy (BIS), (16)

the sample occurs to be in the range of the primary energies of the electron beam (see later).

FINAL STATES IN BIS

The final state resulting from a bremsstrahlung transition is an excited state of the solid in which an otherwise vacant state is occupied. In metallic samples, the potential of this additional electron is screened by the creation of a positive hole resulting from the repulsion of charges (the reverse situation is encountered in photoemission final states). As long as the occupied orbital is a Bloch state extending over the whole crystal, the screening can be practically ignored and the experimental results can be interpreted in terms of single-particle eigenvalues of the undisturbed ground state. This interpretation is no longer valid for localized final states for which the contribution of the other valence electrons is important. This situation has to be analysed within the framework of a many-body formalism, as has been shown for the 4f levels in rare earth metals (17,18). The bremsstrahlung process is very fast so that, in the sudden approximation, not only the adiabatic final state is realized but also a whole spectrum of excited final states. This contribution is certainly not very important for extended final states not far above E_F, but should in principle be observable for sharp structures in BIS spectra. The abrupt occurrence of a local potential due to the occupation of a localized orbital is known to produce a singular response of a Fermi gas. This problem has been extensively studied for photoemission spectra of core levels (10) which show asymmetric lines due to the excitation of low-energy electron-hole pairs in the final state. Similar excitations are clearly observed in BIS studies of rare earths (19) and actinides (20). The lifetime of the Bremsstrahlung final state determines the ultimate resolution of the method. It is evident that a state at E_F is practically infinitely-long-lived and that no lifetime broadening affects Fermi edges. Any increase of the final state energy above E_F opens new decay channels contributing a shortening of the lifetime. A reliable calculation of these mechanisms starting from the density of states and the matrix elements for each type of process appears to be intractable and one must be satisfied with the qualitative statment that the lifetime broadening increases with kinetic energy. The empty density of states does not usually contain very sharp structure at sizable energies above E_F so that this broadening is not a serious handicap for the method and can be allowed for (19). In the exceptional case of Eu which has a localized 4f state 8.6 eV above E_F, the lifetime contribution to the line width is found to be approximately 3 eV (19). An unusual and interesting situation occurs when the primary energy of the incident beam corresponds to the excitation energy of a core level into a localized state above E_F. The electrons can then make a transition to the discrete state corresponding to the double occupation of the localized orbital in the presence of a core hole or can decay directly into the states of the continuum (normal BIS process). The interference of these two competing processes gives rise to a Fano-type resonance which has been observed in the x-ray emission spectrum of Ce (21). For the purpose of studying the density of states above E_F, primary energies in the vicinity of any threshold or discrete excitation should be avoided.

EXPERIMENTAL PROBLEMS

The measurement of the Bremsstrahlung emission can be performed by classical methods and the instrumental details (22) will not be given here. It is evident that at high photon energies it

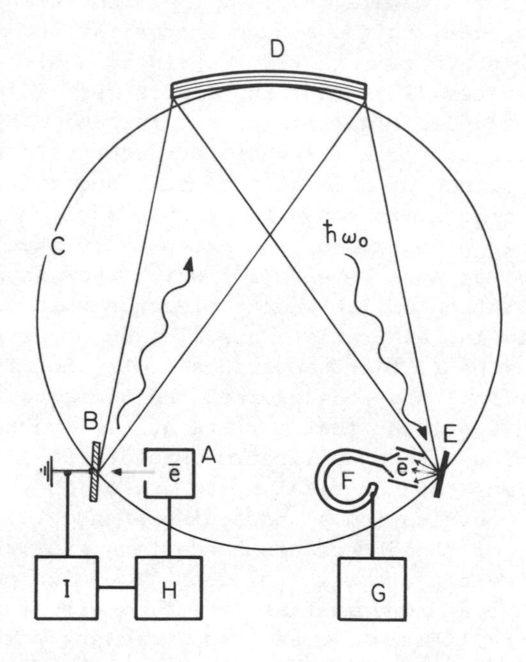

Fig. 3 Diagram of a XPS instrument (22). A- electron gun; B- sample; C- Roland circle of the monochromator; D- spherically bent quartz monocrystal; E- converter; F- channel electron multiplier; G- counting unit; H- step unit; I- high-voltage power supply

is much easier to develop an instrument for BIS rather than for SWL since the x-ray monochromator has only to select one fixed wave length. A schematic diagram of a BIS apparatus is shown in Fig. 3. The electron gun (A) has to be designed to produce an intense electron beam which is as monoenergetic as possible. The energy spread due to the cathode can be minimized by using an indirect heated thermionic emitter of very low work function (an oxide cathode for example). Precaution must be taken to avoid any contamination of the sample by the evaporation of the cathode. The broadening of the energy distribution of the electrons along their path between cathode and sample (Boersch broadening) (23,24,22) is a very serious difficulty. This disturbing effect arising from the mutual electron interaction within the beam can largely be avoided by choosing a suitable electron gun geometry and a sufficiently low emission current (22). This last condition is one of the factors limiting the sensitivity of BIS. UHV and clean vacuum system walls are necessary to prevent the contamination of the sample (B) in the presence of an electron beam. Any type of monochromator having a good resolution and a large collecting angle can be used to select one energy from the radiation emitted by the sample. A monochromator of Johann's type with a spherically bent quartz crystal has been found to be very convenient for an energy of 1.5 keV (22). Soft x-rays are easily detected by the secondary electrons that they produce in a converter (E) (for example CsI). A channel electron multiplier (F) can be used to obtain pulses which are accumulated in a counting unit (G). The instrumental line width (FWHM) of the described apparatus (22) is 0.43 eV for a beam current of 0.3 mA. Many hours are usually necessary to obtain spectra with satisfactory statistics. An apparatus for BIS in the UV range has been recently developed (25). It involves rather different experimental difficulties than those met in the soft x-ray range, in particular in the production of monoenergetic electrons of low energy ($E_k \lesssim 10$ eV).

APPLICATIONS OF BIS

All ground state properties of solids are to a large extent determined by the occupied electronic states so that their study and characterization is a fundamental aim of solid state science. In recent years many electron spectroscopies and in particular all kinds of photoemissions have been used with great success to achieve this purpose. A more complete description of the outermost electronic states including also the vacant states appears to be helpful in many situations: verification of theoretical predictions and calculations, interpretation of results obtained by other spectroscopies (photoemission, optics, ...), description of any kind of electronic excitation, investigation of the stability of electronic configurations (phase transitions), determination of the degree of localization of electronic states, anticipation of the properties of neighbouring elements ... A few examples illustrating some applications of BIS as vacant state spectroscopy will be briefly discussed here. Fig. 4 shows the BIS spectra of the two neighbour elements Pd and Ag in a large energy range (35 eV) above E_F (26). It is obvious that the peak pinned at E_F in the Pd spectrum accounts for the fraction of state remaining unoccupied in the 4d band of this metal. The large cross-section for this symmetry explains the high intensity of the peak which is broader than expected. This can be explained by lifetime broadening, hybridization with states of other symmetries and many-body effects. At higher energies the spectrum shows clearly structures superimposed on a background and extending surprisingly far above E_F. The BIS

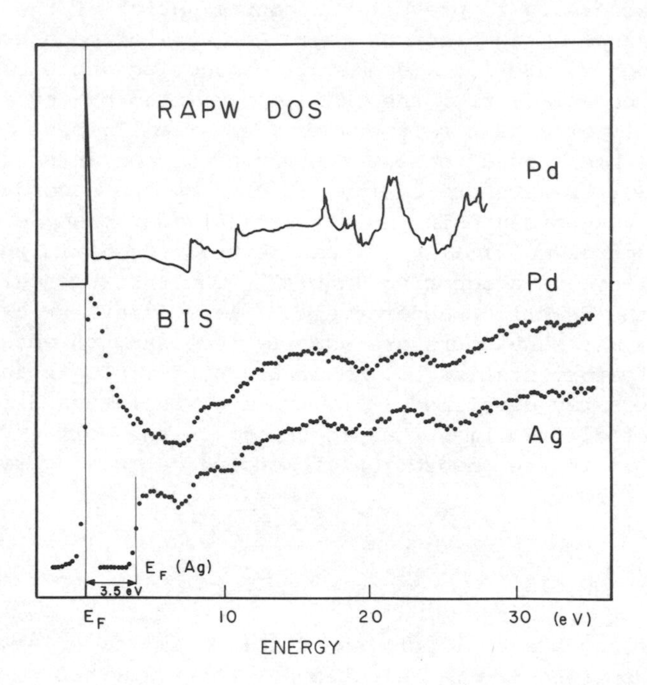

Fig. 4 Comparison of the BIS spectra of Pd and Ag (26) with a calculated DOS of Pd (27)

spectrum of Ag appears to be a nearly perfect replica of the spectrum of Pd, provided that it is shifted by 3.5 eV to account for the filling of the valence band by an additional electron. This observation demonstrates that the pecularities of the potential originating from the atoms forming the solid are still important for electronic states in this energy range. On the top of Fig. 4, a density of states of Pd calculated by the RAPW method (27) has been aligned with the BIS spectra. If one takes into account an important lifetime broadening, the agreement with experiment is clear.

The occupied part of the d band in transition metals has been extensively studied by photoemission and it is interesting to have the corresponding information about the empty part. Fig. 5 shows the available BIS spectra of the 5d transition metals obtained by different authors (Lu, Au, 19; Hf, 28; Ta, W, 29; Ir, Pt, 30). In Lu the 5d band is nearly empty and has a

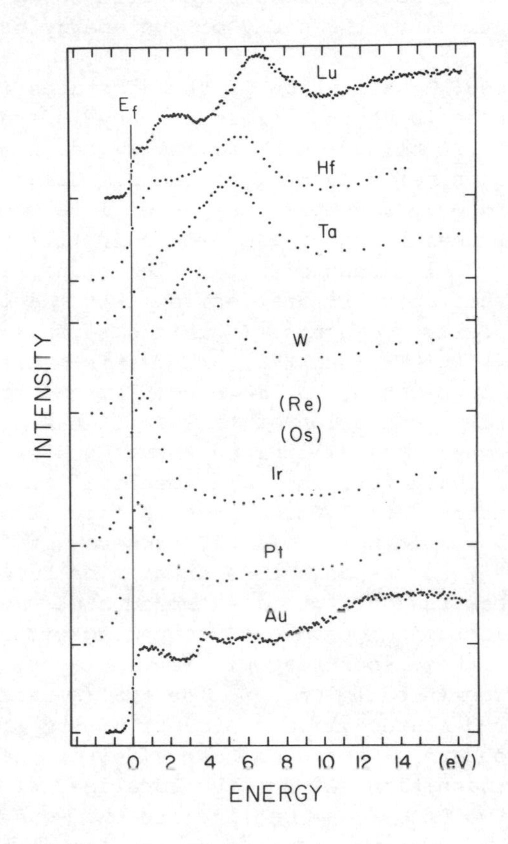

Fig. 5 Systematic comparison of the available BIS spectra of the 5d-transition metals obtained by different authors (see text)

width of approximately 9 eV. It shows a double-peak structure predicted by simple theoretical considerations. In fact the overall shape and dimension of this BIS spectrum is not very different from the XPS spectrum of the occupied d band of Au. In the two next elements Hf and Ta the two peaks move toward E_F as a result of the filling of the band. For W the first peak seems to be occupied and the position of E_F corresponds approximately to the minimum in the spectrum of Lu. Unfortunately the poor resolution of these spectra does not allow a more detailed analysis. At the end of the series (Ir, Pt) the second peak has approached E_F and is partly filled; as expected it disappears completely in Au which has a full 5d band. There is little doubt that systematic

BIS (and XPS) studies at high resolution will bring further valuable contributions in the study of the energy bands.

The exceptional localization of the 4f states in rare earth metals and compounds is at the origin of many interesting properties. No attempt will be made here to review the numerous electron spectroscopic studies devoted to the different aspects of these elements and only a simple example will be presented here to illustrate the usefulness of BIS. The half-filled 4f shell of Gd is occupied by 7 electrons with parallel spins, leaving 7 empty states for the other spin orientation, so that the initial state ($^8S_{7/2}$) is spherically symmetric. In this case the accessible final states (^7F_J) in BIS (and XPS) are 7 spin-orbit components (J = O - 6). The interesting but more complicated situation occuring for an arbitrary population of this shell has been studied and analysed elsewhere (19,31). Since the 4f states in solids keep their atomic character, they are localized on each atoms and do not form extended states. The change of population of this shell ($f^n \rightarrow f^{n+1}$) resulting from a BIS process (or from a XPS process, $f^n \rightarrow f^{n-1}$) corresponds to a transition between discrete energy eigenvalues, in contrast to extended states which have continuous one-electron eigenvalues accounting within Koopmans' approximation for these spectroscopic final states. This situation is schematically depicted in Fig. 6. The XPS (A) and BIS (C) spectra of the extended states yield a picture of the occupied and empty states which are sketched on both sides of the top of the figure. The XPS transition (A) for the localized 4f level can be decomposed into two steps: one can imagine that one part of the photon energy Δ_- is used to excite a f electron to the lowest state at E_F (I) and that an electron at E_F is then excited by the remaining energy (II). The energy Δ_- involved in the step I is then given by the position of the 4f peak related to the observed Fermi energy (B). A similar decomposition of the BIS process (C) yields the energy Δ_+ for the promotion of an electron at E_F into a $4f^{n+1}$ state. Assuming that the same photon energy is used in both cases, the two kinds of processes are identical at E_F (B) so that experimental intensities and energies can be considered to be equal at E_F. The joined XPS-BIS spectrum of Gd corresponding to this situation is shown in Fig. 7. The energy difference ($\Delta_- + \Delta_+$) between the two peaks is the Coulomb correlation energy U for this localized shell embedded in the metal (18). This is easily recognized in Fig. 6 since U is defined as minimum energy required for the transition $4f^n + 4f^n \rightarrow 4f^{n+1} + 4f^{n-1}$ between two non-interacting atoms.

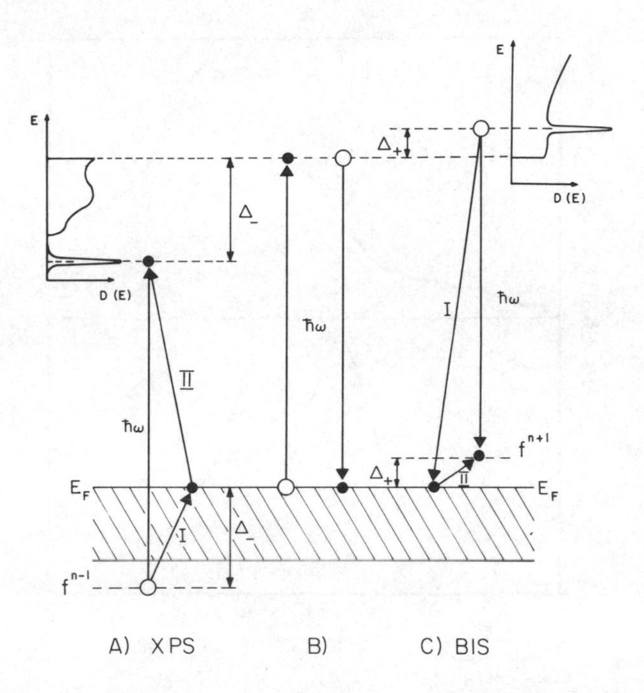

Fig. 6 Schematic representation of the XPS and BIS processes in rare earths

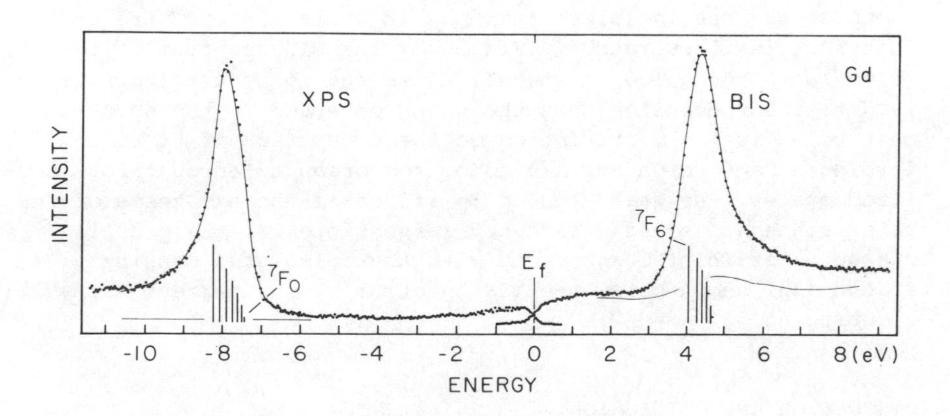

Fig. 7 Combined XPS and BIS spectra of Gd (19)

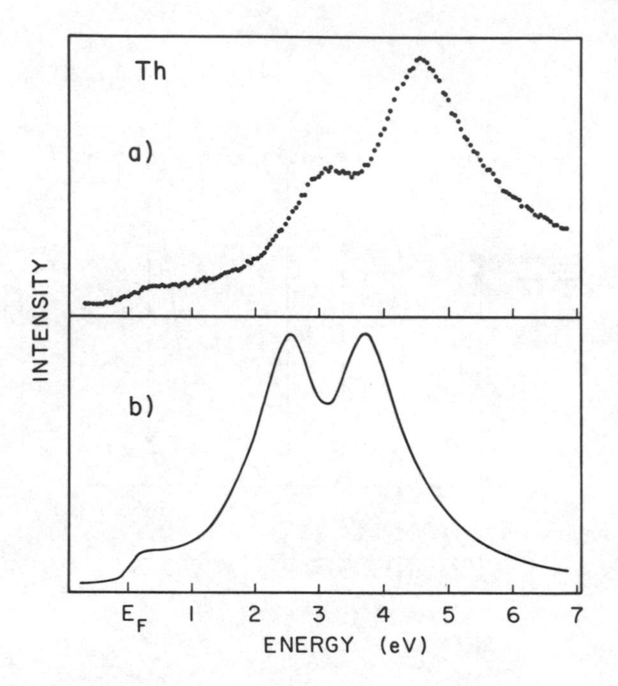

Fig. 8 BIS spectrum of Th (32) compared to a calculated DOS (33)

The 5f states at the beginning of the actinide series are known to be less localized than the 4f states in the rare earths. This fact is illustrated in Fig. 8 by the BIS spectrum of Th (32) which has no occupied 5f states. Cross-section considerations lead to the conclusion that the peaks observed in the spectrum must be exclusively attributed to the occupation of 5f states forming a band which appears to be too broad to account for localized states. The weak step at E_F indicates the existence of the other states of s and d symmetry responsible for the metallic character of Th. The agreement with the calculated density of states (33) shown below the BIS spectrum is not perfect but still encouraging.

DISCUSSION AND CONCLUSION

The empty states can be investigated in a wide energy range

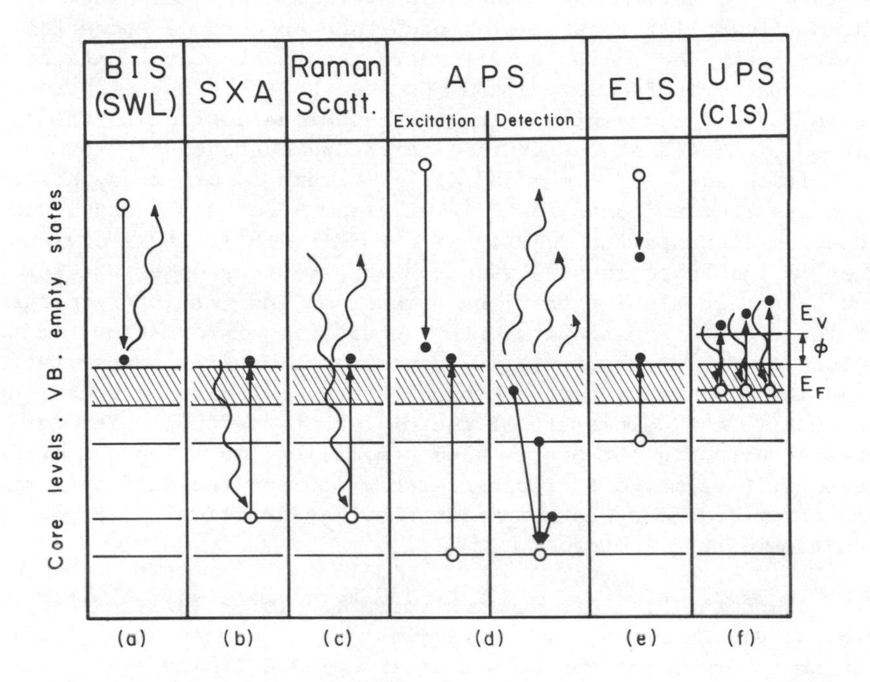

Fig. 9 Schematic energy diagram showing the transitions involved in empty state spectroscopies. (a), BIS: bremsstrahlung isochromat spectroscopy, SWL: short wave length limit. (b), SXA: soft x-ray absorption. (c), Raman scattering. (d), APS: appearance potential spectroscopy. (e), ELS: energy loss spectroscopy. (f), UPS (CIS): UV-photoemission at constant initial state

by different methods; they necessarily involve the occupation of these levels which are vacant in the initial ground state. The electronic transitions induced in the most usual techniques used for this purpose are schematically represented in Fig. 9. In all cases, with the exception of BIS, the electron excited into a vacant state is originating from a level below E_F. The hole produced in this process can modify the local density of states so profoundly that the results obtained by these spectroscopies can no longer be analyzed in terms of the initial density of the ground state. This perturbation resulting from the population modification in the final state becomes particularly important in the case of the occupation of relatively well localized orbitals which have a very strong interaction with the created deep hole. The most complicated situation occurs in Appearance Potential Spectroscopy (APS, Fig. 9 (d)) where the final state contains two additional occupied levels and one deep hole. It has been recently shown (34), for example, that this technique applied to the excitation of the 3d states in rare earths accounts mostly for the transition $3d^{10}\ 4f^{n} + e \rightarrow 3d^{9}\ 4f^{n+2}$ whereas a soft x-ray absorption spectrum accounts for $3d^{10}\ 4f^{n} + \hbar\omega \rightarrow 3d^{9}\ 4f^{n+1}$ and a BIS spectrum accounts for $3d^{10}\ 4f^{n} + e \rightarrow 3d^{10}\ 4f^{n+1}$. It is obvious that in this case these 3 vacant state spectroscopies are not equivalent at all but that the comparison and the interpretation of these different results bring interesting informations on the nature of the electron states in these solids (35). Other factors than the perturbation of the final state must also be taken into account in the comparison of the different spectroscopies represented in Fig. 9. Matrix elements, selection rules and surface sensitivities may differ considerably between the different techniques or even for one technique they may drastically vary as a function of the energy.

Among the high-energy spectroscopies (a) - (e) shown in Fig. 9, BIS is certainly the simplest and the most straightforward method to probe the vacant states in solids. It involves processes which are in many respects perfectly symmetrical to the high-energy photoemission processes so that the spectra obtained by these two techniques yield complementary informations of the same nature on occupied and empty DOS. For photons in the UV range, this symmetry between BIS and photoemission is destroyed by the fact that the momentum conservation becomes important and that the matrix elements have a much stronger energy dependence. UPS in the constant-initial-state mode ((CIS) Fig. 9. (f)) is a very powerful tool to investigate the unoccupied levels. If, in addition to the kinetic energy, the direction of the emitted electron can be selected, this experiment performed on single crystals can yield in favorable cases the dispersion curves $E(\vec{k})$. The only serious limitation is that the interesting energy range between E_F and the vacuum level E_V is not accessible. This drawback would not be encountered in an angular resolved BIS measurement at low energy and constant initial state (Fig. 2 (a)). It would allow the determination of $E(\vec{k})$ straight from E_F, provided that the photon energy is low enough to ensure the $\vec{k}$ vector conservation régime. The realization of an apparatus for measurements of this kind appears to be very difficult since the requirements of very low primary energy and good angular resolution can only be fulfilled at the expense of the intensity. The usual increase of the photoelectric cross sections at low photon energies might be sufficiently large to compensate this loss of intensity and to make this experiment feasible. There is little doubt that, in the near future, BIS will give rise to many developments and will find a wide range of applications.

REFERENCES

1. Duane, W., and Hunt, F.L.: 1915, Phys. Rev. 6, p. 166.
2. Ohlin, P.: 1942, Ark. Mat. Astro. Fys. A 29, p. 3.
3. Nijboer, B.R.: 1946, Physica (Utrecht) 12, p. 461.
4. Albert, L.: 1956, Z. Phys. 143, p. 513.
5. Edelmann, F., and Ulmer, K.: 1967, Z. Phys. 205, p. 476, and refs. therein.
6. Merz, H., and Ulmer, K.: 1968, Z. Phys. 210, p. 92 and: 1968, Z. Phys. 212, p. 435, and refs. therein.
7. Riehle, F.: 1980, Phys. stat. sol. b 98, p. 245.
8. Nagel, D.J.: 1973, in "Band Structure Spectroscopy of Metals and Alloys" (Fabian and Watson, ed.) Academic Press, p. 457.
9. Turtle, R.R., and Liefeld, R.J.: 1973, Phys. Rev. B 7, p. 3411.
10. Wertheim, G.K., and Citrin, P.H.: 1978, Topics in Applied Physics 26, (M. Cardona and L. Ley, ed.) Springer-Verlag, Berlin, p. 197.
11. Best, P.E., and Chu, C.C.: 1977, Phys. Rev. B 15, p. 5160.
12. Tseng, H.K., and Pratt, R.H.: 1971, Phys. Rev. A 3, p. 100.
13. Hahn, Y., and Rule, D.W.: 1977, J. Phys. B: Atom. Molec. Phys. 10, p. 2689.
14. Tseng, H.K., Pratt, R.H., and Lee, C.M.: 1979, Phys. Rev. A 19, p. 187.
15. Chu, C.C., and Best, P.E.: 1979, Phys. Rev. B 19, p. 3414.
16. Böhm, G., and Heyd, R.: 1970, Phys. stat. sol. 39, p. K 19.
17. Herbst, J.F., Watson, R.E., and Wilkins, J.W.: 1978, Phys. Rev. B 17, p. 3089, and refs. therein.
18. Lang, J.K., and Baer, Y., and Cox, P.A.: 1979, Phys. Rev. Lett. 42, p. 74.
19. Lang, J.K., Baer, Y., and Cox, P.A.: 1980, to appear in J. Phys. F: Metal Phys.

20. Baer, Y.: 1980, to appear in Physica.

21. Riehle, F.: 1977, Dissertation, Universität Fridericiana, Karlsruhe.

22. Lang, J.K., and Baer, Y.: 1979, Rev. Sci. Instrum. 50, p. 221.

23. Boersch, H.: 1954, Z. Phys. 139, p. 115.

24. Zimmermann, B.: 1970, Adv. Electron. Electron. Phys. 29, p. 257.

25. Denninger, G., Dose, V., and Scheidt, H.: 1979, Appl. Phys. 18, p. 375.

26. Lang, J.K., and Baer, Y., unpublished results.

27. Christensen, N.E.: 1976, Phys. Rev. B 14, p. 3446.

28. Merz, H.: 1970, Phys. Lett. 33 A, p. 53.

29. Merz, H., and Ulmer, K.: 1966, Phys. Lett. 22, p. 251.

30. Rempp, H.: 1974, Z. Phys. 267, p. 181.

31. Cox, P.A., Lang, J.K., and Baer, Y.: 1980, to appear in J. Phys. F: Metal Phys.

32. Baer, Y., and Lang, J.K.: 1980, Phys. Rev. B 21, p. 2060.

33. Skriver, H.L., and Jan, J.P.: 1980, Phys. Rev. B 21, p. 1489.

34. Harte, W.E., and Szcezepanek, P.S.: 1979, Phys. Rev. Lett. 42, p. 1172.

35. Kanski, J. and Nilsson, P.O.: 1979, Phys. Rev. Lett. 43, p. 1185.

LUMINESCENCE IN INORGANIC SOLIDS

R.G. Denning

Inorganic Chemistry Laboratory, South Parks Road,
Oxford OX1 3QR, U.K.

Abstract. The theory of non-radiative relaxation is outlined for isolated molecules and impurities in continuous lattices. The principal factors controlling the rate of relaxation are discussed. The theory of resonant and non-resonant energy transfer is outlined. Dynamical problems associated with the mobility of electronic excitation energy are reviewed, with some examples.

INTRODUCTION

In contrast to most of the techniques described in this volume, optically excited luminescence is not a surface sensitive technique. The characteristic optical absorption lengths ensure that emission occurs from the bulk. Furthermore the small momentum carried by an optical photon denies any direct access to the dispersion of the optical excitation energy with the wave-vector k. Nevertheless the low energy of the optical photon conveys some especially valuable characteristics. In XPES and X-ray emission the linewidths are determined by the lifetime of the core hole, which is a composite of Auger and radiative relaxation mechanisms. Linewidths of the order of 0.1 to 5 eV are typical, corresponding to lifetimes of $\sim 10^{-15}$sec. For a given radiative transition moment the lifetime is inversely proportional to ν^3 so that, other things being equal, visible emission will result in lifetimes 10^6 longer than X-ray emission. Frequently visible luminescence is associated with transitions having low transition moments so that it is not uncommon to find luminescence lifetimes covering the timescale from 10^{-9} to 1 sec. This timescale is easy to observe directly. During this time there is ample opportunity for the electronic excited state to sample its surroundings, both

P. Day (ed.), Emission and Scattering Techniques, 171–189.

in terms of perturbing vibrations, which can lead to its relaxation, and by the mobility of the excitation through the lattice. It is the time-dependent aspect of the luminescence as opposed to its spectral properties which I will described in this chapter. Spectroscopically the luminescence spectrum provides little more information than that available from an absorption measurement, but changes in the spectrum may provide valuable proof of the mobility of the excitation in the lattice.

To limit the range of the discussion this article will only cover emission from isolated molecules or from dilute impurities in continuous lattices. In other words at the first level of approximation the electronic wavefunctions are considered to be localised on an individual centre or molecule, while the vibrational motion may be localised or collective. In this regime the excited state is described as a tightly bound Frenkel exciton. There are three modes of relaxation, radiative, non-radiative and by the transfer to an adjacent centre-energy transfer. Non-radiative decay may simply create vibrational energy or may result in photochemical products.

There are obvious technological objectives in analysing these processes. The quantum efficiency of the emission process must be high in both lasers and phosphors so that non-radiative decay should be minimised. The influence of energy transfer is more subtle. Neodymium provides important solid state lasers both in phosphate glasses and in yttrium aluminium garnet. Both lasers are pumped by flashlamps but are poor converters of the flashlamp output into laser photons because of the low absorption cross-section of neodymium ion in the visible. An attractive solution involves the addition of a strong visible absorber capable of transferring energy to the neodymium excited state at the appropriate rate. There are, however, also disadvantages with energy transfer. The mobility of the excitation may lead to its efficient annihilation at a defect or impurity. This can be avoided by the use of dilute materials which inhibit the mobility. Similarly, if a narrow band laser is used as a pump the mobility of the excitation can degrade the gain of the laser because the inversion at the initial frequency is depleted by being spread across the inhomogeneous width of the line. Finally, at a time when isotopically selective photochemistry is becoming viable for the commercial separation of isotopes, it is clear that mobility between isotopic species can effectively forbid any separation in the solid state.

NON-RADIATIVE DECAY

The theory of non-radiative decay falls into two parts, that appropriate to systems characterised by localised vibrations and that used in the case of collective modes or phonons. Initially we examine the relaxation in molecular solids, but first we must characterise the non-radiative process by a rate or lifetime. Since

$$1/\tau_D = 1/\tau_{NR} + 1/\tau_R$$

the non-radiative lifetime, τ_{NR}, is accessible from a measure of the observed decay time τ_D, and the radiative decay time, τ_R. In principle the easiest way to obtain τ_{NR} is to measure the quantum efficiency of the emission but in practice this is a notoriously difficult quantity to measure. More frequently $1/\tau_R$ is obtained from the integrated area in absorption, which gives the Einstein coefficients for the transition. In cases where the absorption and emission spectra are not mirror images such a procedure will fail. In this case it is the practice to use the fact that the no-phonon, or pure electronic transition is common to both absorption and emission measurements. The absolute intensity of the emission in the no-phonon transition is then obtained from the absorption spectrum, and is used to obtain the total intensity in the integrated emission spectrum (corrections having been applied for the wavelength variation of the detector sensitivity). Care must be taken that there is no self-absorption of the no-phonon emission line.

The outline treatment of the non-radiative decay for inorganic molecules which follows is taken from Robbins and Thomson [1] and uses the approach developed for organic systems due to Englman and Jortner [2]. Fermi's Golden Rule for the time evolution of an initial state $|a\rangle$ into a final state $|b\rangle$ forms the starting point, the rate being given by

$$W_N = (2\pi/\hbar)|M_{ab}|^2\rho \quad (1)$$

where ρ is the density of final states and

$$M_{ab} = \langle b|\tilde{H}|a\rangle \quad (2)$$

Here H is the Hamiltonian operator from which we select only the dominant terms:

$$\tilde{H}_e^o(q,Q_o) = T_e(q) + V(q,Q_o) \quad (3)$$

Here $\tilde{H}_e^o(q,Q_o)$ describes the electronic energy of the system evaluated in terms of a set of electronic coordinates q, and equilibrium nuclear coordinates Q_o, and is composed of a kinetic energy part $T_e(q)$ and a potential energy part $V(q,Q_o)$. To investigate the effect of vibrations $\tilde{H}(q,Q_o)$ is expanded in a Taylor series about Q_o from which the leading term is

$$\tilde{H}_e(q,Q) = \left(\frac{\partial V}{\partial Q_i}\right)Q_i \tag{4}$$

where Q_i represents a displacement coordinate describing the ith normal mode with respect to the Q_o configuration. Within this approximation eqn. (2) becomes

$$M_{ab} = \Sigma_i V^i_{ab} F^i \tag{5}$$

In this equation V^i_{ab} is an electronic factor, F^i is a vibrational factor and

$$V^i_{ab} = \langle \phi_a{}^o | \frac{dV}{dQ_i} | \phi_b{}^o \rangle \tag{6}$$

$$F^i = \langle n+1 | Q_i | n_i \rangle \Pi_{j \neq i} \langle m_j | n'_j \rangle \tag{7}$$

In eqn. (6) Q_i is the mode responsible for the admixture of the ground and excited state wavefunctions and is called the 'promoting mode'. The occupancy of this mode changes by one quantum in the relaxation process as represented by the first quantity in eqn. (7), and may correspond to the creation of a quantum of this mode in the electronic ground state or its annihilation in the electronic excited state. In the latter instance the process will be thermally activated. The second quantity in eqn. (7) represents the overlap between the vibrational wavefunctions of the ground and excited (primed) electronic states with respect to the remainder of the modes. Often this overlap will only be large for a single mode whose displacement represents the change in equilibrium nuclear configuration in the electronic excited state. Such a mode is described as the 'accepting mode'. If one quantum of the promoting mode is created in the electronic ground state, and we assume only the 'j' accepting mode is excited, energy conservation in the relaxation process requires that $\Delta E = (m-n')h\omega_j + h\omega_i$ where ΔE is the degradation in electronic energy and m and n' are the occupation numbers of the accepting mode in the ground and excited states respectively. At low temperature n' = 0 so that eqn. (7) is proportional to a vibrational overlap integral given by eqn. (8)

$$S_j = \langle p_j | 0'_j \rangle \tag{8}$$

where p_j is the number of quanta of the $'j'^{th}$ mode in the electronic ground state needed to 'bridge the gap' to the electronic excited state. Such an integral is familiar as a Franck-Condon factor. If the displacement along the accepting mode coordinates is zero, then to a first approximation this integral will vanish on account of the orthogonality of the vibrational wavefunctions. In the small displacement (or weak coupling) limit the integral can be evaluated analytically and can be written as

$$S \propto \left[\frac{\{\Delta Q_j\}^2}{p_j}\right]^{p_j} \tag{9}$$

Here ΔQ_j represents the displacement in the 'j' accepting mode. For small displacements $\Delta Q_j << 1$ so that the integral is inversely proportional to p_j, dropping sharply as p_j increases. This dependence upon p_j implies that only the highest frequency vibrations are likely to act as efficient accepting modes, providing that they correspond to the sense of the equilibrium distortion of the excited state.

In isolated transition metal complexes the accepting mode will usually be the totally symmetric stretching mode, but the sensitivity to p_j is so great that weakly-coupled modes of higher frequency may dominate the non-radiative rate. One may expect that modes involving hydrogen atoms will be particularly effective. Robbins and Thomson [1] show that the non-radiative rate in Cr^{3+} complexes correlates with the number of hydrogen atoms bound to the coordinating atoms. Heller [3] has observed that the lifetime of lanthanide luminescence is proportional to the hydrogen atom concentration in H_2O/D_2O mixtures. This sensitivity has provided a general technique for determining the number of water molecules bound to lanthanides in a variety of chemical environments [4]. The necessity of avoiding high-frequency accepting modes means that lasers based on solutions of the rare-earth ions operate best in aprotic solvents [5].

Turning to the electronic factor, V^i_{ab}, it would be expected that $\frac{dV}{dQ_i}$ will in general be large when $V(q,Q_o)$ is large, so that non-radiative rates should decrease as the orbit-lattice (ligand) interaction decreases. This dependence explains the almost universal observation of luminescence amongst lanthanide elements and its rarity amongst d-block transition metals. Beyond this simple observation, one may also examine the symmetry properties of V^i_{ab}. Thomson and Robbins point out that for octahedral complexes of the transition metals the gerade symmetry of the ground and excited electronic states implies that the only internal modes contributing to V^i_{ab} have A_{1g}, E_g and T_{2g} symmetry. For internal modes $\langle \phi_a^o \left| \frac{dV}{dQ_{T_{1g}}} \right| \phi_b^o \rangle = 0$.

On this basis the non-radiative relaxation may prove to be symmetry-forbidden. The Table shows how, for certain d electron configurations, only modes of T_{1g} symmetry link the ground state with the first excited state and that, by and large, these configurations correlate well with luminescent species. It seems clear that

TABLE Symmetry Control of 1st Excited State Luminescence

No. of d electrons	1	2	3	4	5	6 (Low Spin)	6 (High Spin)	7	8	9
Ground State	$^2T_{2g}$	$^3T_{1g}$	$^4A_{2g}$	^5E_g	$^6A_{1g}$	$^1A_{1g}$	$^5T_{2g}$	$^4T_{1g}$	$^3A_{2g}$	E_g
Connecting Mode*	Γ_g	Γ_g	T_{1g}	Γ_g	NIL	T_{1g}	Γ_g	Γ_g	T_{1g}	Γ_g
1st Excited State	^2E_g	$^3T_{2g}$	$^4T_{2g}$, ^2E	$^5T_{2g}$	$^4T_{2g}$	$^1T_{1g}$	^5E_g	$^4T_{2g}$	$^3T_{2g}$	$^2T_{2g}$
Luminescence	NIL	NIL	Cr^{3+} Complexes ruby, MnF_6^{2-}	NIL	Mn^{2+} Complexes MnF_2 TMMC	$Co(CN)_6^{3-}$; Rh^{3+}, Ir^{3+} complexes	NIL	Co/MgF_2	Ni/MgF_2 Ni/MgO	NIL

*Γ_g includes other internal modes of the octahedron other than T_{1g}

symmetry control is one of the leading factors determining the occurrence of luminescence in transition metal complexes.

Impurities in continuous lattices may be treated in an analogous manner except that the accepting modes are now in the lattice phonons. We follow the treatment of Riseberg and Moos [6]. Painstaking lifetime measurements [7] on a large number of lanthanide-containing crystals have established a simple empirical law relating the non-radiative rate connecting two electronic states and the energy gap, ΔE, between them:

$$W_N = \alpha e^{-\beta \Delta E} \tag{10}$$

α and β are dependent on the properties of the lattice.

As before the local modes surrounding the impurity will provide the electronic relaxation process but it will be natural to express the perturbation in terms of the collective lattice phonons and their population. If a single phonon is created in the relaxation then by analogy with eqns. (1)-(6) we may write

$$W_{12} = \sum_{\bar{k}} (2\pi/\hbar^2) \left| \langle a | V^{\bar{k}}_{ab} | b \rangle \langle n(\bar{k}) + 1 | Q^{\vec{k}} | n(\bar{k}) \rangle \right|^2 \delta[\omega(\vec{k}) - \Delta E/\hbar] \tag{11}$$

Here the summation is over all final states satisfying the delta function. $V^{\bar{k}}_{ab}$ is the orbit-lattice electronic interaction energy associated with a particular local mode projected from the phonon of wavevector $\vec{k}$ and $Q^{\vec{k}}$ is the equivalent phonon creation operator. The condition in the delta function implies that only phonons whose cut-off frequency exceeds the electronic energy defect will be able to effect the relaxation. In general this mechanism is very efficient and thermal equilibrium is established rapidly between levels satisfying this criterion.

For electronic energy defects in excess of the phonon cut-off frequency higher order processes involving the emission of more than one phonon must be involved. Eqn. (11) must then be expanded to

$$W_{ab} = \sum_{\vec{k}} \sum_{\vec{k}'} (2\pi/\hbar^2) \left| \frac{V^{\vec{k}}_{ac} V^{\vec{k}'}_{bc}}{\Delta E_{bc}} Q^{\vec{k}} . Q^{\vec{k}'} \right|^2 \delta[\omega_{\vec{k}} + \omega_{\vec{k}'} - (\Delta E/\hbar)] \tag{12}$$

where $|c\rangle$ represents a third electronic state included in the mechanism by the second order perturbation and ΔE_{bc} is the separation of this state from $|b\rangle$. As the electronic energy increases higher orders of perturbation theory must be included so that the perturbation Hamiltonian for a general multiple-phonon process takes the form

$$H_{ab} = \frac{V_{aj}.V_{jk} \; ..... \; V_{bl}.Q_i Q_j \; ..... \; Q_n}{\Delta E_{jk} \; ..... \; \Delta E_{bl}} \qquad (13)$$

Eqn. (13) may be written alternatively for a 'p' phonon process as

$$H_{ab} = A\,\varepsilon^p \quad \text{with } \varepsilon = \frac{V.Q}{\Delta E} \qquad (14)$$

In general the strain energy V.Q is much smaller than the separation between electronic states so that $\varepsilon \ll 1$. Under these circumstances we can write

$$W_{ab} \propto H_{ab}^{\;2} \propto (\varepsilon^2)^p = A\,\varepsilon^{2\Delta E/\hbar\omega_m} = A\exp(2\Delta E/\hbar\omega_m \ln\varepsilon)$$

But with $\varepsilon \ll 1$, $\ln\varepsilon$ is -ve so

$$W_{ab} = A\exp(-2\Delta E/\hbar\omega_m |\ln\varepsilon|) \qquad (15)$$

This expression was first derived by Riseberg and Moos [6]. Such is the sensitivity of W_{ab} to the number of phonons 'p' that it may be assumed that only those of the highest frequency ω_m, which minimize p, will be effective. So p may be replaced by $\Delta E/\hbar\omega_m$ where ΔE is the energy defect. Eqn. (15) expresses the relationship between the non-radiative rate and the energy defect, and confirms the empirical eqn. (10).

The form of eqn. (13) also allows an assessment of the temperature dependence of the rate, since

$$W_{ab} \propto H_{ab}^{\;2} \propto \frac{V_{ab}^{\;2p}\,Q^{2p}}{\Delta E^{2(p-1)}}$$

The term Q^{2p} consists of a series of phonon creation matrix elements $\langle n_i + 1|Q_i|n\rangle^{2p}$. When the Q_i are expressed in terms of the phonons, the phonon creation operator gives terms in $\left\{\sqrt{\frac{n_i + 1}{2}}\right\}^{2p} \propto (n_i + 1)^p$.

Replacing n_i by the Bose-Einstein thermal average occupation number gives

$$(\bar{n} + 1)^p = \left\{\frac{e^{h\omega/kT}}{e^{h\omega/kT} - 1}\right\}^p \qquad (16)$$

for the temperature sensitivity of the non-radiative rate. Some illustrative examples of this temperature dependence are given in

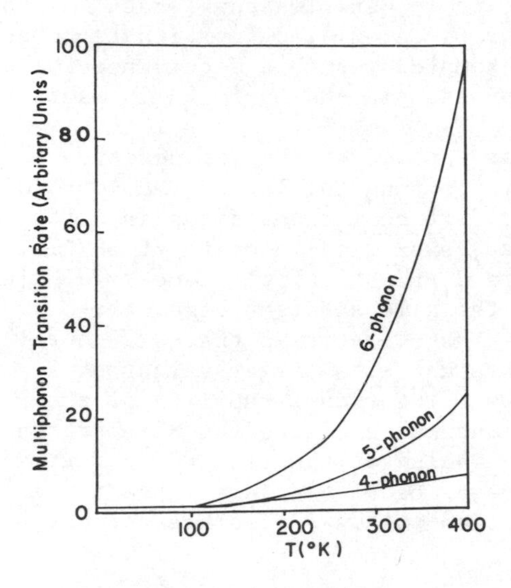

Figure 1. Theoretical temperature dependence of non-radiative rate as a function of the number of phonons created in the decay.
(reproduced with permission from Moos, H.W., 1970, J.Luminescence, 1, p.106)

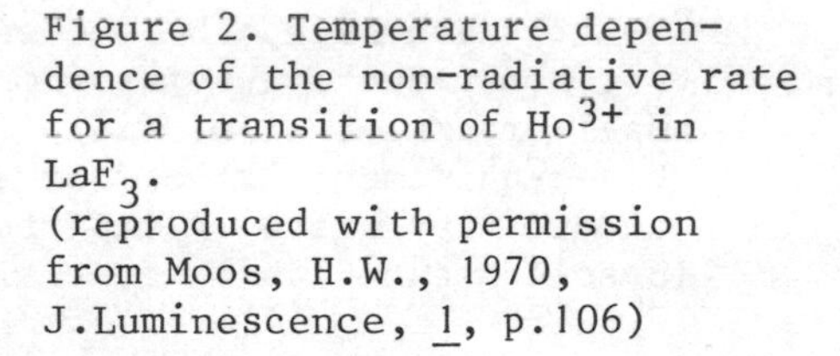

Figure 2. Temperature dependence of the non-radiative rate for a transition of Ho^{3+} in LaF_3.
(reproduced with permission from Moos, H.W., 1970, J.Luminescence, 1, p.106)

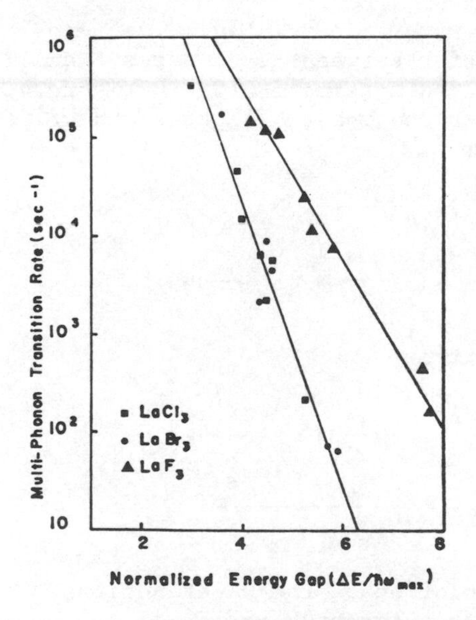

Figure 3. Non-radiative decay rates for a number of rare-earth excited states in LaF_3, $LaCl_3$ and $LaBr_3$ as a function of the number of phonons created in the decay.
(reproduced with permission from Moos, H.W., 1970, J.Luminescence, 1, p.106)

figure 1. The strong dependence on 'p' enables the 'order' of the multiphonon relaxation process to be calculated from its temperature dependence. Figure 2 shows an example of such a dependence for p=5 for a particular non-radiative process in the Dy^{3+} ion in $LaBr_3$.

The success of this model as applied to the luminescence of rare earths can also be shown by plotting the log of the non-radiative rate for a large number of different transitions in different rare earths versus the energy gap, expressed in units of $\Delta E/\hbar\omega_m$. These results collected in figure 3 are strikingly successful [6]. Data for $LaCl_3$ and $LaBr_3$ lie on the same straight line, these lattices being isomorphous. The faster rates in the, structurally different, LaF_3 lattice are consistent with larger values of V_{ab} in the field of the fluoride ions. The transferability of the model between different transitions was explained by Riseberg and Moos as being a result of the averaging of the electronic matrix-elements of the strain as a consequence of the very large number of different high-order perturbation pathways possible for any one non-radiative transition.

To summarize, the success of the theory leads to the conclusion that non-radiative relaxation will be minimized by (i) ligands with small crystal fields, (ii) ligands of large mass. Unfortunately this requirement is not compatible with the property of high thermal conductivity which is important to prevent refraction difficulties in solid state laser materials.

ENERGY TRANSFER

Initially consider the transfer between two centres A and B; the wavefunctions may be expressed diagrammatically in figure 4. We begin by neglecting overlap between the two centres and write the Hamiltonian operator as a sum

$$\hat{H} = H_A + H_B + H_{AB}; \quad H_{AB} = e^2/r_{12}$$

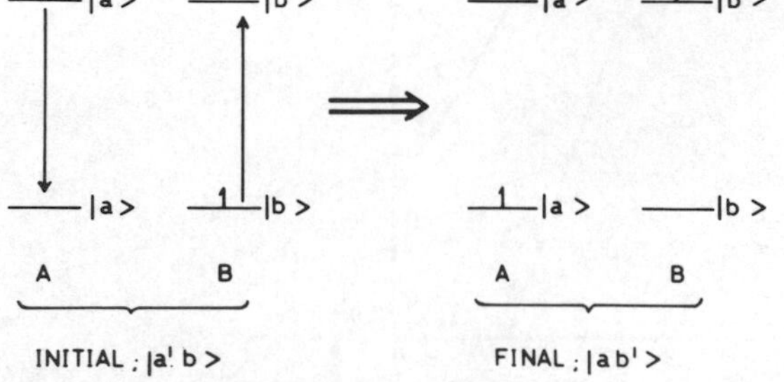

Figure 4. Schematic evolution of the wavefunctions describing the energy transfer process.

including the Coulombic interaction energy of electrons on the two centres. Treated as a perturbation we find a secular determinant

$$\begin{array}{c|cc|}
 & |a'b\rangle & |ab'\rangle \\
|a'b\rangle & E_{A'} + E_B + J_{AB'} - E & X_{AB} \\
|ab'\rangle & X_{AB} & E_A + E_{B'} + J_{AB'} - E
\end{array} = 0; \text{ where, } J_{AB'} = \langle a'b|H_{AB}|a'b\rangle \text{ and } X_{AB} = \langle a'b|H_{AB}|ab'\rangle \quad (17)$$

X_{AB} is the quantity of primary importance and is called the <u>Excitation Exchange Integral</u>. It should be carefully noted that no <u>electron exchange</u> has been permitted in the zero overlap case.

Solving eqn. (17) when A is equivalent to B by symmetry gives two roots:

$$E = E_{diag} \pm X_{AB} \text{ and the wavefunctions } \psi_1 = \frac{1}{\sqrt{2}}(a'b + ab')$$

$$\psi_2 = \frac{1}{\sqrt{2}}(a'b - ab')$$

Now consider an optical transition involving the coupled wavefunctions. We write:

$$\langle\psi_g|e\bar{r}|\psi_1\rangle \approx \langle a|e\bar{r}|a'\rangle + \langle b|e\bar{r}|b'\rangle \quad (18)$$

$$\text{but } \langle\psi_g|e\bar{r}|\psi_2\rangle \approx \langle a|e\bar{r}|a'\rangle - \langle b|e\bar{r}|b'\rangle \quad (19)$$

If A and B are translationally equivalent then (19) will vanish and only one of the two excited states is optically allowed. The extension of this argument to a continuous lattice is straightforward since the translation symmetry forces the choice of 'symmetry-adapted' Bloch functions with the form (familiar from earlier chapters in this volume):

$$\psi(\vec{k}) = \frac{1}{\sqrt{2}} \sum_{n=1}^{N} \exp(i\vec{k}.\vec{r}_n)\phi_1\phi_2 \ldots \phi_n^e \ldots \phi_N \quad (20)$$

each such function describing an excitation wave or Exciton. It is easy to see by comparing eqns. (18), (19), (20) that only for $\vec{k} = 0$ are all translationally equivalent transition moments mutually in phase. In all other cases the transition probability will be zero as required by the conservation of $\vec{k}$. Non-translationally related interacting centres can give rise to Davydov Splitting.

If overlap is permitted between the wavefunctions of the centres A and B, antisymmetrisation is required so that $|ab'\rangle$ is replaced by

$$\frac{1}{\sqrt{2}}(ab' - b'a)$$

The excitation exchange or transfer integral, X_{AB}, now takes the form

$$X_{AB} = \langle a'b|H_{AB}|ab'\rangle - \langle a'b|H_{AB}|b'a\rangle$$

The first term is identical to eqn. (17) and we now call this the Coulomb Transfer Integral, while the second stems from the overlap and electron exchange and to avoid confusion of nomenclature is called the Overlap Transfer Integral.

We concentrate on the Coulomb Transfer Integral since the other integral is seldom tractable. The interaction energy

$$H_{AB} = e^2/r_{12} = \frac{e^2\alpha r_a r_b}{R^3} + \frac{e^2\alpha' r_a N_b}{R^4} + \frac{e^2\alpha'' N_a N_b}{R^5} + \text{etc} \quad ...(21)$$

can be expanded in terms of dipole-dipole, dipole-quadrupole and quadrupole-quadrupole parts. In eqn. (21) the α represent angular factors describing the relative orientation of dipole moments r_i or quadrupole moments N_i separated by a distance R. It is, of course, the Coulomb Transfer Integral, X_{AB}, which connects the initial and final states of the system so that the energy transfer rate is proportional to $|X_{AB}|^2$ or to R^{-6}, R^{-8} or R^{-10} depending on which mechanism is the most important. A straightforward approach to the computation of the rate, explained in detail by Dexter [8], can be illustrated by the observation that in the dipole-dipole case the transfer integral can be represented by

$$\langle a'b|H_{AB}|ab'\rangle = \frac{e^2\alpha}{R^3}\langle a'|r_a|a\rangle\langle b|r_b|b'\rangle \quad (22)$$

The transition moments are easily obtained from the absorption spectra. Eqn. (22) implies that fast transfer can accompany intense dipole-allowed transitions. However, weak optical transitions, allowed by the quadrupole mechanism, but forbidden by the dipole mechanism, can also be very fast at short internuclear distances. The latter mechanism can easily predominate in the f-f transitions of lanthanide ions or in the d-d transitions of transition metal ions.

Eqn. (22) presses the point that, if energy is to be conserved, both A and B must have non-zero transition moments at the same optical energy. Indeed we should include the whole span in the optical energy range in which spectral overlap occurs between the donor emission and the acceptor absorption since this represents the span of final states open to the system. The concept of spectral overlap should be treated with great care. It only applies to the overlap between truly homogeneously broadened lines, since only then is the spectrum a proper guide to the density of final states. When discrete vibrational or phonon sidebands appear in the optical spectrum it is better to consider the problem explicitly, including the electronic-vibrational interaction in the interaction Hamiltonian [9]. The concept of spectral overlap is, however, particularly useful in that it helps to complete the analogy between the virtual photon coupling of the Coulombic interaction and real photon emission and reabsorption, which is generally of small importance as a transfer mechanism except in large samples. Spectral overlap is also helpful in rationalising the temperature dependence of energy transfer between identical ions. If the excited state of the ions relaxes strongly in a particular nuclear coordinate, the emission spectrum and absorption spectrum will only overlap at the zero-phonon line at low temperatures. Under these circumstances the transfer integral is attenuated by a Franck-Condon factor and the slow transfer is said to be due to self-trapping of the excitation. At high temperatures, more spectral overlap occurs, larger Franck-Condon factors are encountered, and the transfer of the excitation is thermally enhanced. Whether this self-trapping occurs depends, of course, on the time-scale of the local vibrational relaxation. Should the transfer integral be much larger than the reciprocal of the relaxation time, no trapping will occur and minimal relaxation will occur in the vicinity of the excitation, which is then described as a large polaron. In weakly coupled systems the excitation is a self-trapped small polaron.

We now turn to the complex question of the measurement of the rate of transfer. We can distinguish two types of transfer (a) between sites which are distinguishable spectroscopically or in efficiency of emission, (b) between identical sites. In the first class the donor A should be dilute so that no A-A transfer occurs. The rate of decay of the emission from A is then investigated as a function of increasing acceptor, B, concentration. B does not have to be an emitter, but in proving that B is an energy transfer acceptor it is invaluable if it is. In general the decay of A, in the presence of B, is non-exponential reflecting inhomogeneity in the A population as a result of a statistical distribution in the distances of B centres. Those A which are close to B decay rapidly, leaving a population of A with larger separations from B.

The degree of non-exponential character then reflects the 'range' dependence of the transfer since a very short range dependence would result in rapid depletion of a small portion of the A population followed by normal radiative decay of the remainder. A longer range dependence will be reflected in a continuing change in the rate with time. This problem has been treated mathematically and can be used as a test of the range-dependence and thus of the transfer mechanism [10].

Figure 5 shows an example of a case in which statistical implications are avoided by ensuring that the donor centres are dilute in a lattice composed entirely of acceptors. Here the donor is a 0.5% isotopic variant incorporated in the lattice of $Cs_2UO_2Cl_4$. The energy transfer is demonstrated by the time evolution of the emission spectrum following a short exciting pulse from a tunable dye laser which primarily excites the donor species. At 4.2K back-transfer is negligible.

The measurement of transfer rates between identical species A is more difficult. Essentially all methods rely on timing the excitation migrating over a definite number of A lattice sites. The most common approach is to measure that rate of diffusion to a trap species B, of known concentration, it being assumed that the trapping step is fast compared to the diffusion. Measurements are made at different concentrations of B. An alternative is to use one excitation to trap another in a process known as biexciton annihilation. The immediate outcome is a double excitation on one centre. In favourable cases this can lead to emission at a new, and often higher, frequency or may provide a pathway for non-radiative decay. In either case the event is detectable and, if the initial concentration of excitons is known, a measurement of the

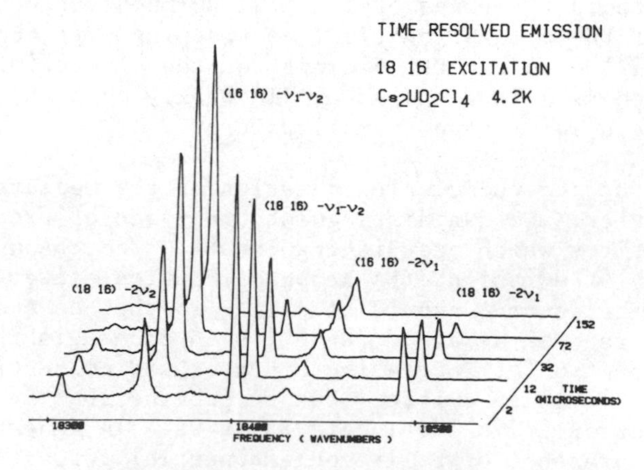

Figure 5. Time resolved emission spectra of $Cs_2UO_2Cl_4$ at 4.2K followed selective excitation on an absorption band of $U^{18}O^{16}O^{2+}$ present at 0.5% abundance.

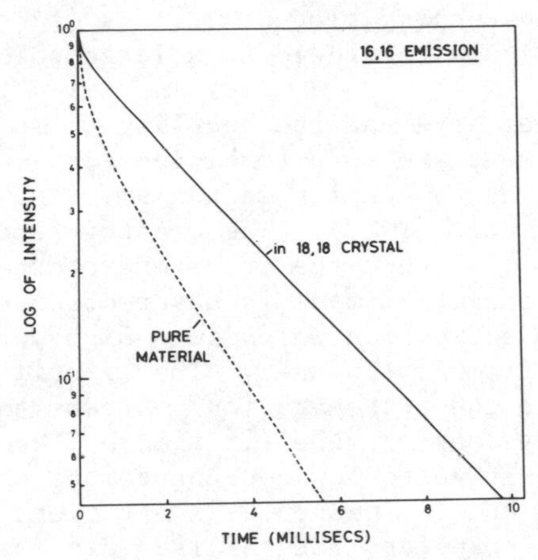

Figure 6. Decay of luminescence of $U^{16}O_2^{2+}$ in natural abundance $Cs_2UO_2Cl_4$ and as a trap in bulk material containing $U^{18}O_2^{2+}$

migration rate is possible. It is also possible to test whether migration is the rate controlling step. If migration is fast compared to trapping the donors 'A' will behave as a homogeneous set, all with the same effective distance from the trap, and their decay will be exponential, but if migration is rate-controlling the decay will be non-exponential. Much effort has been devoted to understanding the dynamics of this process [12].

The role of migration is brought out well in Figure 6, which shows the luminescence of the same species UO_2^{2+} in two different environments. Linear decay is observed when the UO_2^{2+} is trapped and immobile in an isotopic host lattice, but non-linear decay with a shorter lifetime is observed when the same species is itself the host [11].

Finally a range of examples will illustrate the magnitude of the transfer integral and the hopping times where they are known. The first class are those systems which are overlap-coupled. It is generally possible to identify these in cases where all the multipole-transition moments for the Coulombic mechanism are known to be too small to be significant. Examples are transitions which are strongly spin-forbidden such as those in crystals of aromatic molecules. Here the triplet-radiative lifetime can be as long as seconds but the transfer integrals are as large as $1 cm^{-1}$ [13]. In the case of singlet excitation in aromatic crystals the dipole-dipole mechanism may also be important, and hopping times are thought to be about 10^{-11} s [14]. Since this time is long compared to the period of a vibration, $\sim 10^{-13}$ s, these excitations should be thought of as small polarons. If, on the other hand, earlier estimates of the hopping time as 10^{-13} s [14] are accepted then the excitation would be viewed as a large polaron.

By a similar argument the coupling in Mn^{2+} compounds, in which the d-d transitions are spin-forbidden, is sometimes large enough to be dependent on an overlap mechanism. Examples are found in MnF_2 and $[Me_4N]MnCl_3$ (TMMC). The exciton bandwidth in MnF_2 is about 1 cm^{-1} [15] so that the transfer rates are rapid. The radiative lifetime of 33 msec is never observed on account of the trapping of the excitation which in a high purity crystal at 4.2K can give a lifetime of 120 μsec. The lifetime is, however, extremely sensitive to the concentration of traps and to an increase in temperature. By contrast the lifetime of the excitation in TMMC is not very sensitive to low concentrations of impurities nor to temperature and the luminescence is efficient even at room temperature [16]. The transfer integral is again dominated by overlap but its magnitude is not known. A guide is the ground state magnetic exchange energy of 4.3 cm^{-1}. The difference between MnF_2 and TMMC is attributable to their dimensionality, the latter compound being strongly one-dimensional. The significance of the dimensionality is that the excitation, when confined to a linear path, more frequently overwrites previous steps in a random walk than if 'branching' is permitted, so that a smaller proportion of new lattice sites is explored per unit time. According to Montroll [18] the number of hops $\langle n \rangle$ needed to find a trap of concentration C is $(1/C^2)$ in one dimension, but is proportional to (1/C) in two or three dimensions. For example, at a trap concentration of 10^{-5}, 10^{10} hops will be required to find a trap in a one-dimensional lattice as opposed to about 10^5 hops in other lattices. Application of this principle to TMMC gives a hopping time of 10^{-12}s, and also explains why traps are singularly inefficient at quenching the luminescence. Self-trapping in these Mn^{2+} systems probably involves a Jahn-Teller type distortion [16].

Much smaller transfer integrals of Coulombic origin are found in lanthanide lattices. A good illustration comes from the beautiful biexciton annihilation experiments of Mcfarlane et al. [18] in pure $TbPO_4$. Here the radiative lifetime is 3 msec but the actual lifetime is 74 μsec. The difference in the decay at high and low exciton densities is shown in Figure 7, the non-linear behaviour at high densities being due to biexciton annihilation, which is also detectable by anti-Stokes emission. The intensity of the latter emission is proportional to the square of the laser energy (which determines the exciton density). At 45 mJ the beam cross-section may be used to calculate an exciton density of 2×10^{19}/cc while the cation density is 10^{21}/cc giving an exciton concentration of 1 in 100. The initial diffusion of the excitons leads to a diffusion length of about 30 Å and a hopping time of about 10^{-7} s, which in turn reflects the small size of the transfer integral.

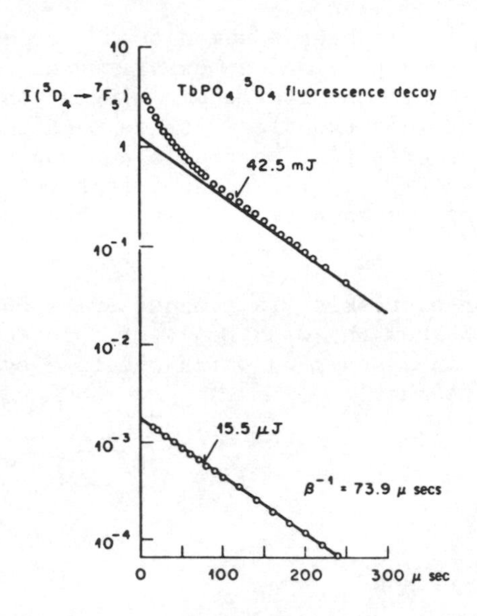

Figure 7. Decay of luminescence in $TbPO_4$, showing second order kinetics due to biexciton annihilation at high laser intensities.
(reproduced with permission from Diggle, P.C., Gehring, K.A., McFarlane, R.M., 1976, Sol.State Commun. 18, p.391)

Finally, the ability to detect the mobility of the exciton suggests that it should be possible to test for Anderson localisation (see chapter by P.A. Cox in this volume) of the excitation. In this situation the 'diagonal' disorder in the lattice is reflected in the inhomogeneous width W of the exciton absorption. The exciton bandwidth is given by $2X_{ab}z$ where X_{ab} is the transfer integral and z is the coordination number of the donor site with respect to acceptor sites. Localisation is predicted if $W > 2X_{ab}z\alpha$ where α is a constant lying in the range $1 \rightarrow 2.7$ [18]. Jortner [20] has interpreted the results of Kopelman [21] in these terms. Kopelman measured the luminescence of iostopic traps in naphthalene crystals. By different degrees of deuteration, 'traps' and 'supertraps' (of greater depth) are created. Absorption occurs in the host. Because the concentration of traps greatly exceeds that of supertraps, which lie to lower energy, the population of supertraps occurs by migration through the random and dilute ($\sim$ 5%) lattice of traps. A critical concentration of traps is found below which migration to supertraps is terminated. Jortner interprets this to mean that X_{ab} varies with the dilution so that at the critical dilution the trap exciton bandwidth drops below the inhomogeneous width and localisation occurs [20]. The argument has been criticised by Kopelman who prefers a percolation model [22].

For some time it has been assumed that single-ion to pair transfer in dilute ruby provided an experimental verification of Anderson localisation in the single-ion single-ion transport [23]. It is now known that this transfer rate is much too slow to satisfy the localisation criteria [24] so that a new explanation must be found for these results. It is probably fair to say that a good experimental test of Anderson type exciton localisation remains to be found.

This selection of results is a necessarily superficial selection designed to show the wide range of interaction energies and dynamical problems in systems on which the time evolution of luminescence provides insight.

References

[1] Robbins, D.J., Thomson, A.J., 1973, Mol.Phys. 25, p.1103.
[2] Englman, R., Jortner, J., 1970, Mol.Phys. 18, p.145.
[3] Heller, A., 1969, J.Amer.Chem.Soc., 88, p.2058.
[4] Horrocks, W. DeW., Sudnik, D.R., 1979, J.Amer.Chem.Soc. 101, p.334.
[5] Heller, A., 1966, Appl.Phys.Lett. 9, p.106; Lempicki, A., Heller, A., 1966, Appl.Phys.Lett. 9, p.108.
[6] Riseberg, L.A., Moos, H.W., 1968, Phys.Rev. 174, p.429; Moos, H.W., 1970, J.Luminescence, 1, p.106.
[7] Weber, M.J., 1968, Phys.Rev. 171, p.283.
[8] Dexter, D.L., 1953, J. Chem.Phys. 21, p.836.
[9] Orbach, R., "Optical Properties of Ions in Crystals", Ed. Crosswhite, H.M., and Moos, H.W., (Interscience, N.Y., 1967) p.445.
[10] Inokuti, M., Hirayama, F., 1965, J.Chem.Phys. 43, p.1978.
[11] Denning, R.G., Ironside, C.N., Thorne, J.R.G., Woodwark, D.R., to be published.
[12] Huber, D.L., 1979, Phys.Rev. B20, p.2307.
[13] Hanson, D.M., 1970, J.Chem.Phys. 52, p.3409.
[14] Powell, R.C., Zoos, Z.G., 1975, J.Luminescence, 11, p.1.
[15] Dietz. R.E., Misetich, A., "Localised Excitations in Solids" (Plenum, N.Y. 1968) p.366.
[16] Yamamoto, H., McClure, D.S., Marzzacco, C., Waldman, M., 1977, Chem.Phys., 22, p.79.
[17] Montroll, E., 1969, J.Phys.Soc.Japan, 26, Suppl., p.6.
[18] Diggle, P.C., Gehring, K.A., McFarlane, R.M., 1976, Sol.State Commun. 18, p.391.
[19] Anderson, P.W., 1958, Phys.Rev. 109, p.1492.
[20] Klaftner, J., Jortner, J., 1977, Chem.Phys.Lett. 49, p.410.
[21] Kopelman, R., Monberg, E.M., Ochs, F.W., 1977, Chem.Phys. 21, p.373.
[22] Monberg, E.M., Kopelman, R., 1978, Chem.Phys.Lett., 58, p.497.
[23] Koo, J., Walker, L.R., Geschwind, S., 1975, Phys.Rev.Lett., 35, p.1669.
[24] Jessop, P.E., Szabo, A., 1980, Phys.Rev.Lett., 45, p.1712. Chu, S., Gibbs, H.M., McCall, S.L., Passner, A., 1980, Phys.Rev.Lett., 45, p.1715.

LOW-ENERGY ELECTRON DIFFRACTION

G. Rovida

Istituto di Chimica Fisica, Università di Firenze

Low-energy electron diffraction (LEED) is the most widely applied technique for investigating the surface structure of solids. After a brief discussion of the nomenclature used in surface crystallography, the fundamental aspects of LEED and its main results are described.

INTRODUCTION

The knowledge of the surface structure of solids is fundamental in order to clarify on an atomic scale the mechanisms of complex phenomena, like adsorption, heterogeneous catalysis and corrosion, and to improve the performance of many electronic devices. The low-energy electron diffraction (LEED) is at present the most widely used experimental technique for the study of the surface structure of solids. Since the early 1960s (when the progress in technology made the LEED experiments much easier in comparison with the first experiments of Davisson and Germer in 1927), its use has rapidly increased giving rise to a new field of research which can be defined as "surface crystallography". Other experimental techniques can also give information about the surface structure, but, up to now, LEED is still the most widely applied. We can say that surface crystallography is related to LEED in the same manner as classical crystallography is related to X-ray diffraction.
We shall start with a brief discussion of some geometrical aspects of surface structures and of the nomenclature used in surface crystallography. Then the fundamental aspects of LEED will be describ-

P. Day (ed.), Emission and Scattering Techniques, 191–212.

ed. Finally, the main results of this experimental technique will be briefly reviewed and discussed, in order to illustrate its applications as well as its present limits.

1. GEOMETRICAL ASPECTS AND NOMENCLATURE OF SURFACE STRUCTURES

The "surface structure" of a solid can be considered the structure of the first atomic layers at and below the "geometrical" surface, to a depth such that the positions of the atoms do not differ appreciably from those in the bulk. This depth cannot be evaluated a priori, since it may vary depending on the solid considered. However, at least in the case of metals, significant structural differences do not usually extend more than two or three atomic planes inside the solid.

We can imagine an "ideal surface" obtained by cutting an infinite crystal into two semi-infinite crystals along a given plane and assuming that the positions of the atoms near the surface do not change. The surface atoms will be arranged in a lattice which is perfectly periodic in the two dimensions parallel but not in the third dimension normal to the surface plane. For the atomic planes parallel to the surface we can define two vectors $\underline{a}$ and $\underline{b}$ such that any vector $\underline{t}=m\underline{a}+n\underline{b}$ (with m and n integers) corresponds to a translation connecting two equivalent atom positions in the same plane. The two vectors define the two-dimensional (2D) unit cell, or unit mesh, which will correspond to one of the five 2D Bravais lattices, as shown in Fig.1.

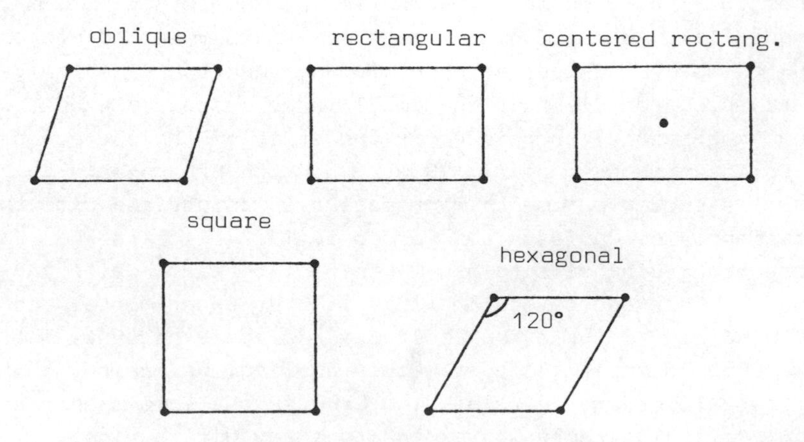

Fig.1-The five Bravais-lattices in two dimensions.

On an atomic plane parallel to the surface, rows of atoms are identified by two Miller indices (hk), which are the 2D equivalent of the Miller indices (hkl) characterizing a given family of atomic planes in 3D crystallography. Inside the unit mesh, the atoms can be arranged according to the 17 plane groups which are possible if we consider symmetry elements operating in two dimensions. If we consider the ensemble of the first atomic planes as a 3D structure limited in the third dimension, and admit also symmetry elements operating in this third dimension (excluding, of course, those requiring periodicity in the direction perpendicular to the surface) we have 80 space groups with 2D periodicity (1,2).
In the case of the "ideal surface" considered, all the atomic planes parallel to the surface would have the same periodicity (unit mesh) and the same spacings as in the bulk. In the case of a real surface, the atoms near the surface will generally relax to new equilibrium positions, so that we must expect (as verified in some cases) both a change of the periodicity along the surface plane (i.e., variation of the unit mesh) (reconstruction of the surface structure), and a change in the spacing of the outermost planes. With better reason, such changes are to be expected in the case of adsorption of atoms or molecules on the surface. In both the case of a reconstructed clean surface and of a surface with the presence of an adsorbate, the new surface structure will generally correspond to a new unit mesh with primitive vectors $\underline{a}_s$ and $\underline{b}_s$, which, in matrix notation, can be expressed in terms of the ideal or bulk mesh:

$$\begin{pmatrix} \underline{a}_s \\ \underline{b}_s \end{pmatrix} = \begin{pmatrix} g_{11} & g_{12} \\ g_{21} & g_{22} \end{pmatrix} \begin{pmatrix} \underline{a} \\ \underline{b} \end{pmatrix} = G \begin{pmatrix} \underline{a} \\ \underline{b} \end{pmatrix}$$

where matrix G can be used to identify the surface unit mesh. For example, the structure in Fig. 2a (which can be assumed to derive from adsorption of oxygen atoms onto a square array of metal atoms, like the (100) plane of a f.c.c. metal) can be indicated as $\begin{pmatrix} 1 & 1 \\ 1 & -1 \end{pmatrix}$, while that in Fig. 2b as $\begin{pmatrix} 2 & 0 \\ 0 & 2 \end{pmatrix}$. This notation is the most complete, and can be used in all cases. Moreover, the determinant of G directly gives the ratio between the area of the surface mesh and that of the ideal mesh: this allows the surface coverage to be derived immediately, in the case of an adsorption structure, if the number of adsorbed atoms per unit mesh is known. Another kind of notation, which is in more common use, can be applied when the angles between the primitive vectors of the surface and bulk unit

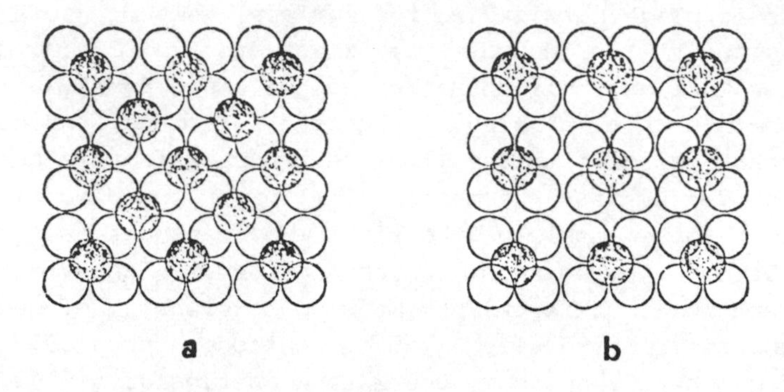

Fig.2-Examples of simple surface structures.

meshes are the same (1). In this case the new mesh is identified by

$$\left(\frac{a_s}{a} \times \frac{b_s}{b}\right) \alpha$$

where α is the angle of rotation between the two meshes (omitted if zero). In the case of the structures shown in Fig.2a and 2b, they can be indicated as $(\sqrt{2}\times\sqrt{2})45$ and (2x2) respectively. In some cases a centered unit mesh is chosen for simplicity; in this case the letter c is placed before the parentheses: the structure of Fig.2a can also be indicated as c(2x2), while that of Fig.2b is better defined as p(2x2) (primitive unit mesh). In the case of structures formed by adsorbates, the symbol of the atom or molecule is placed after the expression. For instance, if the structures in Fig.2 correspond to adsorption of oxygen on the (100) face of nickel, they can be respectively indicated as Ni(100)-c(2x2)-O (or Ni(100)-$(\sqrt{2}\times\sqrt{2})$45-O) and Ni(100)-p(2x2)-O.
As in the case of 3D structures, for 2D structures as well the concept of reciprocal lattice is used to describe better the diffraction mechanism, as will be shown later. The 2D reciprocal lattice is formed by an array of points which is periodic in the two dimensions parallel to the surface plane; these points, with respect to one of them chosen as the origin, define vectors

$$\underline{q}_{hk} = h\underline{a}^{*} + k\underline{b}^{*}$$

where k and k coincide with the Miller indices of the atomic rows perpendicular to the vector $\underline{q}_{hk}$ in the direct lattice; $\underline{a}^{*}$ and $\underline{b}^{*}$ are the primitive vectors identifying the reciprocal unit mesh.

The latter are related to the primitive vectors of the direct lattice by the relations

$$\underline{a} \cdot \underline{a}^* = \underline{b} \cdot \underline{b}^* = 2\pi$$

$$\underline{a} \cdot \underline{b}^* = \underline{a}^* \cdot \underline{b} = 0$$

that is, $\underline{a}^*$ has a module which is inversely proportional to the distance of the rows parallel to $\underline{b}$ and is perpendicular to $\underline{b}$; the same for $\underline{b}^*$ in relation to $\underline{a}$.
In the case of a surface structure with periodicity differing from that of the bulk, the surface reciprocal lattice can be described in terms of the reciprocal lattice corresponding to the 2D periodicity of the parallel planes in the bulk. Indicating with $\underline{a}_s^*$ and $\underline{b}_s^*$ the primitive vectors of the surface reciprocal lattice, we have

$$\begin{pmatrix} \underline{a}_s^* \\ \underline{b}_s^* \end{pmatrix} = \begin{pmatrix} g_{11}^* & g_{12}^* \\ g_{21}^* & g_{22}^* \end{pmatrix} \begin{pmatrix} \underline{a}^* \\ \underline{b}^* \end{pmatrix} = G^* \begin{pmatrix} \underline{a}^* \\ \underline{b}^* \end{pmatrix}$$

It can be shown that the matrix G^* is the inverse transposed of matrix G of the direct lattice.

2. LOW-ENERGY ELECTRON DIFFRACTION

2.1. Experimental conditions

In LEED, a collimated beam of monoenergetic electrons is used with energy usually in the range 10-300 eV. The sample on which the electrons impinge must be a single crystal cut along the face to be studied: this must have a diameter of several mm , since the electron beam diameter may vary between 1 and 3 mm depending on the primary energy.
When the beam of the primary electrons strikes the sample surface, only a fraction of them is backscattered elastically. Most of them suffer a loss of energy due to inelastic events, so that the distribution in energy of the electrons coming from the surface is typically as shown in Fig.3. At low energy, the large maximum is due to the true secondary electrons emitted by the solid; the peak at E_p corresponds to the electrons scattered elastically, while the small peaks at lower energies are due to losses of discrete amounts of energy due to excitations of interband transitions or

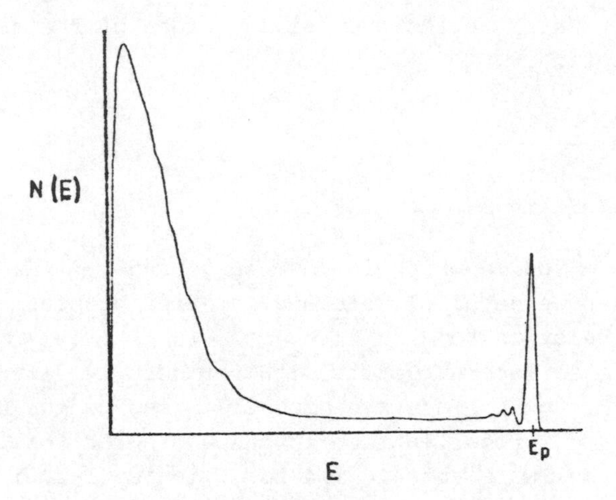

Fig.3-Typical energy distribution of secondary electrons (E_p = energy of the primary electron beam).

plasma oscillations in the solid. In LEED, only the angular distribution of the electrons scattered elastically is studied, so that all the electrons with energy different from that of the primary beam must be eliminated.

To detect the scattered electrons, a Faraday cup can be used, which must be moved in the whole range of angles in front of the sample surface. More commonly, a fluorescent screen is placed in front of the sample: each diffracted beam gives rise to a spot on the screen, so that the entire diffraction pattern can be directly observed through a window of the vacuum chamber; the intensity of each diffraction spot can be measured from outside using a spot photometer. The use of the fluorescent screen is particularly useful when studying surface structures which are only partially ordered (in this case the diffraction pattern consists of streaks or diffuse spots which would escape an analysis made with a Faraday cup) and when the pattern changes rapidly during the experiment (as in the case of adsorption studies).

In a display type LEED apparatus (Fig.4), the electrons coming from the sample are first filtered in order to eliminate all the electrons with energy different from that of the primary beam; this is accomplished by a set of at least two grids, the first being held at the same potential of the sample, the second at a potential near to that of the cathode emitting the primary electrons: the latter grid has the function of repelling the inelastic electrons. The elastically scattered electrons are then accelera-

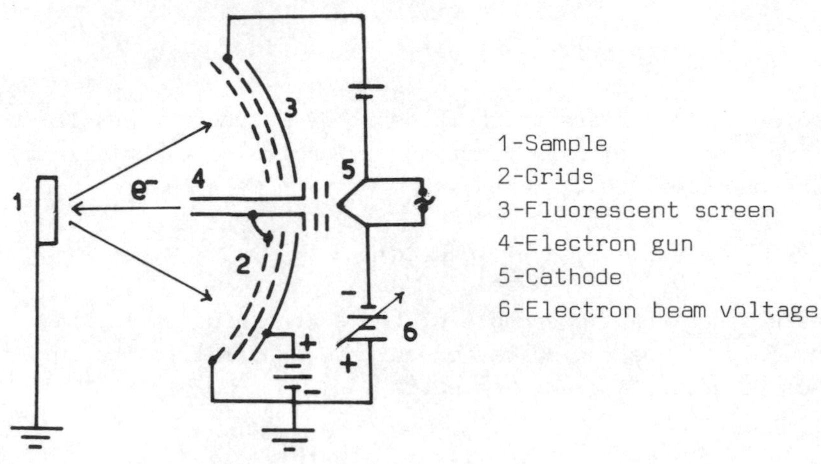

Fig.4-Schematic LEED apparatus.

ted by a strong potential (of the order of several KeV) onto the fluorescent screen. The grids and the screen are usually of spherical shape with the sample in the geometrical center, so that the radial trajectories of the electrons are not distorted by the applied electric fields.

Of course, the LEED optics must be placed in an apparatus in which ultrahigh vacuum (10^{-9}-10^{-10} Torr) can be maintained, as for all other techniques for surface studies, in order to prepare and maintain a clean surface. The sample surface is cleaned in situ, in general by repeated cycles of ion bombardments and annealings at high temperatures: the latter are needed in order to restore the long range order in the surface structure, which has been damaged by ion bombardment.

2.2. The LEED pattern

In LEED we are concerned only with the elastically scattered electrons. Assuming that the incident electrons impinge on a perfectly ordered surface, the interaction will be as follows. The incident beam can be described by a wavefunction

$$\psi_i = A\exp(i\underline{K}_o \cdot \underline{r}) \qquad (1)$$

where $|\underline{K}_o|^2=2E$ (using atomic units so that $\hbar^2=m_e=e^2=1$) while the scattered electron's wavefunction, due to the lattice periodicity along the two dimensions parallel to the surface plane, can be written in the Bloch form

$$\psi_s = \exp(i\underline{K}_{o//}\cdot\underline{\rho})\ u_s(\underline{r}) \tag{2}$$

where $\underline{K}_{o//}$ and $\underline{\rho}$ are the components of $\underline{K}_o$ and $\underline{r}$ parallel to the surface and $u_s(\underline{r})$ is a function having the same periodicity of the surface plane, so that it can be written as a Fourier expansion

$$u_s(\underline{r}) = \Sigma_{\underline{q}}\ \alpha_{\underline{q}}(z)\ \exp(i\underline{q}\cdot\underline{\rho}) \tag{3}$$

where the $\underline{q}$ are the vectors of the 2D reciprocal lattice; $\alpha_{\underline{q}}(z)$ are only a function of the coordinate z normal to the surface. Thus, substituting eq.3 in eq.2

$$\psi_s = \Sigma_{\underline{q}}\ \alpha_{\underline{q}}(z)\ \exp\left[i(\underline{K}_{o//} + \underline{q})\cdot\underline{\rho}\right] \tag{4}$$

Outside the crystal, since the potential is zero, it can be shown that

$$\frac{d^2\alpha_{\underline{q}}(z)}{d\,z^2} + \left[2E - |\underline{K}_{o//} + \underline{q}|^2\right]\alpha_{\underline{q}}(z) = 0 \tag{5}$$

so that the $\alpha_{\underline{q}}(z)$ are of the form

$$\alpha_{\underline{q}}(z) = A_{\underline{q}}\exp\left[\pm i(2E - |\underline{K}_{o//} + \underline{q}|^2)^{1/2} z\right] \tag{6}$$

Chosing the minus sign, since we are dealing with waves coming back from the sample,

$$\psi_s = \Sigma_{\underline{q}}\ A_{\underline{q}}\exp\{i[(\underline{K}_{o//}+\underline{q})\cdot\underline{\rho} - (2E - |\underline{K}_{o//}+\underline{q}|^2)^{1/2} z]\} \tag{7}$$

Thus the scattered wave is formed by a series of plane waves corresponding to discrete beams. Each beam corresponds to a vector of the reciprocal lattice with indices (hk) and has a wave vector component on the surface given by $\underline{K}_{o//} + \underline{q}$. Thus, the angles of the diffracted beams are related to the dimensions and symmetry of the unit mesh, but the geometry of the diffraction pattern does not give information about the positions of the atoms inside the unit mesh; this is contained in the amplitudes $A_{\underline{q}}$ of the scattered beams.

From the above considerations, the LEED pattern is composed of a set of beams, each corresponding to a vector $\underline{q}_{hk}$ of the 2D reciprocal lattice, and identified by the two Miller indices (hk).

The allowed wave vectors of the scattered beams are defined by the relations

$$|\underline{K}| = |\underline{K}_o| \qquad \underline{K}_{//} = \underline{K}_{o//} + \underline{q} \tag{8}$$

This can be directly illustrated by the so called Ewald construction (as is also done in X-ray crystallography) shown in Fig.5 for the simple case of normal incidence of the primary beam.

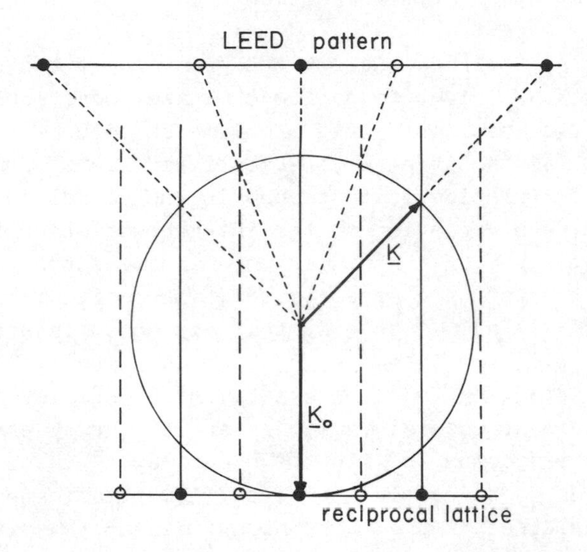

Fig.5-Ewald construction for LEED (normal incidence). Full circles: 2D reciprocal lattice points of the clean surface. Open circles: additional points due to the presence of an adsorbate.

Here, a vertical section of the 2D reciprocal lattice is shown; the directions of the scattered beams are defined by the intersections between the sphere of radius K and the vertical lines corresponding to each lattice point. It can be realized that the LEED pattern displayed on the fluorescent screen is a kind of projection of the 2D reciprocal lattice. So that, starting from the pattern of the clean surface, the adsorption of atoms or molecules in ordered layers causes a change in the surface periodicity and new beams appear in the pattern.
Each beam, corresponding to a given $\underline{q}_{hk}$, can be considered as arising from in-phase scattering by the atomic rows with Miller indices (hk); in fact, the conditions 8 correspond to the relation:

$$d_{hk}(\sin\theta_{hk} - \sin\theta_o) = 2\pi/K \quad (9)$$

where θ_o and θ_{hk} are the angles of the incident and scattered beam respectively, with respect to the direction normal to the surface. So that it is possible to derive the distance between atomic rows and the dimensions of the unit mesh from the measurement of the diffraction angles.

2.3. Determination of atomic positions

As shown before, the knowledge of the geometry of the scattering process is not sufficient to derive the exact positions of the atoms within the unit mesh. This problem can only be resolved by an analysis of the intensities of the scattered beams. Usually, the experimental information is contained in the so called I/E curves. For each beam, the variation of the integrated intensity as a function of the energy E of the primary beam is measured (in general, by a spot photometer or a Faraday cup). The most common range of energies is 20-200 eV. Up to now, there is no satisfactory theoretical treatment of the intensities which can directly give the atomic structure starting from the experimental data. Usually, starting from a given structural model, intensity curves are calculated for each beam and compared with the experimental ones. The structure is considered as resolved if a sufficient agreement is reached. The theoretical treatment of the intensities of the diffracted beams for low-energy electrons is much more complex in comparison with X-rays, for which the so called kinematic theory is sufficient. This is due to the very low atomic cross sections for X-ray scattering, so that the scattered wave intensity is a very small fraction of the incident one. In this case the coherent superposition of single scattering events from each atom is sufficient to describe the interaction between the X-ray photons and the crystal. On the other hand, in the case of LEED the scattering cross section is of many orders of magnitude higher, so that multiple scattering effects are of importance, and a dynamical treatment is required, which takes into account, for each incoming electron, a number of scattering events, only limited by the free path of the electron in the solid. Such theoretical treatment was developed in the years around 1970 and later has been applied successfully to the determination of the structure of clean surfaces as well as of adsorbed layers (3,4).
In the following, the basic concepts of the theory will be outlined in order to show the main approximations made and the parame-

ters which are introduced in the treatment.
The problem of calculating the LEED intensities is in some respects similar to that of calculating the band structure of the solid. The main differences are that for LEED the energy range is different and that there is the incoming electron in excess over the N electrons of the solid. The complete treatment should thus resolve the Schrödinger equation for a system of N+1 electrons and give, for each beam and each energy, the coefficients $A_{\underline{g}}$ in equation 7. Moreover, the movement of the nuclei should also be considered as well as the excitation of electronic transitions. This complete treatment is actually not possible, so that some simplifications are introduced to reduce the problem to the resolution of a monoelectronic Schrödinger equation.
The first approximation is to consider the electron as moving in a rigid lattice of atoms; the effect of atomic vibrations is however introduced in the theory in a simple way, as will be shown later. The second approximation is that which leads to the Hartree-Fock equation. The third is to substitute the exchange term in the Schrödinger equation with a local operator; the Slater approximation is often used, as usual in band structure calculations. The interaction of the incoming electron with the other electrons leading to excited states (excitation of plasmons or interband transitions) is accounted for by introducing an imaginary term (V_i) in the potential (this corresponds to a gradual decrease in amplitude of the wave along its travel inside the solid).
The resulting simplified Schrödinger equation is then resolved by writing the wavefunction inside the crystal as a Fourier series like that of equation 4 . A system of equations is thus obtained for the functions $\alpha_{\underline{g}}(z)$

$$\frac{d^2\alpha_{\underline{g}}(z)}{dz^2} + (2E - |\underline{K}_{o//} + \underline{g}|^2)\alpha_{\underline{g}}(z) = \Sigma_{\underline{g}'} V_{\underline{g}-\underline{g}'}(z)\ \alpha_{\underline{g}'}(z) \qquad (10)$$

where

$$V_{\underline{g}}(z) = \frac{1}{\sigma} \int V(r)\ \exp(-i\underline{g}\cdot\underline{\rho})\ d\underline{\rho}$$

is the Fourier transform of the potential; σ is the area of the surface unit mesh. To resolve the system of equations 10 the most commonly applied method is that in which the crystal is considered as formed by atomic layers parallel to the surface and separated by empty spaces where the potential is constant (inner potential, V_o).

Inside each layer, the so called "muffin tin" approximation is used. In the interlayer spacings, the wave function of the electron is written as a superposition of plane waves propagating both in the positive and negative directions with respect to the z axis. Inside each atomic layer, the waves coming from the interlayer spacings are mixed with those coming from each atom in the layer. The waves scattered by a given layer have amplitudes $\beta^{\pm}_{\underline{q}'}$ which are related to those of the incident waves $\alpha^{\pm}_{\underline{q}}$ by equations of the type

$$\beta^{\pm}_{\underline{q}'} = \sum_{\underline{q}} M^{\pm\pm}_{\underline{q}'\underline{q}} \alpha^{\pm}_{\underline{q}}$$

where the signs ± refer to the positive or negative directions of propagation; the elements M are functions of the wave vector, of the atomic positions and of the atomic scattering properties; the latter properties are described in terms of the phase shifts δ_l which are derived from the atomic potential. The waves scattered by a layer are then treated as incident waves on the adjacent layer, and so on. Thus, the problem is to calculate the elements M. This requires very big computer programs. Usually, 7-8 phase shifts and up to 100 beams (different vectors $\underline{q}$) may be required. Taking into account the symmetries can in part reduce the complexity of the calculation. Moreover, other types of treatment can lead to a more rapid convergence.

Without entering the details of the theory (see Ref.3 and 4), it is worth noting that the treatment of the solid layer by layer permits planes with different periodicity or different atomic species to be considered as well: this is particularly important in the case of surface reconstruction or in the presence of an adsorbate. The complexity of the computer programs and the computation time actually limit the evaluation of LEED intensities to very simple structures with small unit mesh (in general, with area no more than 3-4 times that of the support).

The effect of the temperature is introduced by using effective phase shifts (usually complex) which can be derived from those at T=0 K by solving equations in which a new parameter appears, the surface Debye temperature (Θ). The result is a general smoothing of the peaks in the I/E curves, without significant changes in the positions of maxima and minima.

In summary, the parameters entering the calculation are of two types: structural and non-structural. The former are the unit mesh size and the atomic positions. The latter are: i) the real part of the potential (this is derived from atom charge densities);

ii) the imaginary part of the potential (V_i) (this is usually varied to obtain a better fit, although it can be approximately derived from the width of the peaks in the experimental I/E curves); iii) the inner potential (V_o) (this is evaluated as the sum of the Fermi energy and of the work function of the solid, with small changes to improve the agreement between theory and experiment); iv) the surface Debye temperature (θ). Notwithstanding the high number of non-structural parameters, in general the calculated intensities are found to depend critically on the structural parameters and not to be too sensitive to small variations in the non-structural ones; when the latter are chosen in a physically plausible way, small variations in their values lead to a better fit with the experimental results, but in general do not cause the choice of the right structural model to change from one model to another.

2.4. The determination of a structure

The procedure for the determination of the right structural model starts, of course, with the collection of reliable experimental data, usually in the form of I/E curves for different beams and different incidence angles of the primary beam. Problems may derive from the effects of stray magnetic fields, errors in the determination of the incidence angle and azimuth, contact potential difference between the sample and the cathode of the electron gun, subtraction of the background, non uniformity in the fluorescent screen, corrections for non constant primary beam intensity, etc. Surface defects of the sample may also affect the experimental results. In general, it is found that LEED is not so sensitive to steps or other microscopic defects as can be expected: relatively bad surfaces may give good LEED diagrams. In some cases, lack of complete agreement between theory and experiment has been attributed to surface imperfections. The extent, however, to which surface defects may affect the I/E curves has not been well defined as yet. The experimental results are then compared with theoretical curves calculated on the basis of different possible models of the surface structure. The right model is that which gives the best agreement. The problem now arises how to judge such "best agreement". In some cases, the difference between two models is so evident that a visual inspection allows one of them to be ruled out. But not infrequently two different models, for particular values of the various parameters, give comparable agreement.
Several types of "reliability factors" (R) have been proposed in

order to make the choice independent from subjective judgement. The most simple are similar to those used in X-ray crystallography (see, for example, Ref.5)

$$R = \sum_n |I_{n,ex} - cI_{n,th}| \Big/ \sum_n |I_{n,ex}|$$

where $I_{n,ex}$ and $I_{n,th}$ are the experimental and theoretical intensities taken at given energy points (n) of the I/E curves, and c is a scaling factor. Other types also consider the derivative of the curves ($I' = dI/dE$). An example of a more complete R factor is the following (6)

$$R = \frac{1}{\int I_{ex}\, dE} \int W(E)\, |\, I'_{ex} - cI'_{th} \,|\, dE$$

where W(E) is a weighting function depending on the second derivative, and the integration is extended over the energy range available from experimental curves.
For a given model and a given set of experimental curves, R depends on the values of the structural and non-structural parameters. Usually, R factors are plotted as "maps" in which one structural parameter (for example, the distance between the first and second atomic planes) is varied as well as one non-structural parameter, in general V_o ; the latter is chosen since it is found to be the most critical (for a correct model) to determine the exact values of the structural parameters. The minimum in the R-map indicates the best values of the parameters tested. An example of these R-maps is shown in Fig.6 . In general, a structural model is considered to be correct if the value of R is below 0.2 . For example, in Ref.8 R-maps are reported for different structural models, showing that only in the case of one of them the R-map has a minimum corresponding to acceptable values of R.
At present, the atomic positions can be determined with an accuracy of about 0.1 Å in the direction perpendicular to the surface ; this often corresponds to a higher accuracy in bond lengths.

3. APPLICATIONS OF LEED

The number of surfaces already studied has rapidly increased since the beginning of use of LEED, so that only representative results will be reported here. For a recent bibliography on structures investigated, see Ref.4 . For other general information, see Ref.9-12.

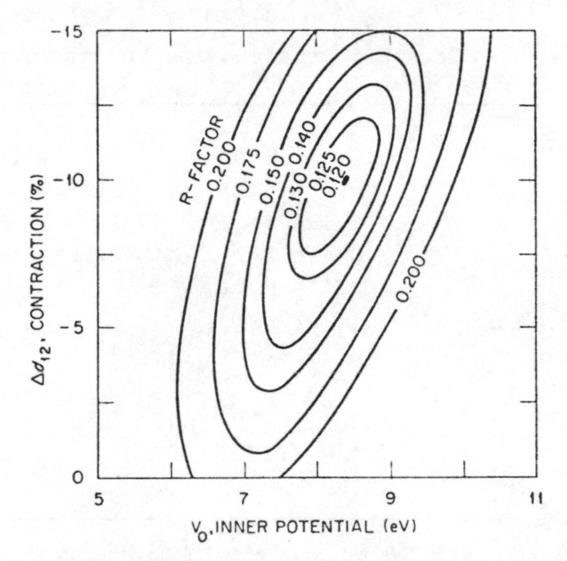

Fig.6-Example of a R-map: Cu(110) surface. d_{12} is the variation of the distance between the first and second atomic planes, relative to the bulk spacing.(From Ref.7)

3.1. Clean surfaces

The study of the structure of the clean surface of a solid is obviously preliminary to any investigation on surface phenomena. At present, due to the availability of high purity single crystals, almost all solid surfaces can be studied by LEED.
One of the most significant results of LEED was the demonstration that in some cases the structure of a clean surface differs significantly from that of the parallel planes in the bulk. A change in the surface unit mesh (reconstruction) was observed in the cases reported in Tab.I .
In general, metals do not show surface reconstruction, with the exception of Au, Pt, Ir, Mo, W and Cr. Tungsten has a c(2x2) superstructure on the (100) face below 300 K, whose structure has recently been studied by LEED intensity analysis and interpreted with small shifts of the surface atoms within the plane of the surface (13). In many cases, a contraction of the first interplanar spacing is derived by LEED analysis; this variation is mainly observed for faces with less compact structure, like (110) faces of f.c.c. metals, for which contractions as high as 10% are reported (7). The interplanar spacing between second and third planes

Table I - Clean surfaces showing reconstruction

Si	(111)-(7x7), (111)-(2x1), (100)-(2x1) (100)-c(4x2), (110)-(5x2)
Ge	(111)-(2x8), (111)-(2x1), (100)-(2x1) (100)-c(4x2), (110)-(2x1)
Te	(0001)-(2x1)
GaAs,GaSb,InSb	(111)-(2x2)
CdS	(0001)-(2x2)
W, Mo, Cr	(100)-c(2x2)
Pt, Ir, Au	(100)-(5x1), (110)-(1x2)
Bi	(11$\bar{2}$0)-(2x10)
Sb	(11$\bar{2}$0)-(6x3)

appears to remain almost unchanged, within the limits of reliability for LEED analysis (a few units per cent). By cutting crystals along high Miller index planes (lower atomic density), faces exhibiting ordered step and kink structures can be obtained (15).These structures have remarkable thermal stability. The reversible changes of surface structure taking place upon adsorption are of great interest in studies of catalytic activity and sintering of small particles (9).
Semiconductors are the solids which show more frequently surface reconstruction. Only in a few cases, however, was the structure determined by LEED intensity analysis, due to the size of the unit meshes and to the number of atomic planes involved. An example is the structure of Si(100)-(2x1) (14).
Ionic crystals (alkali halides, oxides, etc.) have unreconstructed faces. For these compounds, however, only cleavage faces were studied.

3.2. Surfaces with adsorbates

When atoms or molecules are adsorbed on a previously cleaned surface, the LEED diagram immediately shows whether the adsorption layer has long range order or is partially or completely disordered. One of the first and striking results of LEED was the direct evidence of formation of ordered phases with different structures when the surface coverage of a given adsorbate is varied.
When ordered structures are formed, the surface unit mesh can be derived, at least when the relation between the periodicity of the

layer and that of the support is rather simple. Sometimes the identification of the unit mesh is not straightforward, due to the possible equivalent orientations of the layer with respect to the support. A simple example is that of a (2x1) superstructure on a (111) face of a cubic crystal: in this case equivalent ordered domains can exist with orientation differing by 120°. If their dimensions are much smaller than the electron beam diameter (as usually observed), the LEED diagram is the superposition of three (2x1) diagrams rotated 120° with respect to each other, so that it is not possible to distinguish this case from that of a p(2x2) structure. Of course, the interpretation of the LEED pattern may be more complex with larger unit meshes, particularly when the array of adsorbed atoms has an unit mesh which is rotated with respect to that of the support: in this case, multiple diffraction effects may give rise to additional diffracted beams, thus further complicating the interpretation of the pattern geometry (24).

When the surface unit mesh is not too large (typically c(2x2), p(2x2), p(2x1), etc.) the theory can be applied to the analysis of LEED intensities, so that the atomic positions and, consequently, the bond angles and distances can be derived. At present, a great number of simple adsorption structures has been determined. When the adsorbate is monoatomic, it is often found that the adsorbed atoms are located on top of the substrate atoms in sites of highest symmetry, and that the adsorbate-substrate bond distances correspond to the sum of the covalent radii. A typical example is that of the structure of sulfur overlayers on the (100), (110) and (111) faces of Ni (16). An exception is the case of nitrogen adsorption on Ti(0001) (17) in which a (1x1) structure is formed: the N atoms penetrate into the octahedral holes between the first and second atomic planes of the metal, giving a structure which is similar to that of (111) planes of bulk TiN (Fig.7). In Fig.8, several I/E curves calculated for different models are compared with the experimental curve (at the bottom); not only the on-top models can be ruled out, but also different positions of the N atoms below the first Ti plane can be distinguished, since only one of the possible models gives sufficient agreement, as confirmed by the analysis of other diffraction beams.

Only a few structures of molecular adsorbates have been resolved as yet. This is due in part to the difficulty of obtaining reliable intensity measurements, since, in these cases, the adsorbed molecules are easily decomposed by the LEED beam and fast methods of intensity measurement must be employed. The more extensively studied system is the c(2x2) structure of carbon monoxide on Ni(100),

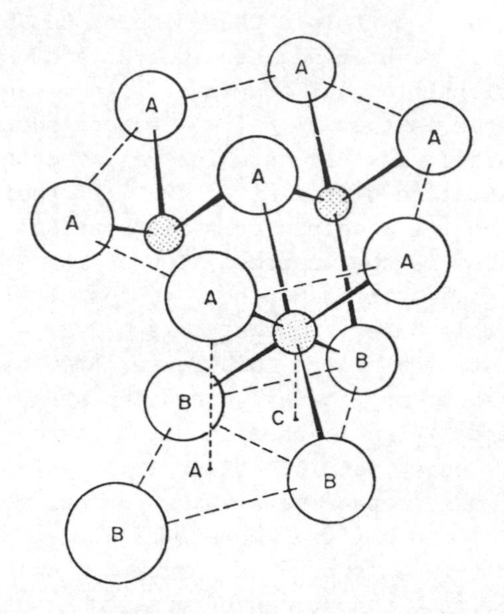

Fig.7-Model of the Ti(0001)-(1x1)-N structure. Empty circles: Ti atoms (A, first plane; B, second plane); shaded small circles: N atoms. (From Ref.17)

for which recent results of different authors are available (18-20) and are compared in Fig.9. Substantial agreement is found, not only for the geometry (CO adsorbed on top of a Ni atom; C-O axis normal to the surface), but also for the bond distances, within the expected reliability of the method. The bond lenghts for adsorbed CO are very near those found in the compound $Ni(CO)_4$ (Ni-C 1.82 Å ; C-O 1.145 Å).
Another case of molecular adsorption is that of acethylene on the Pt(111) face, showing a p(2x2) pattern (21). The model giving at present the best agreement is that shown in Fig.10. The molecule is adsorbed in the form of an ethylidyne species, coordinated to a threefold site, with the C-C axis normal to the surface. The C-H bond distances are assumed, since it was found that the LEED intensities were rather insensitive to H atom positions.
LEED can also be applied to the study of the structure of epitactic layers deposited by evaporation under vacuum on a given support: in these cases, information about the structure of successively deposited layers can be obtained. A recent example of this kind of studies is that of the deposition of Cd on the (0001) face of Ti (22). The completion of successive atomic layers of Cd up to the fourth was followed. For the first layer, the Cd atoms are

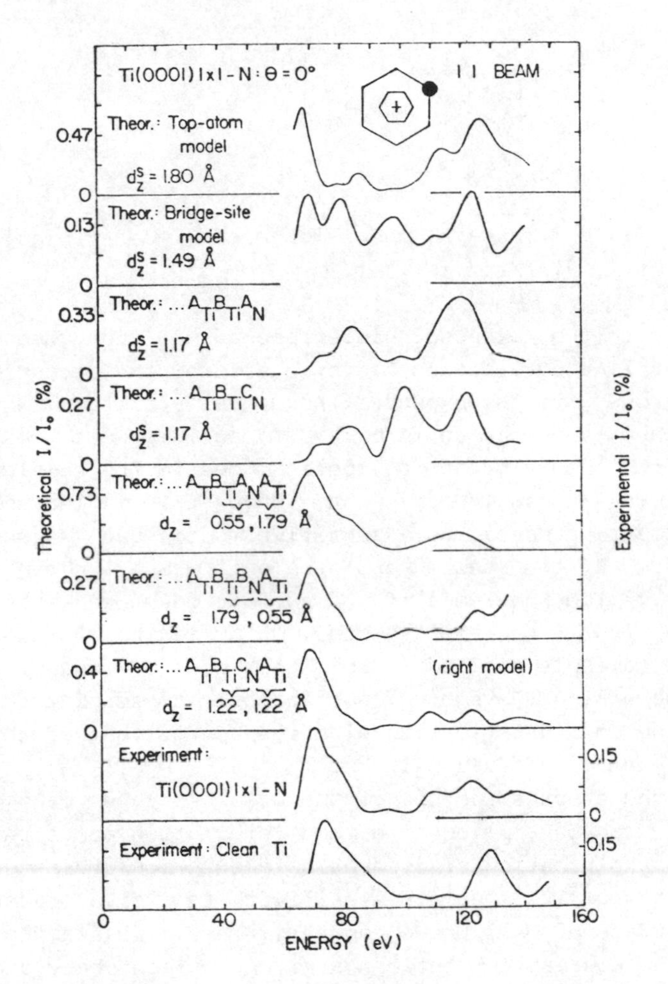

Fig.8-Ti(0001)-(1x1)-N structure: I/E curves for the (11) beam, for different N-atom positions. The letters A,B,C refer to lattice positions in Fig.7; d_z are the distances in the direction normal to the surface. (From Ref.17).

	Ref.18	Ref.19	Ref.20
O–C	1.15	1.10	1.13
C–Ni	1.72	1.80	1.7

Fig.9-Carbon monoxide adsorption on Ni(100): results of different authors (bond distances in Å).

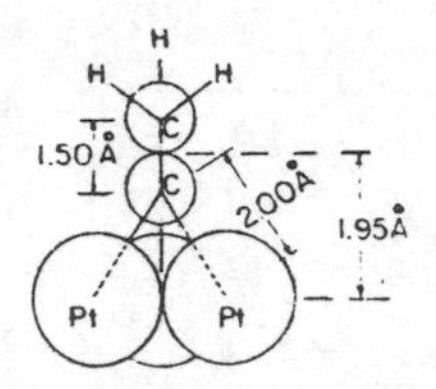

Fig.10-Structural model for acethylene adsorption on Pt(111). (From.Ref.21).

not located in the hexagonal closed packed position that would be occupied by Ti atoms, but in f.c.c. position; the second layer, however, grows with the expected hexagonal structure, with normal interlayer spacings as in bulk Cd metal.

When the LEED theory cannot be applied, due to the complexity of the structure (in usual adsorption studies this is the most frequently occurring case), only tentative models can be proposed, on the basis of the knowledge of the unit mesh size and symmetry, as well as of the knowledge of the surface coverage (this is usually obtained from Auger spectroscopy, which is the auxiliary technique most commonly applied in LEED studies). Of course, in these cases, some ambiguity remains, particularly concerning the positions of the adsorbate lattice with respect to that of the support (see, for example, Ref.23-24).

Even when the exact structure of the adsorbed layer cannot be derived, LEED can be applied to obtain information about the energy as well as the kinetics of adsorption-desorption processes, with the advantage of its specific sensitivity to a given ordered phase on a surface of well known geometry. Thus, for example, the study of the equilibrium between a given surface phase and the gas, as a function of the pressure and temperature, allows the isosteric heat of adsorption to be derived: an example is that of CO adsorption on Ir(111) (23). At 1/3 coverage, CO gives a $(\sqrt{3}\times\sqrt{3})30$ structure: the conditions for the appearence of this structure in the LEED pattern by varying the CO pressure and the temperature, allow the adsorption isostere to be determined corresponding to a coverage near 1/3 (Fig.11). From the plot, an isosteric heat of 39 Kcal/mole was derived.

In favorable cases, the activation energy for desorption can also be derived: as opposed to classical methods, like flash desorption, the variation of the LEED intensity of an extra-beam (characteristic of a well defined adsorption structure) during the desorption process, can give an actvation energy which is related to adsorpt-

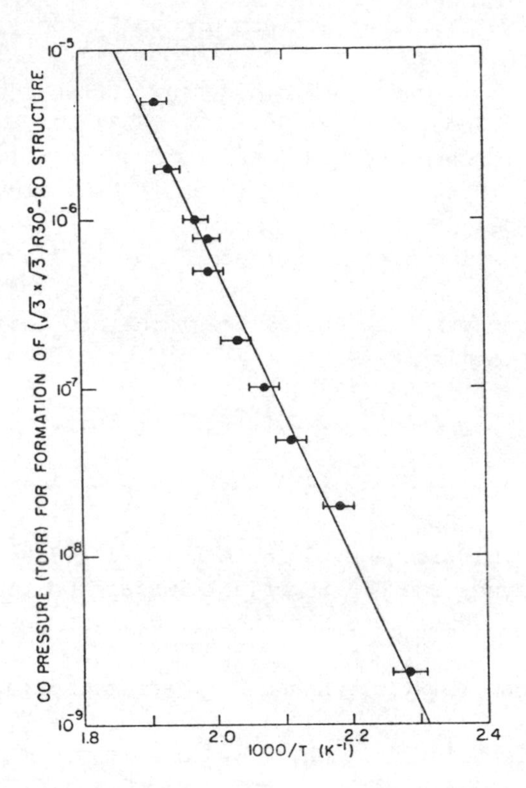

Fig.11-Adsorption isostere for CO adsorption on Ir(111) for a coverage of 1/3. (From Ref.23).

ion sites of an unique and well defined geometry (26). Of course, it must be ascertained that the long range order in the layer is retained at the temperature of desorption.
Another kind of application of LEED is the study of order-disorder transitions in adsorption layers. Recent application of lattice-gas models allowed the adsorbate-adsorbate interaction energy to be derived from the observation of the temperature dependence of the degree of order; this can be obtained by measuring the variation of LEED intensities with the temperature (27-28).

REFERENCES

(1) E.A.Wood, J.Appl.Phys. 35 (1964) 1306.
(2) E.A.Wood, Bell System Techn.J.43 (1964)541.
(3) J.B.Pendry:"Low Energy Electron Diffraction", Academic Press, 1974.

(4) M.A.Van Hove and S.Y.Tong:"Surface Crystallography by LEED", Springer-Verlag, 1979.
(5) M.A. Van Hove, S.Y.Tong and M.H.Elconin, Surface Sci.64(1977)85.
(6) E.Zanazzi and F.Jona, Surface Sci. 62 (1977) 61.
(7) H.L.Davis, J.R.Noonan and L.H.Jenkins, Surface Sci.83(1979)559.
(8) S.Hengrasmee, P.R.Watson, D.C.Frost and K.A.R.Mitchell, Surface Sci. 92 (1980) 71.
(9) G.A.Somorjai: "Principles of Surface Chemistry", Prentice Hall 1972.
(10) G.Ertl and J.Kuppers:"Low Energy Electrons and Surface Chemistry", Verlag Chemie, 1974.
(11) F.Jona, Surface Sci. 68 (1977) 204.
(12) S.Y.Tong, Progr.Surface Sci. 7 ((1975) 1; Inst.Phys.Conf.Ser. 41 (1978) 270.
(13) R.A.Barker, P.J.Estrup, F.Jona and P.M.Marcus, Solid State Comm. 25 (1978) 375.
(14) S.Y.Tong and A.L.Maldonado, Surface Sci. 78 (1978) 459.
(15) B.Lang, R.W.Joyner and G.A.Somorjai, Surface Sci.30(1972)440.
(16) J.E.Demuth, D.W.Jepsen and P.M.Marcus, Phys.Rev.Lett.32(1974) 1182.
(17) H.D.Shih, F.Jona, D.W.Jepsen and P.M.Marcus, Surface Sci. 60 (1976) 445.
(18) M.Passler, A.Ignatiev, F.Jona, D.W.Jepsen and P.M.Marcus, Phys.Rev.Lett. 43 (1979) 360.
(19) S.Andersson and J.B.Pendry, Phys.Rev.Lett. 43 (1979) 363.
(20) S.Y.Tong, A.Maldonado, C.H.li and M.A.Van Hove, Surface Sci. 94 (1980) 73.
(21) L.L.Kesmodel, L.H.Dubois and G.A.Somorjai, J.Chem.Phys. 70 (1979) 2180.
(22) H.D.Shih, F.Jona, D.W.Jepsen and P.M.Marcus, Phys.Rev. B 15 (1977) 5550, 5561.
(23) D.I.Hagen, B.E.Nieuwenhuys, G.Rovida and G.A.Somorjai, Surface Sci. 57 (1976) 632.
(24) U.Bardi and G.Rovida, Proc. IV Int.Conf.Solid Srfaces, Cannes 1980 (Suppl."Le Vide" N.201) p.325.
(25) G.Rovida and F.Pratesi, Surface Sci. 67 (1977) 367.
(26) G.Rovida, J.Phys.Chem. 80 (1976) 150.
(27) G.Doyen, G.Ertl and M.Plancher, J.Chem.Phys. 62 (1975) 2957.
(28) E.D.Williams, S.L.Cunningham and W.H.Weinberg, J.Chem.Phys. 68 (1978) 4688.

EXAFS AND SURFACE EXAFS:
PRINCIPLES, ANALYSIS AND APPLICATIONS

J. Stöhr

Stanford Synchrotron Radiation Laboratory
Stanford University, Stanford, CA 94305

A discussion is given of principles and analysis methods of the Extended X-ray Absorption Fine Structure (EXAFS) technique with emphasis on applications in surface crystallography. The various detection techniques in use such as transmission, fluorescence, electron yield and ion yield will be discussed and compared with respect to their applicability for surface studies. Analysis procedures of the EXAFS data will be outlined. Finally selected examples of surface structure investigations will be presented such as electron yield EXAFS studies of oxygen on Al(111) and ion yield EXAFS studies of oxygen on Mo(100).

I. INTRODUCTION

The present paper is not intended to be a comprehensive review article (1) on the Extended X-ray Absorption Fine Structure (EXAFS) technique but instead a rather elementary tutorial for non-specialists. I shall only cover selected applications of the technique which because of my personal interest are in the field of surface science. Whenever possible I will, however, give extensive reference to aspects not discussed here. Sections II and III illustrate the principles of the EXAFS phenomenon and give some fundamental equations. The detection techniques which allow the application of EXAFS to surface structure investigations are discussed in Section IV. Experimental details are discussed in Section V and data analysis precedures are outlined in Section VI. Section VII discusses applications in surface crystallography using the chemisorption of oxygen on Al(111) and on Mo(100) as examples. The paper is summarized in Section VIII with an outlook on possible future applications of the technique.

P. Day (ed.), Emission and Scattering Techniques, 213–250.

II. THE EXAFS PHENOMENON

In a conventional x-ray absorption experiment (2) the intensity transmitted through a sample is measured as a function of photon energy ($h\nu$). As the photon energy is swept through the threshold of a core electron excitation the transmitted flux is reduced due to loss of photons through creation of photoelectrons. The sudden intensity change with photon energy which occurs when the photon energy is equal to the binding energy (E_B) of a core electron is called the absorption edge. For an isolated absorbing atom (e.g. a noble gas) the absorption cross-section above the absorption edge is a smoothly varying function which is mostly determined by the overlap of the initial state wavefunction of the core electron and the final state wavefunction of the photoelectron. In general this overlap becomes gradually smaller with increasing photon energy.

If the absorbing atom is bonded to any neighbors the final state wavefunction of the escaping photoelectron may be modified by scattering due to the neighbor atom(s). Depending on the wavelength of the photoelectron (which of course depends on its kinetic energy ($E_k = h\nu - E_B$)) the outgoing and backscattered waves can interfere constructively or destructively. Quantum mechanically this photoelectron interference effect in the final state modifies the absorption cross-section and results in modulations of the transmitted x-ray intensity above the absorption edge known as Extended X-ray Absorption Fine Structure (EXAFS) (3)

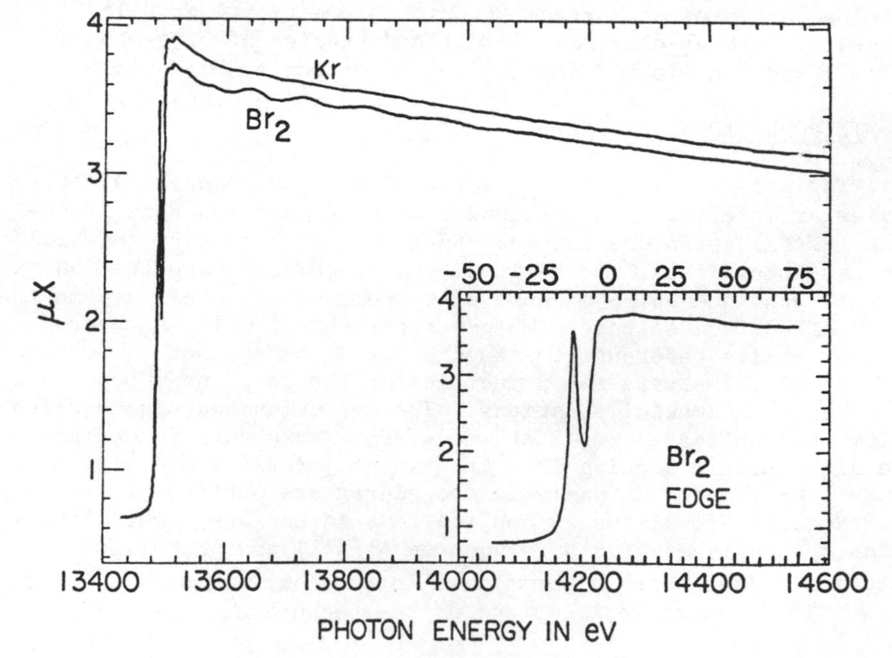

Figure 1

Figure 1 shows the absorption spectra of the gases Kr and Br_2 taken from Ref. (4). The plotted <u>absorption coefficient</u> μ is obtained from the incident (I_o) and transmitted (I) X-ray intensity through a sample of thickness x according to

$$\mu x = \ell n\ (I_o/I) \tag{1}$$

While the atomic Kr spectrum shows a smooth decrease of the absorption coefficient above the K-edge (excitation of 1s electrons) the diatomic Br_2 spectrum exhibits small sinusoidal intensity oscillations, i.e. EXAFS. It is these small oscillations which are the subject of the present paper. In many cases the observed oscillations are considerably larger in amplitude than those shown in Figure 1 and exist for several hundred eV above the absorption edge (1).

Because the EXAFS oscillations arise from an interference effect of the photoelectron wave it is apparent that the phase of the oscillations must depend on the distance R_i from the absorbing to the neighbor atom i involved in the backscattering process. The amplitude of the EXAFS oscillations will among other factors depend on the number of nearest neighbors N_i at a given distance R_i.

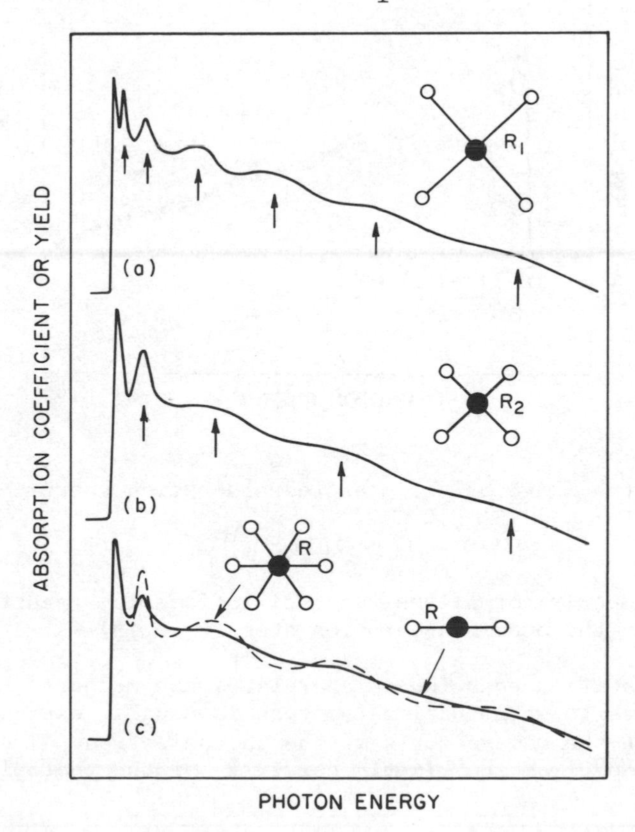

Figure 2

Without detailed analysis of the absorption spectrum one can by comparison of two EXAFS spectra often obtain useful information as suggested by Figure 2.

Note that shorter bonds are characterized by larger-spaced EXAFS oscillations and the EXAFS amplitude increases with the number of neighbor atoms.

III. THE EXAFS EQUATION

In order to analyze the EXAFS oscillations quantitatively we need to establish a unique definition. We are interested in the EXAFS signal per absorbing atom and corresponding to a given core excitation. Hence all background due to other absorption processes in the sample need to be eliminated. As shown in Figure 3 this is done by extrapolation of the pre-absorption-edge background to energies above the edge using a suitable polynomial γ^*. Another smooth polynomial γ_o (usually a spline polynomial consisting of two or three polynomials of second order) is fit through the oscillatory structure above the edge.

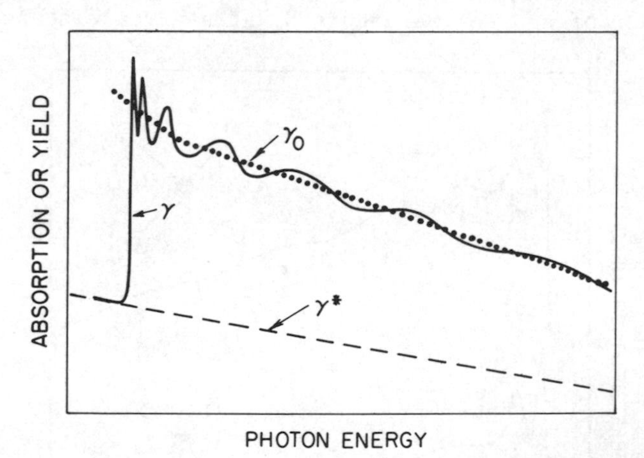

Figure 3

We define the EXAFS of the measured absorption spectrum γ as,

$$\chi(h\nu) = (\gamma-\gamma_o)/(\gamma_o-\gamma^*) \tag{2}$$

Eq. (2) properly normalizes the oscillations in γ about a mean value γ_o to the atomic absorption step $(\gamma_o - \gamma^*)$.

The detailed equation which relates the measured EXAFS oscillations to structural parameters is usually expressed as a function of the wavevector k of the photoelectron. The wavevector k is computed from the kinetic energy E_k of the photoelectron using the free-electron dispersion relation

$$E_k = \hbar^2 k^2/2m \tag{3}$$

If we assume that at a threshold energy E_o which we define as the inflection point of the absorption edge the photoelectron has zero kinetic energy we are lead to equation

$$k = 0.5123\ (h\nu - E_o)^{\frac{1}{2}} \tag{4}$$

where k is in $Å^{-1}$ and $h\nu$ and E_o are in eV units. Using Eq. (4) we can now obtain the measured EXAFS $\chi(h\nu)$ as a function of wavevector. For s-initial states (K or L_1 edges) the EXAFS is given by (5)

$$\chi(k) = -\sum_i A_i(k) \sin[2kR_i + \phi_i(k)] \tag{5}$$

The summation extends over all neighbor shells i separated from the absorbing atom by a distance R_i. $\phi_i(k)$ is the total phaseshift which the photoelectron wave experiences from the absorbing (a) and backscattering (b) atoms (6),

$$\phi_i(k) = \tau_a(k) + \tau_b(k) \tag{6}$$

In the case where the neighbor shell i consists of identical atoms, the total amplitude $A_i(k)$ in Eq. (5) is given by (5)

$$A_i(k) = (N_i^*/kR_i^2)\ F_i(k)\ e^{-2\sigma_i^2 k^2}\ e^{-2R_i/\lambda(k)} \tag{7}$$

The exponential terms in Eq. (7) are the Debye-Waller factor-like term and the damping term due to inelastic scattering (mean free path $\lambda(k)$) of the photoelectrons. For a detailed discussion of these terms the reader is referred to references (7-8).

N_i^* is the effective coordination number of the central atom at distance r_i (ith shell) and is given by (K or L_1 edges only) (9)

$$N_i^* = 3 \sum_j^{N_i} \cos^2\theta_j \tag{8}$$

Here, the sum extends over all neighbor atoms j (total number N_i) in the ith shell and θ_j is the angle between the electric field vector $\vec{E}$ of the x-rays at the central atom site and the vector $\vec{r}_{ij}$ from the central atom to the jth atom in the ith shell. Eq. (8) tells us that a neighbor atom j only contributes to the EXAFS signal if the $\vec{E}$ vector has a sizeable component along its position vector $\vec{r}_{ij}$. For inherently anisotropic systems like surfaces the polarization dependence of the EXAFS signal therefore provides an extremely powerful tool to sort out neighbor atoms which are located in different directions from the central atom. In fact, the $\vec{E}$ vector can be invisioned as a "search light" revealing all neighbors in a given direction. For isotropic systems like polycrystalline or amorphous materials, an average over all angles yields $N_i^* = N_i$ such that the effective coordination number N* is equal to the real coordination number N. For single-crystal materials, or in the case of oriented molecules or atoms on surfaces, Eq. (8) has to be evaluated for an assumed model geometry. Figure 4 shows the dependence of N* on the geometry for a chemisorbed atom in a threehold hollow site (e.g. (111) surface) and a two-fold hollow site (e.g. (110) surface) of the substrate. For higher than twofold symmetry N* does not depend on the azimuthal substrate orientation (ϕ) with respect to the $\vec{E}$ vector but only on the polar (θ) orientation. Note that the equation for the twofold site also describes the fourfold site (a = b) and the twofold bridge site (b = o, a = 2d and a factor 6 instead of 12 because of only 2 neighbors).

(a) Three-Fold Hollow Site

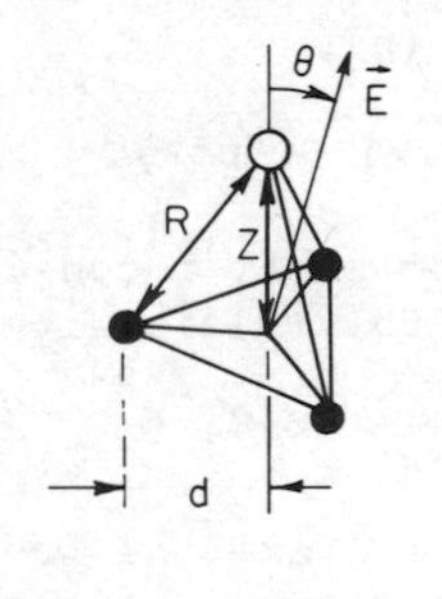

$$N^* = \frac{9}{Z^2+d^2}\left\{Z^2\cos^2\theta + \frac{d^2}{2}\sin^2\theta\right\}$$

(b) Two-Fold Hollow Site

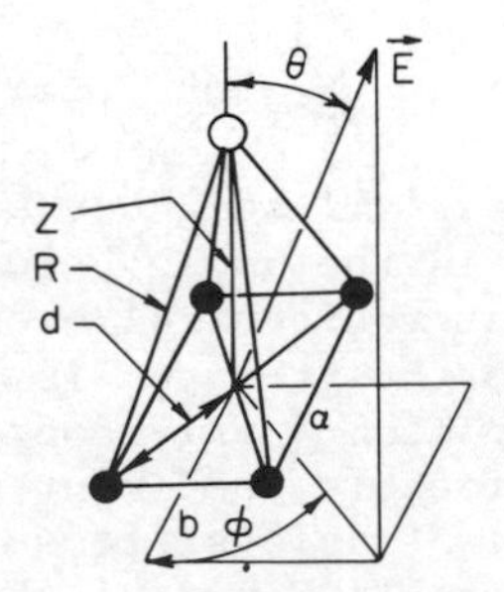

$$N^* = \frac{12}{Z^2+d^2}\left\{Z^2\cos^2\theta + d^2\sin^2\theta\left[\frac{a^2}{4d^2}\cos^2\phi + \frac{b^2}{4d^2}\sin^2\phi\right]\right\}$$

Figure 4

$F_i(k)$ is the backscattering amplitude of the neighbor atoms which depends on their atomic number Z as shown in Figure 5 for C (Z = 6), Si (Z = 14), Ge (Z = 32), Sn (Z = 50) and Pb (Z = 82) (Ref. 6). Without analysis of the detailed size of $F_i(k)$ which is complicated by corrections for the Debye-Waller and mean-free path terms in Eq. (7) one can often obtain valuable information by using the k-dependent maximum or maxima as a fingerprint for the atomic number of the neighbor atom. This concept has been successfully applied for oxygen on GaAs(110) (10).

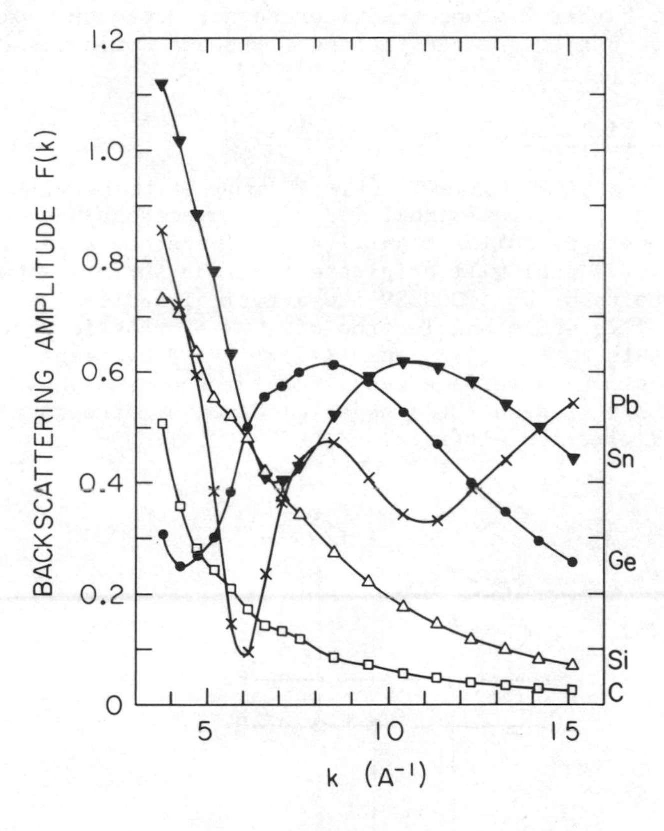

Figure 5

IV. SURFACE EXAFS (SEXAFS)

For surface crystallographic studies by means of EXAFS a detection technique is needed which allows discrimination of the signal originating from surface atoms relative to that from the bulk. As discussed in Section II the absorption coefficient is the probability of making a core excitation, i.e. the probability of producing a core hole. Thus any process whose statistical average is proportional to the annihilation of the core hole can also be used as a measure of EXAFS. When electrons from an initial state A are excited by x-ray absorption the created hole can be filled by a radiative or nonradiative transition from a lower binding energy level B. Radiative transitions result in fluorescent x-ray emission (11) while Auger electrons are produced in the nonradiative process. The relative probability of the two competing channels varies with atomic number Z of the excited atom (12). For K-edges (1s excitation) the Auger and x-ray fluorescence process have equal probabilities for Z = 30. For lower Z atoms the Auger and for higher Z atoms the fluorescence processes dominate, respectively. For $L_{2,3}$ edges the Auger process dominates up to Z = 90 (12).

A. Electron Yield SEXAFS

For surface EXAFS (SEXAFS) (13-15) studies it is advantageous to use the Auger electron signal (9) since electrons have a shorter scattering length in solids than x-rays. Therefore a larger fraction of the signal will originate from the surface atoms. For example, a photon of $h\nu$ ~1000 eV has a typical scattering length of $1/\mu \sim 2000$ Å (16) while an electron of 1000 eV kinetic energy has a mean free path of L ~ 20 Å, only (17). EXAFS by means of Auger electron detection is surface sensitive for the same physical reasons as photoemission (PES) Auger electron spectroscopy (AES) or low energy electron diffraction (LEED). (18).

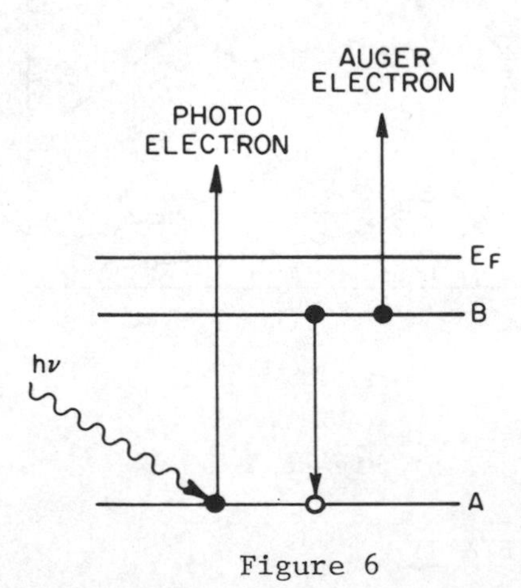

Figure 6

The Auger process is depicted in Figure 6. The elastic (no energy loss) Auger electrons are only a small fraction of the total number of electrons emitted from the sample. In fact, almost all Auger electrons will suffer a loss of energy on their way out of the solid through electron-electron (19), electron-plasmon (20) or electron-phonon (21) scattering. This is true also for the excited photoelectrons such that the total electron yield from the sample is completely dominated by inelastically scattered electrons (22).

The total number of electrons γ created in a depth L of a sample with absorption coefficient μ is proportional to the number of absorbed photons, i.e. $\gamma \sim 1-\exp(-\mu L)$. For $1/\mu >> L$ this relation reduces to $\gamma \sim \mu L$ such that the total electron yield γ is directly proportional to the absorption coefficient μ. At sufficiently high (>50 eV) photon energies the x-ray absorption length $1/\mu$ will always be much larger than a typical electron escape depth L such that the relation $\gamma \sim \mu$ is valid and the total electron yield can be used as a direct measure of the EXAFS (22-25). Experimental details of Auger and total yield measurements are discussed below.

The effective electron escape depth L is a complicated sum of electron scattering lengths depending on the details of the cascading process of the hot primary Auger and photoelectrons inside the sample (see Figure 7). In the soft x-ray region L is of the order of 50-100 Å. Electrons excited deeper inside the sample will have thermallized their energy before reaching the surface. If we measure the electron yield signal from a sample consisting of atoms B only a very small fraction (1-10%) of the signal will originate from the surface layer. Thus we cannot hope to disentangle the EXAFS of atoms B on the surface from that of bulk atoms B.

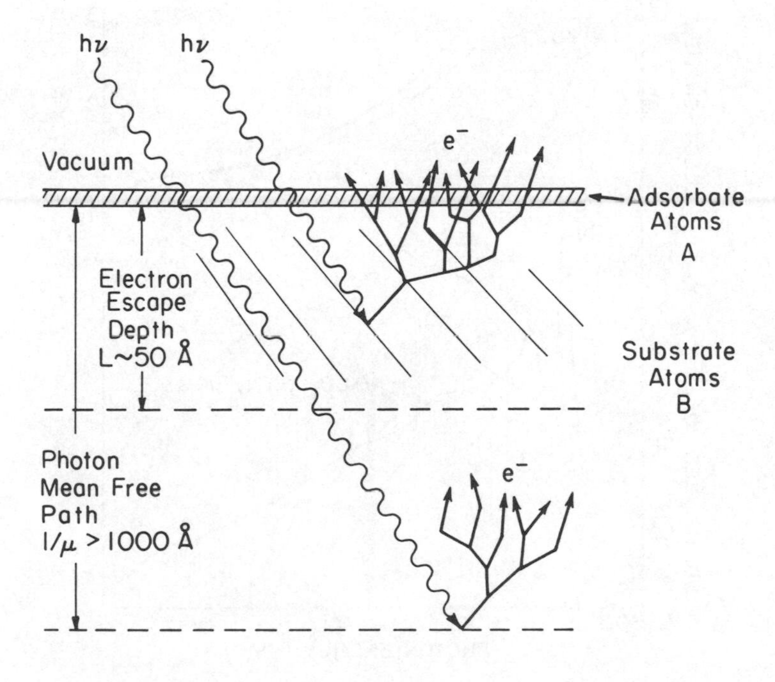

Figure 7

Total yield EXAFS measurements above the absorption edge of a sample will thus give <u>bulk</u> information. However, if we adsorb atoms A on the surface of a substrate consisting of atoms B we can distinguish the EXAFS of atoms A if atoms B do not produce any EXAFS in the energy range of interest. As a rule of thumb an adsorbate coverage of one monolayer produces an edge jump which is about 10% of the (structureless) background signal of the substrate. This is shown in Figure 8a for a sample of about 2 monolayers of oxygen on Ni(100) (26). The small structure in the clean Ni(100) electron yield around 560 eV reflects a modulation in the monochromator transmission function (Au N_3 edge). This structure is divided out in the normalized yield spectrum in Figure 8b which now clearly shows the EXAFS above the O K-edge of the O atoms on the Ni(100) surface.

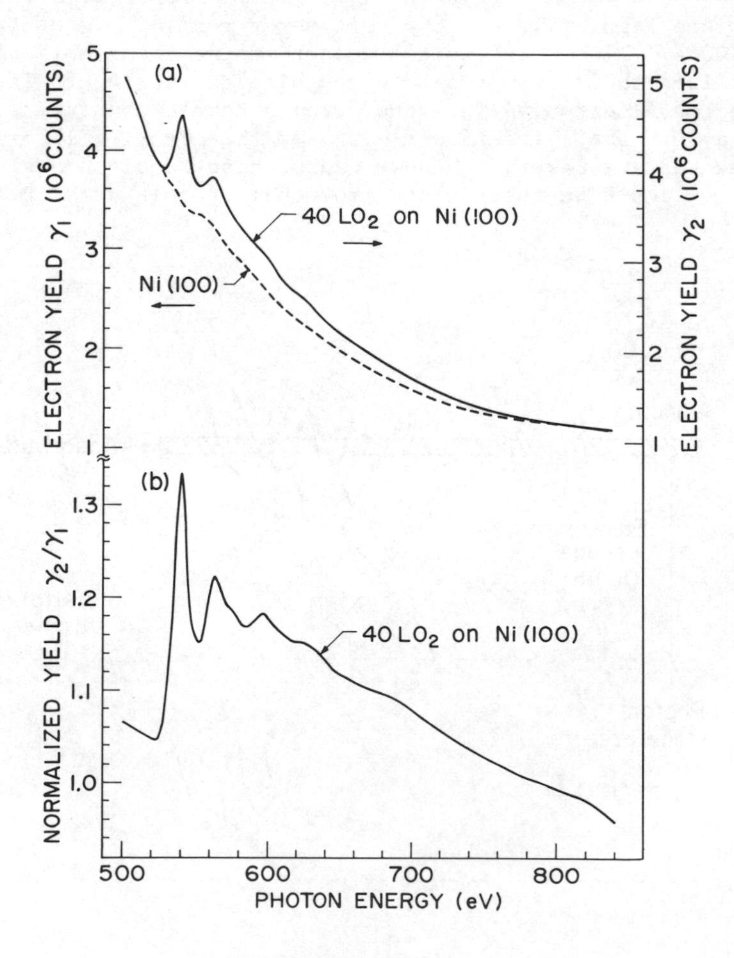

Figure 8

B. Ion Yield SEXAFS

The ion yield originating from photon stimulated desorption (PSD) (27) of atoms adsorbed on surfaces can be used as a signal for SEXAFS measurements (28). As illustrated in Figure 9 a core hole on an adsorbate or a surface substrate atom may be filled by an intra- or inter-atomic Auger process which results in holes in the valence shell breaking the bond of the adsorbate complex. As first suggested by Knotek and Feibelman (29) the desorption process is directly linked to filling the core hole. The PSD ion yield thus directly measures the absorption coefficient of the adsorbate atoms prior to their desorption and of those surface substrate atoms to which the desorbing adsorbate ions were bonded.

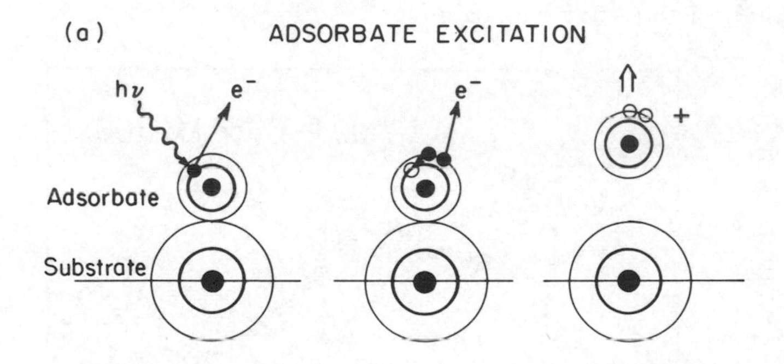

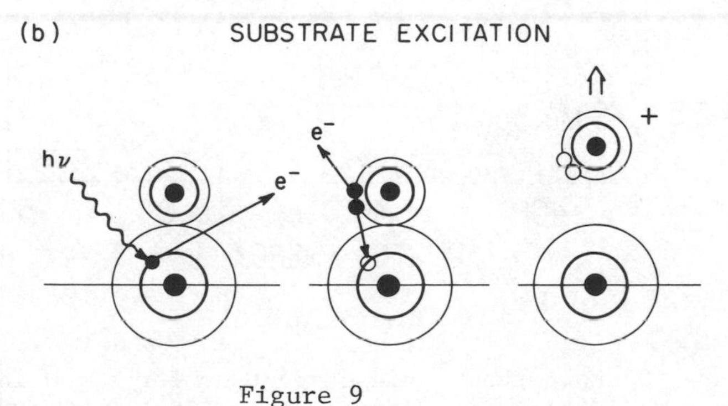

Figure 9

The ion yield γ_{ion} following the excitation of a photoelectron from a core level n is proportional to the partial absorption coefficient $\mu_n(h\nu)$ of this level according to (30)

$$\gamma_{ion} \sim (1-f)\, P_n\, \mu_n(h\nu) \tag{9}$$

Here P_n is the probability for Auger transitions into the core hole in shell n which result in holes in the valence region such that the surface complex is transformed into a repulsive ionic state. P_n is independent of $h\nu$ but depends on the overlap of the <u>valence</u> and <u>core</u> wave functions involved. f is the probability for reneutralization of the repelled ion either causing the recapture of the ion or its desorption as a neutral. The survival probability (1-f) is also $h\nu$ independent but depends strongly on the overlap of valence wavefunctions of the repelled ion and its neighbors which leads to the high site specificity discussed previously (27-32). Because of the vanishing (1-f) term for bulk atoms the PSD signal originates exclusively from surface atoms.

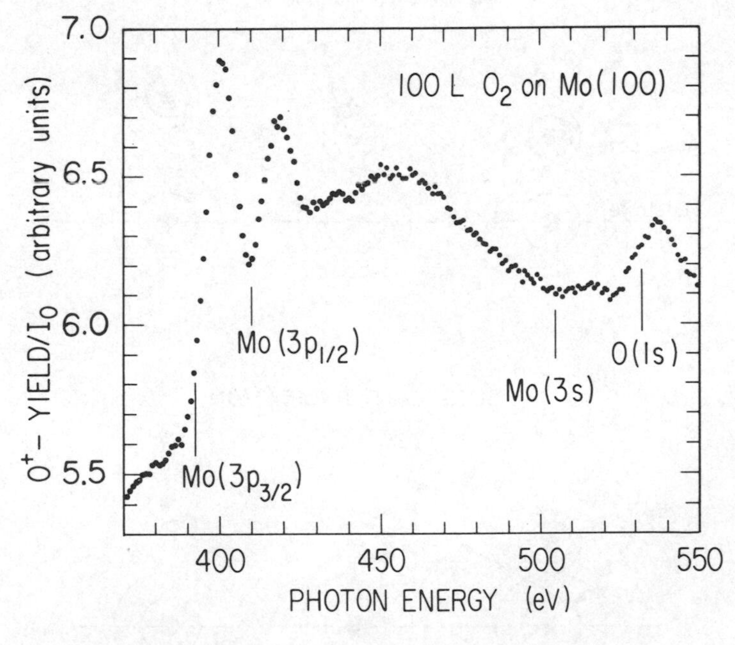

Figure 10

Figure 10 shows the O^+ yield spectrum for a monolayer of oxygen on a Mo(100) surface in the range 370-550 eV (30). Mo has three M-absorption edges between 350 and 550 eV, namely M_3 (393 eV), M_2 (410 eV) and M_1 (505 eV) (33). These edges correspond to excitations of $3p_{3/2}$, $3p_{1/2}$ and 3s electrons, respectively. The O K-edge (532 eV, 1s electrons) also lies in the above spectral range. As expected the ion yield clearly reveals the three absorption edges of Mo and an EXAFS oscillation

(460 eV) above the $M_{2,3}$ edge. The spectrum up to 520 eV resembles the absorption coefficient of clean Mo (30). There are subtle differences, however, due to the fact that the spectrum in Figure 10 corresponds to the absorption coefficient of only those Mo atoms on the surface which have O atoms as nearest neighbors. Above 520 eV the O K-edge is clearly visible demonstrating the importance of desorption processes following _intra_-atomic Auger transitions. In the present case the jump at the O K-edge is relatively small (~4%) owing to the large background from the preceding Mo edges and a relatively high scattered light contribution at the time of the measurement. EXAFS studies above the Mo L_1 edge (2866 eV) are discussed below (28).

V. EXPERIMENTAL DETAILS

Surface EXAFS studies require a high (~10^{10} photons/sec) photon flux incident on the sample. This is necessary to obtain sufficient signal-to-noise ratios (~10^3) in electron yield experiments and sufficient count rates (>10^3 counts/sec) in the low-cross section(<10^{-6} ions/photon) PSD measurements. At present such experiments are only feasible by utilizing monochromatized synchrotron radiation (34). Suitable monochromators for studies in the spectral range 250-20,000 eV are available as discussed elsewhere (35-37).

Electron yield EXAFS measurements are best carried out in the experimental arrangement shown in Figure 11. The monochromatic x-ray beam is first collimated to reduce scattered light. The transmitted beam impinges on a high transmission (~80%) metal grid which can be coated _in situ_ (38). The electron yield signal from this grid amplified by a spiraltron electron multiplier serves as a dynamic intensity monitor. The coating material is chosen not to exhibit any absorption edges in the photon energy range of interest. We have used Cu between 200 and 930 eV and above 2000 eV and carbon at energies above 600 eV. The x-ray beam is then incident on the sample under investigation. The electron yield signal is again amplified by a spiraltron electron multiplier. Two hemispherical grids between the sample and the detector may be used to retard low energy electrons and enhance the surface sensitivity as discussed elsewhere (10). In general the output signal of the electron multiplier is high enough to employ current measurement techniques using a floating battery box and a current amplifier as discussed in Ref. 39. Typical output currents in our measurements were of the order of 10^{-8}-10^{-7}Ampere. Auger yield SEXAFS measurements (13) are carried out using an electron energy analyzer (e.g. a cylindrical mirror analyzer (CMA)). The analyzer window is kept fixed at the kinetic energy of the elastic Auger electrons from the adsorbate (22, 26).

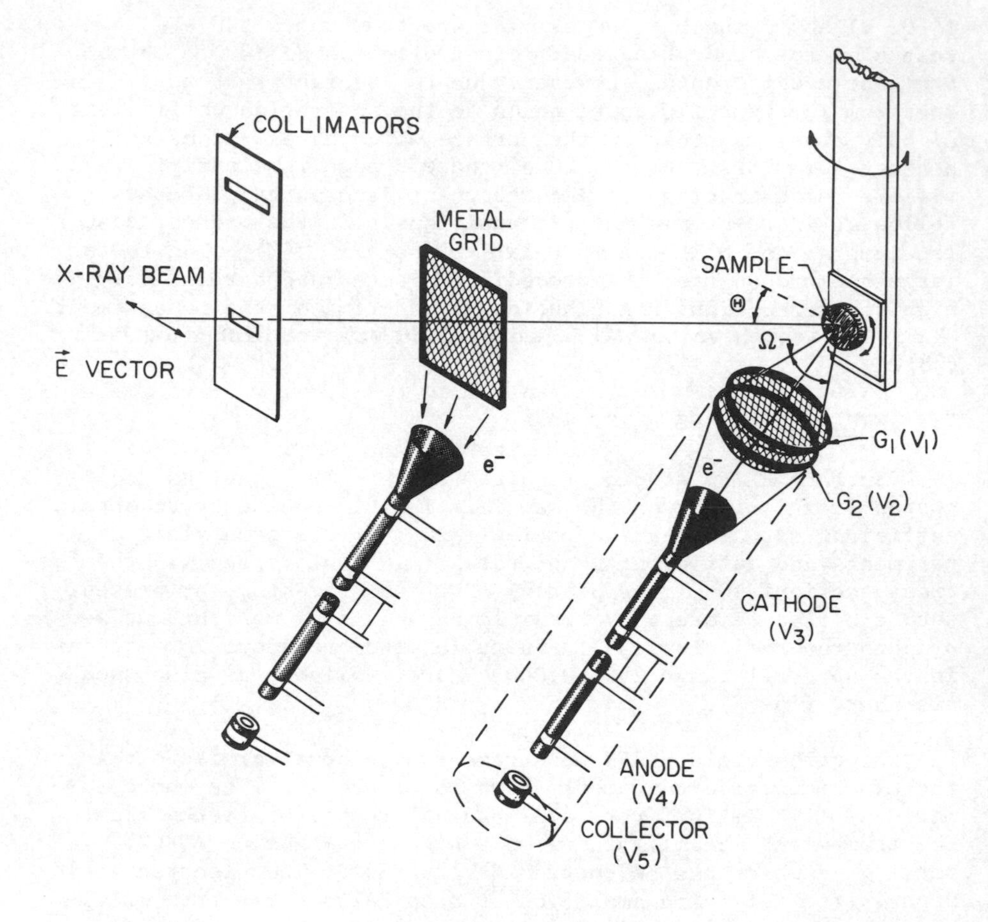

Figure 11

The efficiency of the experimental arrangement in Figure 11 in normalizing out intensity fluctuations is shown in Figure 12. Here we show the electron yield signals of a SiO_2 sample and a Cu coated grid monitor. Both spectra are severely distorted by intensity fluctuations caused by electron beam instabilities of the storage ring SPEAR during the sweep. The glitches are completely normalized out when the signal from the sample (I) is divided by that of the reference monitor (I_o). This indicates a high degree of linearity of the employed detection scheme.

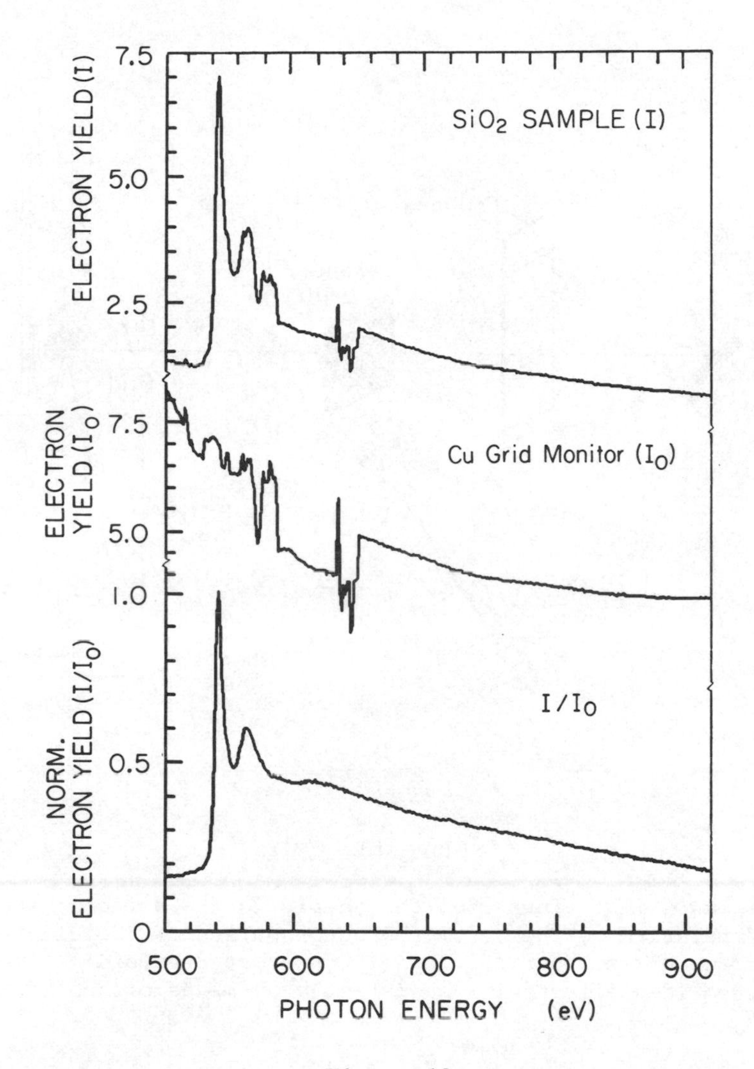

Figure 12

Ion yield EXAFS measurements are performed in a similar arrangement as electron yield studies except that the ion signal from the sample is detected by means of a time-of-flight (TOF) detector (see Figure 13). The positive ions released from the sample (+2.5 kV) are accelerated toward a grid (-1.5 kV) and detected by a channelplate electron multiplier (-2.0 kV) (30).

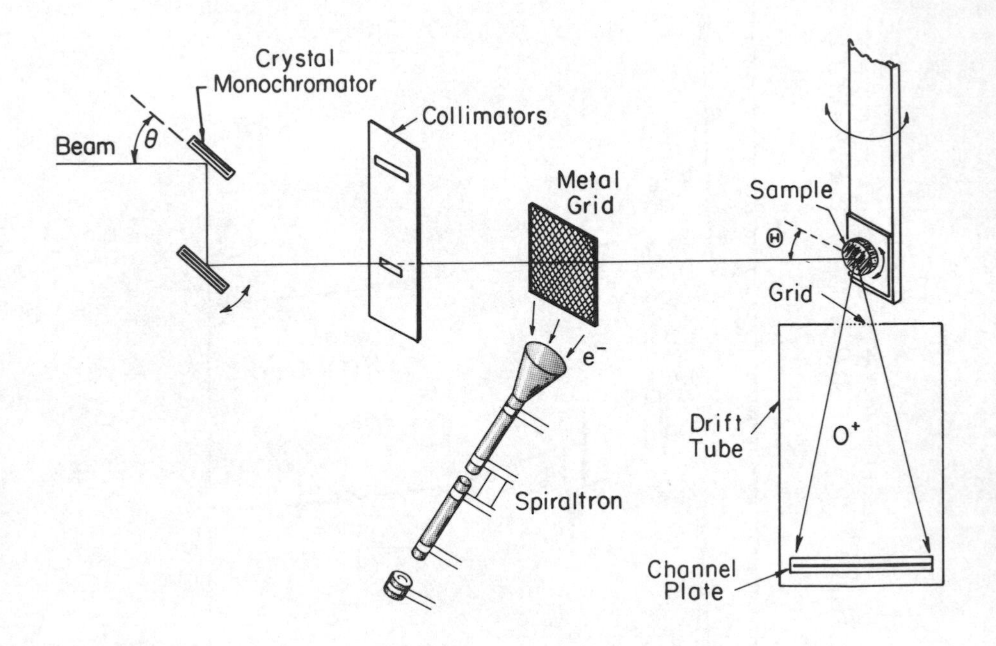

Figure 13

The ion flight times from the sample to the detector are measured relative to the "prompt" photon pulses of the electron storage ring of know periodicity (780 nsec at Stanford). The ions can be identified self consistently by means of the TOF relation.

$$M_i/M_k = (t_i/t_k)^2 \tag{10}$$

where $M_{i,k}$ and $t_{i,k}$ are the mass numbers and the flight times of the ion species i and k respectively. Figure 14 shows the TOF spectrum from a rather surface-contaminated Mo(100) sample (30). The separation of the O^+ (M = 16) and OH^+ (M = 17) peaks illustrates the mass resolution. One channel in Figure 14 corresponds to 0.61 nsec. The ion yield EXAFS spectra are recorded by monitoring the energy dependent intensity variations of a given peak in the TOF spectrum.

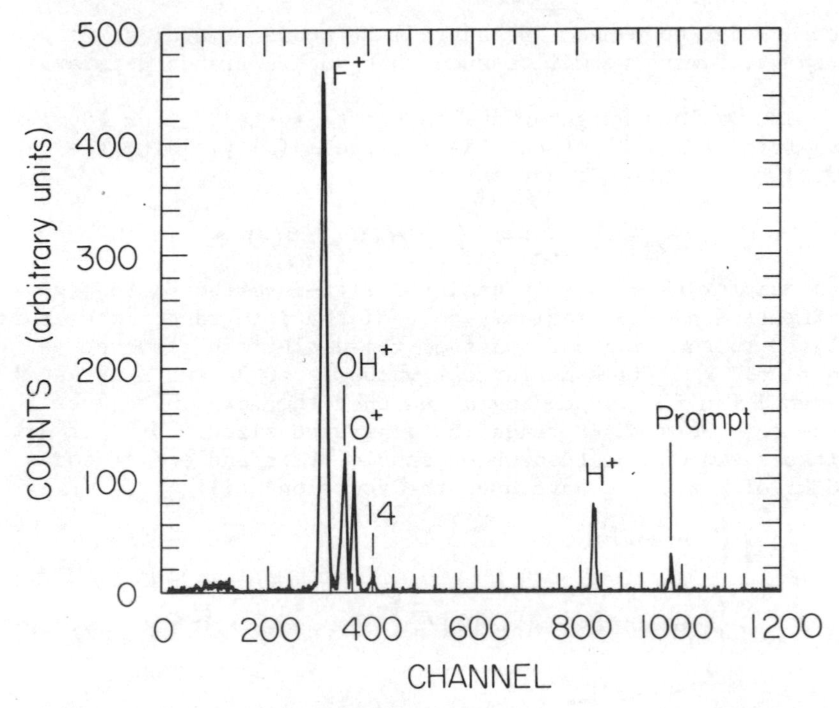

Figure 14

VI. DATA ANALYSIS

Analysis of the measured EXAFS spectra starts with background subtraction as briefly discussed in Section III and illustrated in Figure 3. Normalization of the EXAFS oscillations to the absorption step height (Eq. 2) is important for comparison of the amplitudes since we want to compare the size of the EXAFS oscillations per absorbing atom. Proper subtraction of the background γ^* in Figure 3 can be checked by plotting $\gamma' = \gamma-\gamma^*$. For a given absorption edge the function γ' must exhibit an identical overall shape with increasing photon energy for all EXAFS spectra. This is because the atomic absorption (which is characterized by the function γ_c in Figure 3) should be independent of crystal structure.

The EXAFS as a function of k is given by Eq. (5). The task is to disentangle the signals from all neighbor shells i and then obtain the phase and amplitude of the EXAFS for a given neighbor shell. Evaluation of the phase will yield the distance R_i and the amplitude will give information on number and kind of atoms in the ith shell. The EXAFS analysis can be carried out in a theory- independent fashion by comparison with a model compound ("standard") of known crystal structure. In the following we will assume that such a standard is available.

The various neighbor shells can be sorted out by Fourier transformation (40) of the EXAFS signal $\chi(k)$ yielding the radial distribution function,

$$F(r) = -\frac{1}{\sqrt{2\pi}} \int_0^{k_{max}} \chi(k)\, k^n\, W(k)\, e^{-i2kr}\, dk \tag{11}$$

If a complex transform is used the sign-inverted EXAFS signal is inputted as the imaginary part of the integrand since it is related to the imaginary part of the dielectric response function. k_{max} is some large k value ($\gtrsim 50$ Å) and n is usually chosen $1 \leq n \leq 3$ depending on whether the low- (n = 1) or high- (n = 3) k EXAFS range is to be emphasized. W(k) is a suitable window function which should start and end at nodes k_1 and k_2 of $\chi(k)$. We have used the function (41)

$$W(k) = \begin{cases} \left\{1 - \cos[\pi(k-k_1)/D]\right\}/2 & k_1 \leq k \leq k_1 + D \\ 1 & k_1 + D \leq k \leq k_2 - D \\ \left\{1 + \cos\pi[k - (k_2-D)]/D\right\}/2 & k_2 - D \leq k \leq k_2 \\ 0 & \text{otherwise} \end{cases} \tag{12}$$

where D is chosen to be of the order of $(k_2-k_1)/10$.

The peaks in the magnitude function $|F(r)|$ occur at distances R_i' which are related to the real neighbor shell separation R_i by a phase shift α, i.e. $R_i = R_i' + \alpha_i$. α is determined from a model compound. In many cases the scattering phase $\phi(k)$ (Eq. 6) is linear in k to a good approximation, i.e.

$$\phi(k) = a - bk \tag{13}$$

Then it is easy to show that $\alpha = b/2$. Of the various ways of EXAFS analysis (41-44) the Fourier filtering approach has many advantages and it shall be discussed here. This method gives more accurate results than simply adding a phaseshift correction α to R_i'. The different neighbor shells which correspond to peaks in F(r) can be isolated by a suitable window function W(r) (e.g. Eq. 12) in r space. The inverse transform of the complex function F(r)

$$\chi'(k) = \frac{2}{k^n \sqrt{2\pi}} \int_0^{r_{max}} F(r)\, W(r)\, e^{i2kr}\, dr \tag{14}$$

produces a real and imaginary part

$$\chi'(k) = A(k)\, e^{i\delta(k)} = X(k) + iY(k) \tag{15}$$

The amplitude A(k) and phase (see Eq. 5) $\delta(k) = 2kR_i + \phi_i$ corresponding to the EXAFS of the ith neighbor shell can now be computed,

$$A(k) = [(X(k))^2 + (Y(k))^2]^{\frac{1}{2}} \tag{16}$$

and

$$\delta(k) = \arccos(X/A) \qquad Y > 0 \tag{17a}$$

$$\delta(k) = -\arccos(X/A) \qquad Y < 0 \tag{17b}$$

$$\delta(k) = 0 \qquad X > 0,\ Y = 0 \tag{17c}$$

$$\delta(k) = \pi \qquad X < 0,\ Y = 0 \tag{17d}$$

Comparison of the phase $\delta(k)$ and amplitude A(k) for a structurally unknown system and a model compound consisting of the same central atom and backscatterer pair then yields structural parameters for the former.

A. PHASE ANALYSIS

Let $\delta_s(k) = 2kR_s + \phi(k)$ be the experimentally determined phase for the "standard" with known distance R_s between a given atom- pair. For the unknown system consisting of the same atom pair we denote the determined phase by $\delta_x(k) = 2kR_x + \phi(k)$. In both cases the scattering phaseshift $\phi(k)$ (Eq. 6) is the same except for a small difference due to possible changes in bonding. This concept of phaseshift transferability (45) has its physical origin in the fact that the core-electron potential dominates the scattering process. Only at low-k values (typically $k \lesssim 4$ Å^{-1}) does the potential of the valence electrons have a non-negligible contribution. We allow for differences in bonding by introducing an "inner potential" Δ (41). The momentum of the photoelectron is then redefined as (compare Eq. (4))

$$k' = 0.5123\,(h\nu - E_o - \Delta)^{\frac{1}{2}} \tag{18}$$

or

$$k' = (k^2 - 0.2625\Delta)^{\frac{1}{2}} \tag{19}$$

where Δ is in units of eV and k' in Å^{-1}. Comparison of the two phases yields the unknown bond length

$$R_x = (2kR_s + \delta_x - \delta_s)/2k' \tag{20}$$

Note that δ_x and δ_s were both derived for $\Delta = 0$. We can now plot R_x as a function of k for various values of Δ. For the proper choice of Δ (typically $-10\ \text{eV} \leqslant \Delta \leqslant +10\ \text{eV}$) R_x should be a constant for all k values. This procedure determines the distance R_x and Δ in a self-consistent fashion.

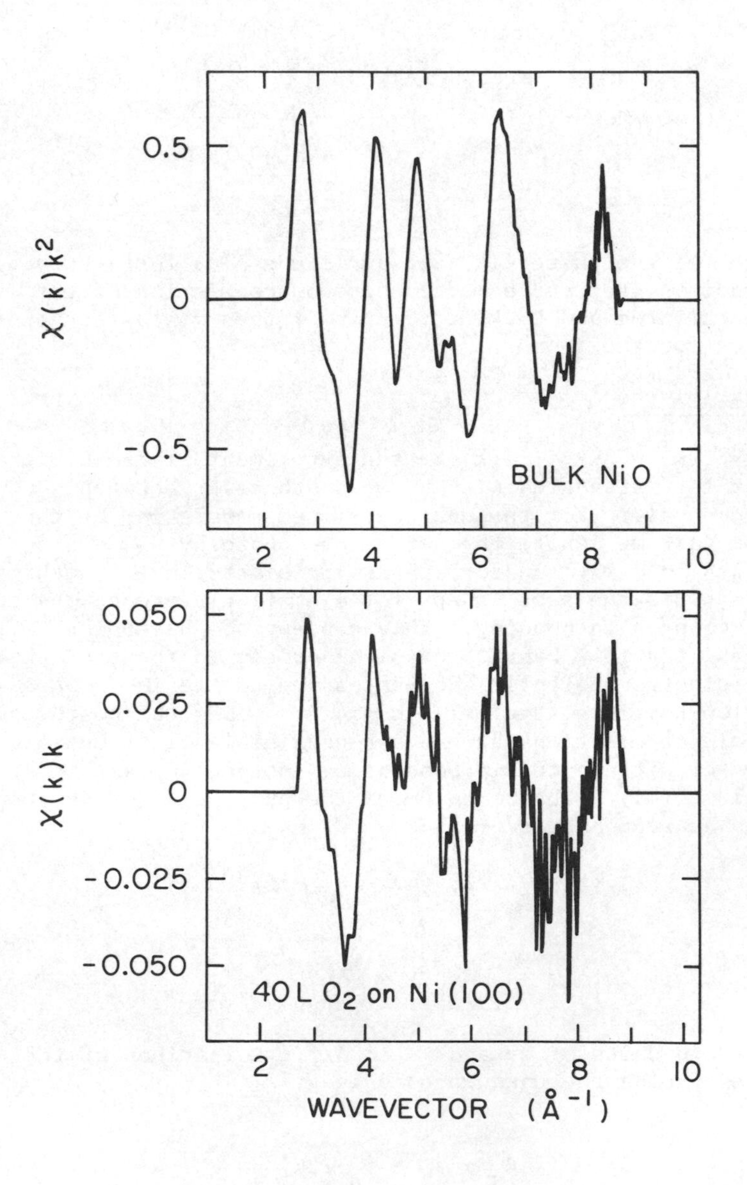

Figure 15

As an example we show in Figure 15 the EXAFS oscillations of a model compound NiO and a sample of 40L (1 Langmuir = 1 x 10^{-6} Torr sec) oxygen on Ni(100). The Fourier transforms |F(r)| for the signals are shown in Figure 16. Both transforms are dominated by a single peak A which is readily indentified from the bulk NiO spectrum as the O-Ni nearest neighbor bond length (2.09 Å).

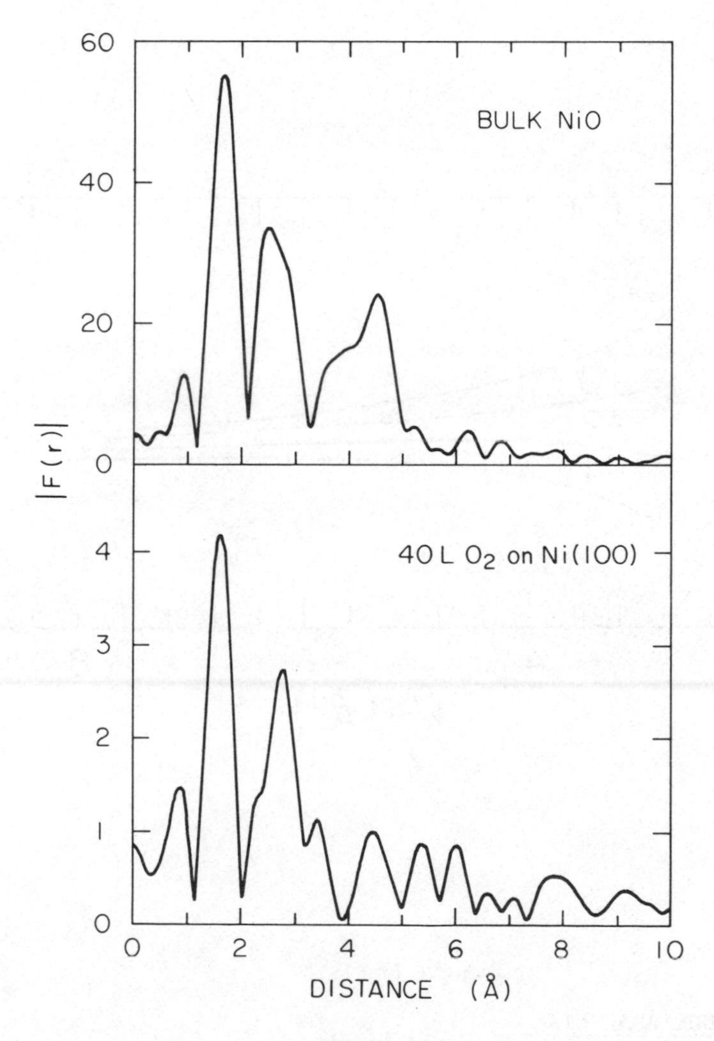

Figure 16

If peaks A are filtered out and back-transformed we can obtain the phase-function δ(k) for the two cases and plot the unknown distance for the O on Ni(100) case according to Eq. (20). This is done in Figure 17 for various values Δ. For Δ = -2 eV we obtain the O-Ni distance (2.04 ± 0.01) Å where the error bar reflects the variation of the Δ = -2 eV curve over the measured k range. This illustrates the great accuracy of EXAFS for distance determinations.

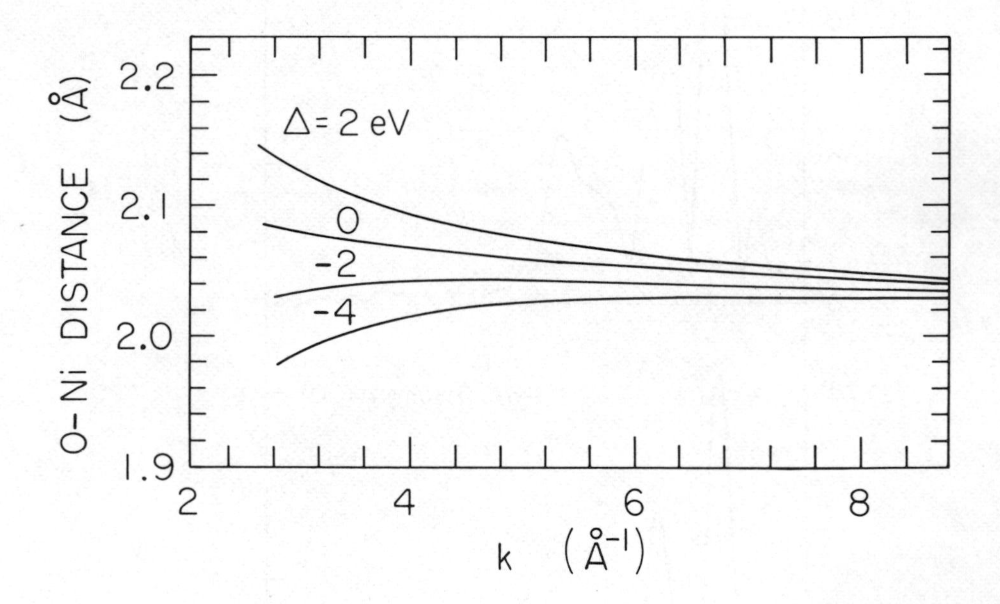

Figure 17

B. AMPLITUDE ANALYSIS

The amplitude function A(k) for a given neighbor shell as determined from the inverse Fourier transform (Eq. 16) depends on the terms listed in Eq. (7). Of all these terms, however, only the backscattering amplitude F(k) has resonance behaviour (6) as a function of k as shown in Figure 5. Thus the overall shape of A(k) often is sufficient to give an indication of the atomic number Z of the neighbor atoms. For a quantitative analysis of Z the Debye-Waller and electron-loss terms need to be corrected for. Analysis of F(k) does not allow discrimination of atomic numbers $\Delta Z \lesssim 5$.

Determination of the coordination number requires comparison with a model system, a procedure which relies on amplitude transferability. This concept, however, does not work as well as for the phase (46, 47). Both the Debye-Waller and inelastic loss terms in Eq. (7) are inherently determined by the chemically sensitive valence electrons, even at larger k (> 4 $Å^{-1}$) values. That is why even in favorable cases coordination numbers can only be determined to an accuracy of about 10% (46, 47).

Comparison of amplitudes in bulk systems is usually done by plotting $\ln[A_s(k)/A_x(k)]$ as a function of k^2 (48). Here $A_s(k)$ is the amplitude function corresponding to a given neighbor shell of a standard and $A_x(k)$ is that of the system under investigation (cp. Eq. 16). From Eq. 7 we obtain

$$\ln(A_s(k)/A_x(k)) = \ln(N_s^* R_x^2/N_x^* R_s^2) + 2k^2(\sigma_x^2 - \sigma_s^2) + 2(R_x/\lambda_x - R_s/\lambda_s) \tag{21}$$

For an isotropic electron mean free peak $\lambda_x \sim \lambda_s \sim \lambda$ the last term is negligible because $2|R_x - R_s| \approx 0.1 Å << \lambda \approx 5 Å$. Eq. 21 also assumes that the backscattering amplitude $F_x(k) = F_s(k)$. A linear plot of $\ln[A_s(k)/A_x(k)]$ versus k^2 yields the unknown coordination number N_x^* from the ordinate intercept at $k \to 0$ and σ_x from the slope.

The use of Eq. 21 for a bulk standard and an unknown surface structure is problematic because the loss term $2(R_x/\lambda_x - R_s/\lambda_s)$ may no longer be negligible owing to an inherently anisotropic mean free path λ_x at the surface (49). This problem is overcome by comparison of the amplitudes of SEXAFS measurements recorded at different angles of x-ray incidence with respect to the surface (8,13). In this case both the Debye-Waller and electron-loss terms should be indentical and a plot of $\ln(A(\theta_1)/A(\theta_2))$ versus k^2 should give a constant for all k values. Alternatively one can simply plot the amplitude ratio versus k yielding directly the ratio of effective coordination numbers

$$A(\theta_1)/A(\theta_2) = N^*(\theta_1)/N^*(\theta_2) \tag{22}$$

Often the experimentally determined ratio of $N^*(\theta_1)/N^*(\theta_2)$ is sufficient to distinguish between various model geometries for which the coordination numbers can be calculated (Figure 4).

VII APPLICATIONS IN SURFACE CRYSTALLOGRAPHY

A. Electron Yield SEXAFS: O on Al(111)

The interaction of oxygen with aluminum surfaces is a textbook example of protective oxide formation at metal surfaces. The properties of the surface oxide layer have important industrial implications. From an academic point of view the oxygen-aluminum system is of interest because the bonding is of simple s-p character thus enabling *ab initio* theoretical model calculations. In the past such calculations have in fact been carried out using a variety of theoretical techniques (50) for the earliest stages of oxidation, i.e. the chemisorption of oxygen on clean Al surfaces.

Experimentally the oxygen on Al(111) system has been investigated by numerous techniques. Photoemission studies (51) have revealed two distinct phases for the initial sorption of oxygen on Al(111) as evident from different shifts of the Al 2p core line. At small oxygen exposures $\lesssim$100L (1L = 10^{-6} Torr sec) an ordered (1 x 1) configuration was found corresponding to an Al 2p chemical shift of 1.4 eV. Photoemission (52) and LEED (53-55) studies showed that oxygen chemisorbs in the three-fold hollow site on the Al(111) surface with no Al atom in the second Al layer underneath. The site was also favored by theoretical calculations (50). Work function measurements (56) imply that around monolayer coverage (~100 - 150L O_2) the O atoms are outside the surface. The second phase which increases with oxygen coverage or temperature corresponds to an Al 2p shift of 2.7 eV which is the same value as for bulk Al_2O_3 relative to clean Al. Thus this phase which often accompanies the chemisorbed phase has been associated with bulk-like oxide formation where O atoms penetrate through the surface Al layer.

Exposure of the clean Al(111) surface to 50L O_2 produces a photoemission spectrum in the region of the Al 2p level shown by the broken line in Figure 18. The Al 2p level, whose spin-orbit splitting is barely visible with the resolution used here, is accompanied by a shoulder 1.4 eV higher in binding energy: this is the shift characteristic of the chemisorbed oxygen state (51). Upon heating to 200°C for ten minutes, the photoemission spectrum given in Figure 1 by the solid line is obtained. This spectrum, taken with improved analyzer resolution, shows clearly a peak shifted by 2.7 eV from the Al 2p level, and no peak shifted by 1.4 eV. The heating converts the chemisorbed state into a phase exhibiting the same binding energy as bulk Al_2O_3.

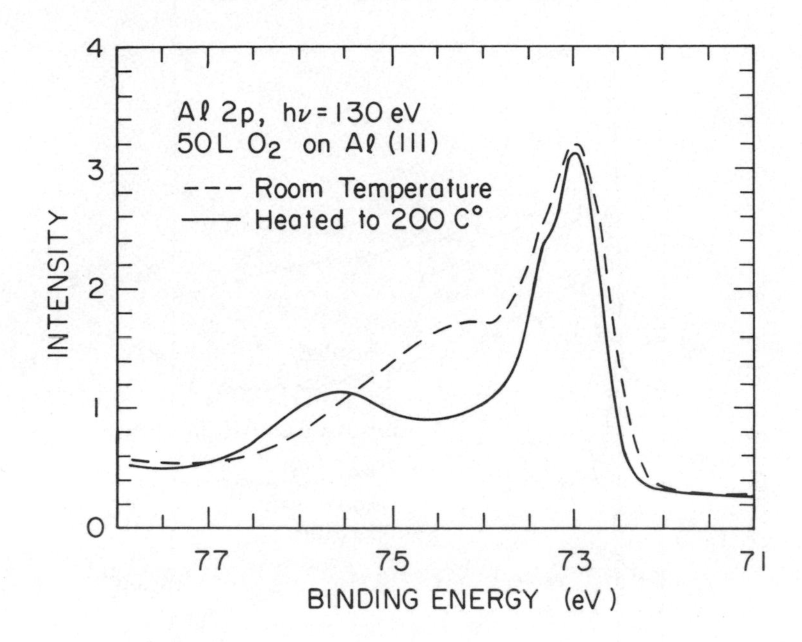

Figure 18

Figure 19 depicts the EXAFS spectra (58) above the oxygen K-edge for the two oxygen coverages corresponding to the photoemission spectra of Figure 18 and also for our standard, a sample of corundum, α - Al_2O_3 which has a well known bond length of R_{O-Al} = 1.92 Å (57). The EXAFS oscillations are nearly sinusoidal in all three spectra indicating that the O - Al nearest-neighbor scattering dominates over more distant neighbor-shells. The dashed lines join the peaks due to O - Al nearest neighbor scattering and show just by considering the frequency of the oscillations and without further analysis (cp. Figure 2) that the O - Al nearest-neighbor distance is longer for α - Al_2O_3 than for the oxygen-exposed Al(111) surfaces, and that there is essentially no change in bond length between the chemisorbed and the heated surface phases. Fourier transformation of the EXAFS spectra and correction by the O - Al scattering phaseshift derived from the bulk Al_2O_3 gives values of R = (1.76 ± 0.03)Å for the O - Al distance in the <u>chemisorbed</u> phase and R = (1.75 ± 0.03)Å for the <u>heated</u> phase.

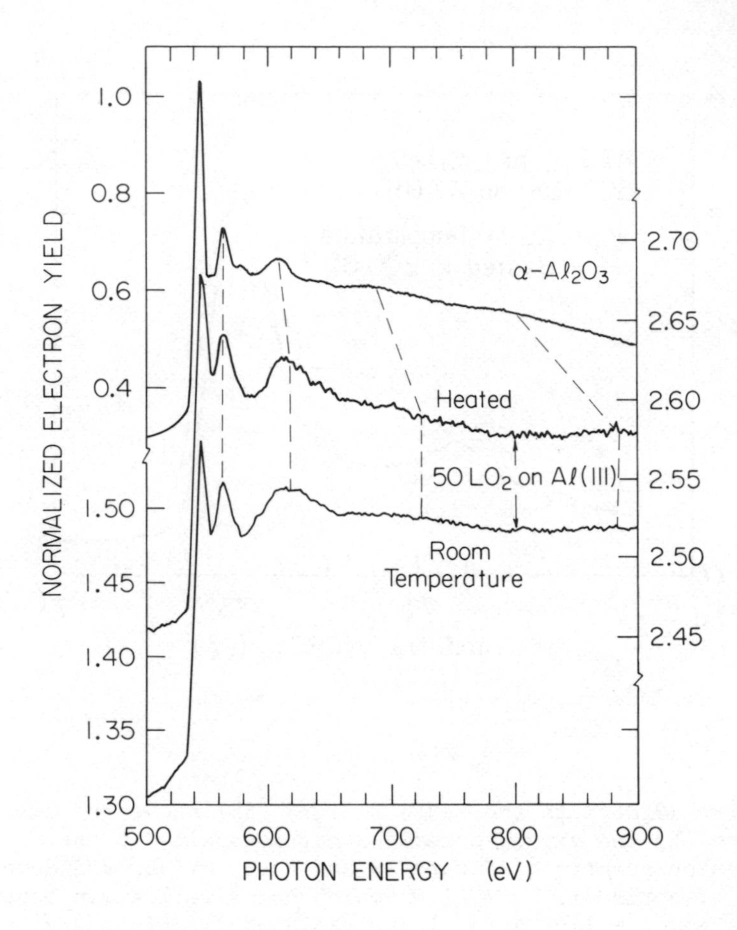

Figure 19

For the chemisorbed state we find strong support of the (1 x 1) oxygen overlayer structure suggested previously from LEED and depicted in Figure 20. Note that, although we determine the O - Al nearest-neighbor-distance to a high degree of accuracy with surface EXAFS, our measurements did not determine the bonding site. In principle this is possible by observation of the O-Al second-nearest-neighbor distance or the number of nearest Al neighbors but the extraction of both of these parameters, although routine in bulk EXAFS, is at present still difficult for surface EXAFS measurements on low-Z atoms. As is usually found with chemisorption systems, the adsorbate (O) atoms are located in positions which are a continuation of the fcc stacking of the bulk (Al) lattice. However, the interplanar spacing (0.6 ± 0.1)Å derived here is considerably reduced from the interplanar separation of 2.33 Å for bulk Al.

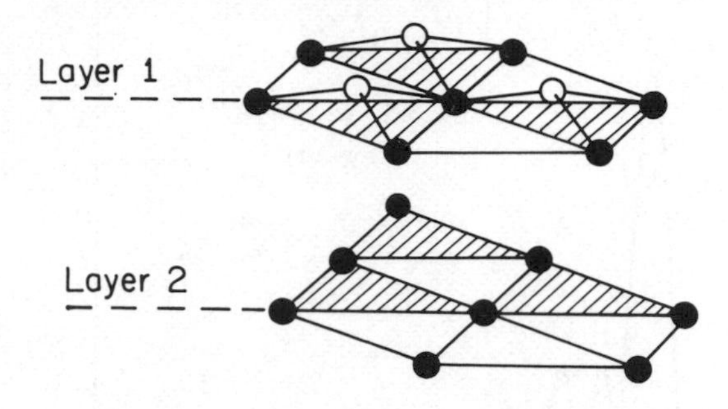

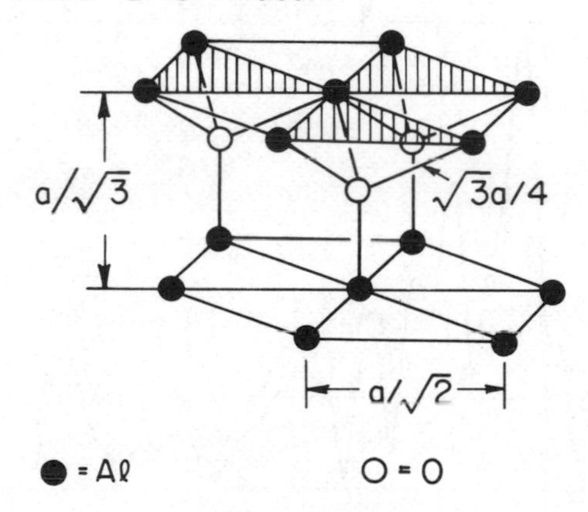

Figure 20

The in-plane O-O distance can also be obtained (49, 59) by using the polarization dependence of the EXAFS (Eq. 8). Figure 21 shows the Fourier transforms for spectra taken around monolayer coverage at grazing ($\theta = 11^{o}$) and a 45° angle of x-ray incidence. For $\theta = 45^{o}$ the electric field vector has a sizeable component parallel to the Al(111) surface and the in-plane O-O distance can be observed (peak B) in addition to the O - Al distance (peak A). Peak B yields an O - O distance of (2.90 ± 0.05)Å. This value is in excellent agreement with the expected value of 2.86 Å for an ideal (1 x 1) overlayer on Al(111).

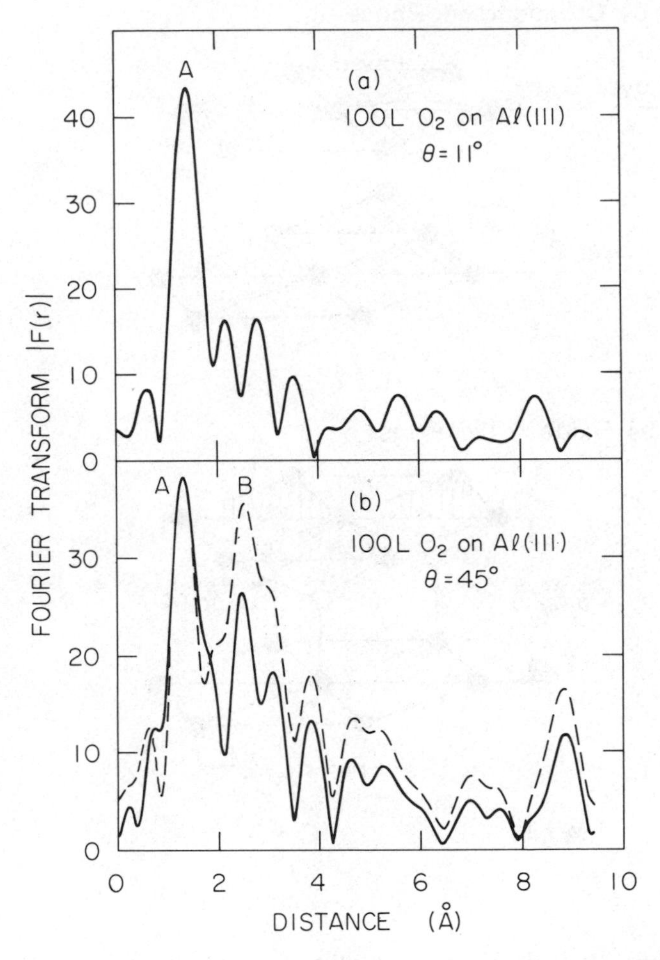

Figure 21

For the initial <u>oxide phase</u> we find the same O - Al bond length, within experimental error. This strongly suggests the geometry shown in Figure 20 for this phase, in which an O atom is bonded to four Al atoms as in corundum and γ-alumina and each surface Al atom is bonded to three O atoms. The ideal O - Al bond length for the shown tetrahedral interstitial position is 1.75 Å which is exactly equal to our experimental value. This site can be occupied simultaneously with the chemisorption site without distortion of the lattice, and both chemisorbed and oxide-like phases have previously been observed together (51). For both phases the O atoms form a (1 x 1) configuration with respect to the clean Al(111) surface. This can explain why a sharp (1 x 1) LEED pattern persists after relatively high exposures (<1000L O_2) corresponding to more than monolayer coverage. The constant Al 2p shift of 2.7 eV and yet the large O - Al bond length change (0.17 Å) between the low-coverage oxide phase on Al(111) and bulk corundum can be reconciled by a change in the number of oxygen neighbors per Al atom as discussed elsewhere (58).

The SEXAFS results for the O - Al bond length R = (1.76 ± 0.03)Å in the chemisorbed phase are in gross disagreement with three independent LEED investigations (53-55) of the same system. The LEED results suggest R ⩾ 2.12 Å. Reevaluation of the experimental LEED data suggests that they can also be explained by the short bond length derived by SEXAFS and that the previously derived large R value might, in fact, be due to a multiple coincidence problem (61). This is only one of many examples which demonstrates the general need to study the same system with a variety of experimental techniques.

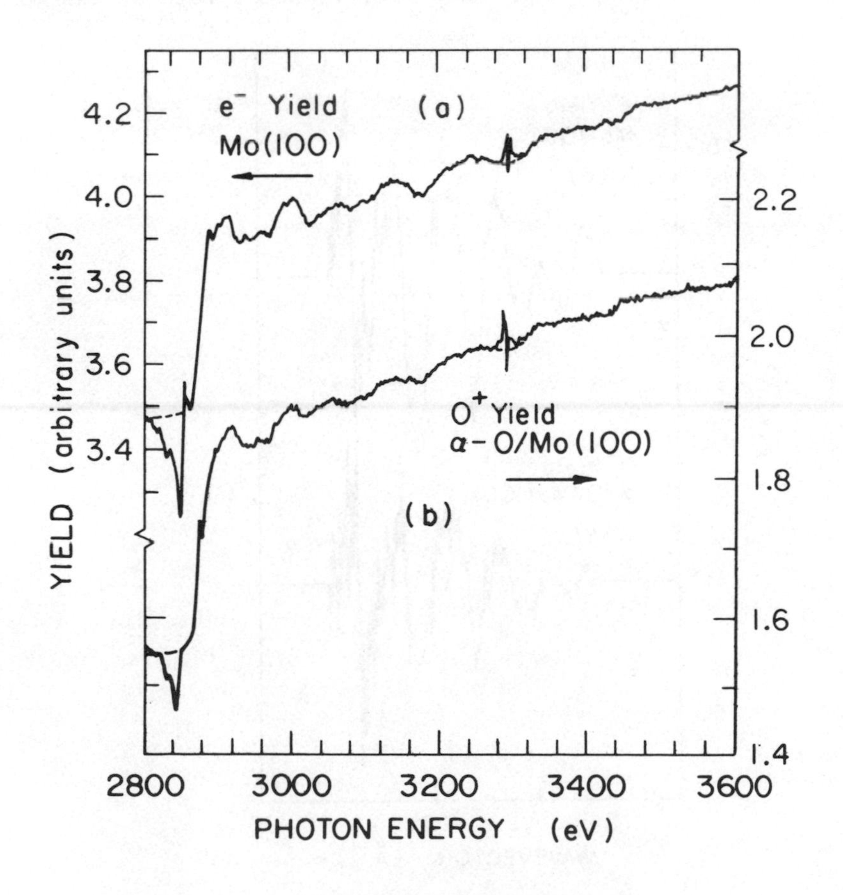

Figure 22

B. Ion Yield SEXAFS: O on Mo(100)

Below we shall discuss some first results of ion yield SEXAFS measurements (28). The oxygen on Mo(100) system was chosen because previous ESD studies (62) revealed a high cross-section for O^+ desorption at high oxygen coverage. The high desorption cross-section is attributed to a minority oxygen species (i.f. α - oxygen) which forms around monolayer coverage. SEXAFS spectra were recorded above the L_1 (2s) absorption edge of the Mo substrate. We also recorded the electron yield EXAFS spectrum of clean Mo(100) as a reference.

Figure 22 compares the total electron yield Mo L_1 EXAFS for a clean Mo(100) sample with the O^+ yield signal obtained from a Mo(100) surface exposed to 100 L oxygen. In both cases the x-ray beam was incident on the sample at a grazing angle of $\sim 20^o$. Both spectra show a sharp rise in count rate at the Mo

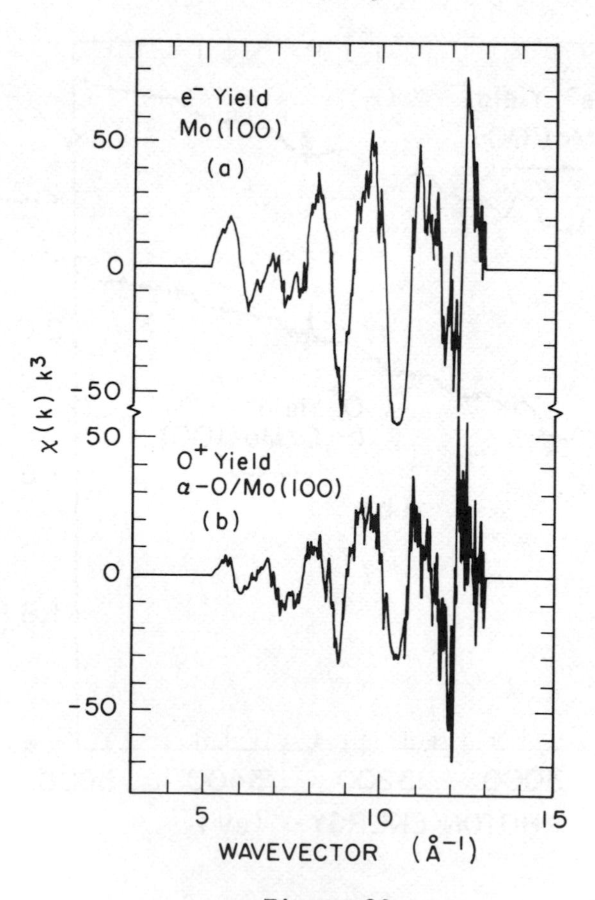

Figure 23

L_1 absorption edge around 2870 eV and EXAFS oscillations are clearly visible up to 700 eV above the edge. The oscillations above threshold are closely the same in frequency with an overall reduction in amplitude for the O^+ yield spectrum. The structures around 2840 eV and 3290 eV are caused by multiple Bragg reflection in the Mo(100) single crystal and in the Ge(111) monochromator crystals, respectively. Both structures can be removed as indicated by the dashed lines in Figure 22.

The EXAFS oscillations $\chi(k)k^3$ obtained from Figure 22 after background subtraction and normalization to the L_1 edge jump are shown in Figure 23. Within statistics the two signals are identical in frequency with a reduction in amplitude by a factor of 2 for the O^+ yield spectrum. The Fourier transforms for the two signals using the EXAFS range 7.8 - 13.0 $Å^{-1}$ are shown in Figure 24. Both transforms are dominated by a single peak A. The peaks in Figure 24 fall within 0.03 Å of each other.

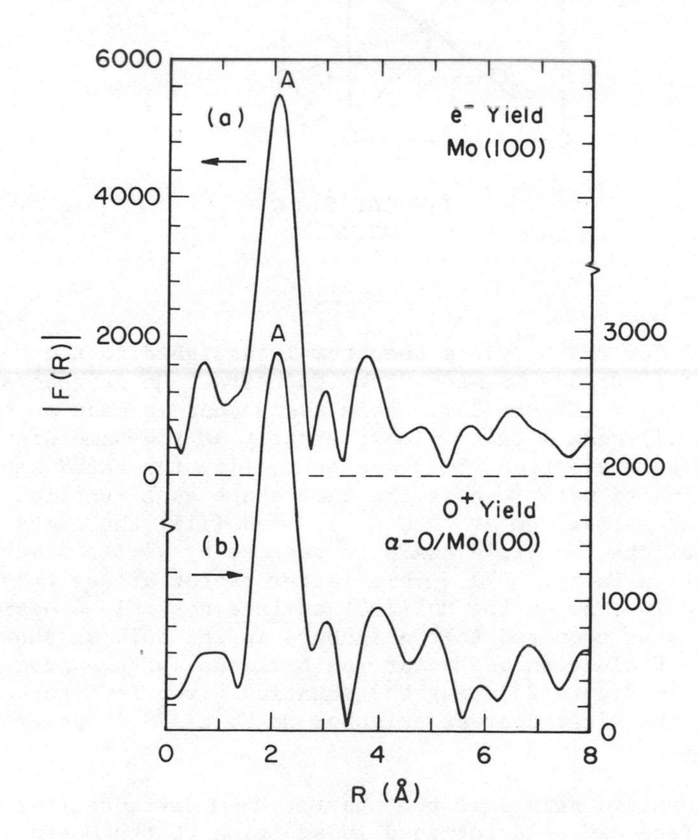

Figure 24

The electron yield EXAFS amplitude corresponding to peak A is bigger by almost exactly a factor of 2(± 10%). From the total electron yield EXAFS spectrum it is clear that peak A in Figure 24 corresponds to the Mo-Mo nearest neighbor distance R = 2.73 Å in bcc Mo metal as shown in Figure 25. Because the electron yield signal originates from atoms within 50-100 Å of the (100) surface the corresponding EXAFS spectrum is representative of bulk Mo atoms with only a small (< 10%) surface contribution.

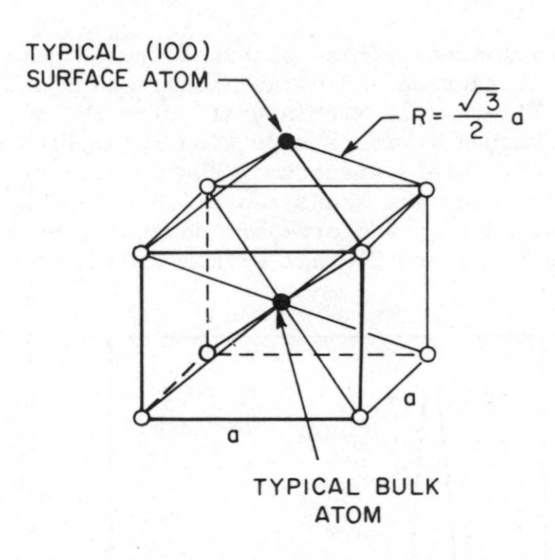

Figure 25

Peak A for the O^+ yield spectrum is assigned to the Mo-Mo distance of a surface Mo atom with its nearest Mo neighbors in the second layer (Figure 25). This assignment is made on the basis that (i) peak A falls almost exactly at the same distance as for the bulk electron yield spectrum, (ii) the EXAFS amplitude corresponding to peak A shows the same shape as a function of k for both cases (maximum at ~ 10 $Å^{-1}$), and (iii) the overall amplitude of the ion yield EXAFS is reduced by almost axactly (within 10%) a factor of 2. This latter factor arises from the fact that a Mo atom on the Mo (100) surface has only 4 nearest Mo neighbors as compared to 8 neighbors in the bulk as shown in Figure 25. Evaluation of N* for the bulk and surface atom geometries in Figure 25 using the equation given in Figure 4b shows that the first nearest neighbor Mo-Mo EXAFS is polarization independent.

The question arises if the shorter Mo-O distance (~ 2.0 Å) at the surface can be determined in addition to the Mo-Mo distance (2.725 Å). The Mo-O EXAFS will be sizeable only at low-k values because of the rapidly decreasing size of the O backscattering amplitude (compare Figure 5). There are indeed changes on the low-distance side of peak A if the low-k EXAFS region 5-8 $Å^{-1}$ is included in the analysis. However, our present data do not allow a reliable determination of the shorter O-Mo distance because the low-k EXAFS region above the L_1 edge contains a non-negligible contribution from the EXAFS above the L_3 and L_2 edges which lie 346 eV and 241 eV below the L_1 edge, respectively.

The fact that the EXAFS for Mo atoms in bulk Mo and at the α-oxygen on Mo(100) surface are essentially identical shows that the α-oxygen phase forms without reconstruction of the Mo(100) surface. This excludes any local oxide structure formation as might be expected from the large oxide-like Mo 3p chemical shift (2.9 eV) for the oxygen coordinated Mo surface atoms (30). This points to a very similar chemisorption structure as for the high-oxygen coverage phase on W (100) discussed by Madey (63) where the O atoms are believed to be bonded to one W surface atom of the outmost substrate layer. The high desorption rate and large chemical shift for the high coverage (minority) α-oxygen phase on Mo(100) can be explained by a highly ionic chemisorption bond as opposed to a more covalent like O-Mo bond for the low-coverage (majority) phase where oxygen is believed to occupy the fourfold hollow site.

The above measurements establish PSD as a new powerful detection technique for SEXAFS studies. In many cases such studies can offer unique advantages over all other known methods of surface structure determination. It is now possible to obtain structural information of specific chemisorption sites which are present in addition to other (e.g., majority) sites with smaller desorption cross sections. It is also possible to distinguish between sites where a given adsorbate (e.g., O) is present in atomic or molecular form (e.g., by O^+ versus OH^+ or O_2^+ detection). PSD SEXAFS measurements can be carried out above both the substrate as well as the adsorbate absorption edges and offer thus additional information as compared to electron yield SEXAFS (adsorbate edge only) studies.(5,6) Finally, by measuring H^+ desorption structural studies of the interaction of hydrogen with surfaces can be performed, even if the surface is disordered.

VIII CONCLUSIONS

The present paper is intended to provide the reader with a taste of the principles, analysis procedures and applications of EXAFS and SEXAFS studies of solids. It is by no means a complete discussion of the field but gives my personal view with emphasis on surface applications. Both electron and ion yield SEXAFS studies are shown to be powerful tools for surface crystallographic investigations. In addition to the chemisorption systems of interest to the chemist (gas adsorption) electron yield SEXAFS studies will no doubt be applied to solid state physics problems such as Schottky barrier and heterojunction formation. PSD allows the study of chemisorption sites which are not observable by other methods. Future work has to show if the states selected by PSD are possibly those which are playing the key role in catalysis. If they are ion yield SEXAFS measurements will truly uncover new horizons in surface and catalysis science.

ACKNOWLEDGEMENTS

Scientists who have worked on storage rings which operate in a 24-hour-per-day mode know how important it is to have good collaborators. I am indebted to many people for their help, especially to Sean Brennan, Josef Feldhaus, Rolf Jaeger, Leif Johansson and David Norman who were part of the preparation, execution and evaluation of the experiments presented here. In particular, most of the credit for the ion yield SEXAFS work should go to Rolf Jaeger.

The work at SSRL was supported by the National Science Foundation under Contract DMR77-27489 in cooperation with SLAC and the Basic Energy Division of the Department of Energy and the NSF-MRL Program through the Center for Materials Research at Stanford University.

REFERENCES

(1) Stern, E.A.: 1978, Contemp. Phys. 19, p. 289; Eisenberger, P. and Kincaid, B.M.:1978, Science 200, p. 1441; Sandstrom, D.R. and Lytle, F.W.: 1979, Ann. Rev. Phys. Chem. 30, p. 215.

(2) Azaroff, L.: "X-ray Spectroscopy" (McGraw-Hill, 1974).

(3) Sayers, D.E., Lytle, F.W., Stern, E.A.: 1971, Phys. Rev. Lett. 27, p. 1204.

(4) Eisenberger P. and Kincaid, B.M.: 1978, Science 200, p. 1441.

(5) Stern, E.A.: 1974, Phys. Rev. B10, p. 3027; Ashley, C.H. and Doniach, S.: 1975, *ibid.* B11, p. 1279; Lee, P.A. and Pendry, J.B.: 1975, *ibid.* B11, p. 2795.

(6) Teo, Boon Keng and Lee, P.A.: 1979, J. Am. Chem. Soc. 101, p. 2815.

(7) Beni, G. and Platzman, P.M.: 1976, Phys. Rev. B14, p. 1514; Greegor, R.B. and Lytle, F.W.: 1979, Phys. Rev. B20, p. 4902

(8) Heald, S.M. and Stern, E.A.: 1978, Phys. Rev. B17, p. 4049; Stern, E.A., Bunker, B.A. and Heald, S.M.: 1980, Phys. Rev. B21, p. 5521.

(9) Lee, P.A.: 1976, Phys. Rev. B13, p. 5261.

(10) Stöhr, J., Bauer, R.S., McMenamin, J.C., Johansson, L.I. and Brennan, S.: 1979, J. Vac. Sci. Technol. 16, p. 1195.

(11) Jaklevic, J., Kirby, J.A., Klein, M.P., Robertson A.S., Brown, G.S. and Eisenberger, P.: 1977, Solid State Commun. 23, p. 679.

(12) Kostroun, V.O., Chen, M.H. and Crasemann, B.: 1971, Phys. Rev. A3, p. 533; Chen, M.H., Crasemann, B. and Kostroun, V.O.: 1971, Phys. Rev. A4, p. 1.

(13) Citrin, P.H., Eisenberger, P. and Hewitt, R.C.: 1978, Phys. Rev. Lett. 44, p. 309.

(14) Stöhr, J., Denley, D. and Perfetti, P.: 1978, Phys. Rev. B18, p. 4132; Stöhr, J.: 1978, Jpn. J. Appl. Phys. 17-2, p. 217.

(15) Bianconi and Bachrach, R.Z.: 1979, Phys. Rev. Lett. 42, p. 104.

(16) Veigele, W.J.: 1973, At. Data Tables 5, p. 51.

(17) Lindau, I. and Spicer, W.E.: 1974, J. Electron Spectrosc. 3, p. 409.

(18) For examples see: Pendry, J.B.: 1974, *Low Energy Electron Diffraction* (Academic, New York); Duke, C.B., *Surface Effects in Crystal Plasticity*, edited by La Tanision, R.M. and Fourie, J.F. (Noordhoff, Leyden, 1977), p. 165-219; Tong, S.Y., *Electron Diffraction 1927-1977*, edited by Dobson, P.T. et al. Institute of Physics, London, p. 270-280; Jona, F.: 1978, J. Physic. C11, p. 4271.

(19) Ritchie, R.H. and Ashley, J.C.: 1965, J. Phys. Chem. Solids 26, p. 1689; Kane, E.O.: 1967, Phys. Rev. 159, p. 624.

(20) Quinn, J.J.: 1962, Phys. Rev. 126, p. 1453; Kleinmann, L.: 1971, Phys. Rev. B3, p. 2982.

(21) Llacer, J. and Garwin, E.L.: 1969, J. Appl. Phys. 40, p. 2766; Mahan, G.D.: 1973, Phys. Stat. Sol. (b)55, p. 703.

(22) Stöhr, J., Johansson, L.I., Lindau, I. and Pianetta, P.: 1979, Phys. Rev. B20, p. 664.

(23) Gudat, W. and Kunz, C.: 1972, Phys. Rev. Lett. 29, p. 169; Gudat, W.: 1974, Ph.D. Thesis (Hamburg University, Internal Report No. DESY F41-74/10 (unpublished).

(24) Martens, G., Rabe, P., Schwentner, N. and Werner, A.: 1978, J. Phys. C11, p. 3125.

(25) Stöhr, J.: 1979, J. Vac. Sci. Technol. 16, p. 37.

(26) Stöhr, J.: 1978, Jpn. J. Appl. Phys. Suppl. 17-2, p. 217. In this paper a calculated O-Ni phaseshift was used which results in too short an O-Ni bond length. The correct O-Ni bond length is (2.04 ± 0.02)Å.

(27) Knotek, M.L., Jones, V.O. and Rehn, V.: 1979, Phys. Rev. Lett. 43, p. 300.

(28) Jaeger, R., Feldhaus, J., Haase, J., Stöhr, J., Hussain, Z., Menzel, D. and Norman D.: 1980, Phys. Rev. Lett. 45, p. 1870.

(29) Knotek, M.L. and Feibelman, P.J.: 1978, Phys. Rev. Lett. 40, p. 964; Feibelman, P.J. and Knotek, M.L. (1978), Phys. Rev. B18, p. 6531.

(30) Jaeger, R., Stöhr, J., Feldhaus, J., Brennan, S. and Menzel, D.: Phys. Rev. B (to be published).

(31) Franchy, R. and Menzel, D.: 1979, Phys. Rev. Lett. 43, p. 865.

(32) Madey, T.E., Stockbauer, R., van der Veen, J.F. and Eastman, D.E.: 1980, Phys. Rev. Lett. 45, p. 187.

(33) Siegbahn, K. et al: 1967, *Electron Spectroscopy for Chemical Analysis, Molecular and Solid State Structure by Means of Electron Spectroscopy*, Nova Acta R. Soc. Sci. Ups. Serv. IV 20, p. 224.

(34) Perlman, M.L., Rowe, E.M. and Watson, R.E.: 1974, Physics Today, July, p. 30.

(35) Brown, F.C., Bachrach, R.Z. and Lien, N.: 1978, Nucl. Instrum. & Methods 152, p. 73.

(36) Fontaine, A., Lagarde, P., Raoux, D. and Esteva, J.M.: 1979, J. Phys. F9, p. 2143.

(37) Cerino, J., Stöhr, J., Hower, N. and Bachrach, R.Z.: 1980, Nucl. Instr. & Methods 172, p. 227.

(38) Stöhr, J., Jaeger, R., Feldhaus, J., Brennan, S., Norman, D. and Apai, G.: 1980, Appl. Optics 19, xxx

(39) Stöhr, J. and Denley, D. in Proceedings International *Workshop on X-ray Instrumentation for Synchrotron Radiation*, Edit. Winick, H. and Brown, G.S. (Stanford, April 1978) SSRL Report 78/04.

(40) Bergland, G.D.: July 1969, IEEE Spectrum, p. 41.

(41) Lee, P.A. and Beni, G.: 1977, Phys. Rev. B15, p. 2862.

(42) Hayes, T.M., Sen, P.N. and Hunter, S.H.: 1976, J. Phys. C9, p. 4357.

(43) Martens, G., Rabe, P., Schwentner, N. and Werner, A.: 1978, Phys. Rev. B17, p. 1418.

(44) Rabe, P.: 1978, Jpn. J. Appl. Phys. 17-2, p. 22.

(45) Citrin, P.H., Eisenberger, P. and Kincaid, B.M.: 1976, Phys. Rev. Lett. 36, p. 1346.

(46) Lengler, B. and Eisenberger, P.: 1980, Phys. Rev. B21, p. 4507.

(47) Stern, E.A., Bunker, B.A. and Heald, S.M.: 1980, Phys. Rev. B21, p. 5521.

(48) Stern, E.A., Sayers, D.E. and Lytle, F.W.: 1975, Phys. Rev. B11, p. 4836.

(49) Stöhr, J., Johansson, L.I., Brennan, S., Hecht, M. and Miller, J.N.: 1980, Phys. Rev. B22, p. 4052.

(50) Lang, N.D. and Williams, A.R.: 1975, Phys. Rev. Lett. 34, p. 531; and 1978, Phys. Rev. B18, p. 616; Salahub, D.R., Roche, M. and Messmer, R.P.: 1978, Phys. Rev. B18, p. 6495.

(51) Flodström, S.A., Bachrach, R.Z., Bauer, R.S. and Hagström, S.B.M.: 1976, Phys. Rev. Lett. 37, p. 1282.

(52) Eberhardt, W. and Himpsel, F.J.: 1979, Phys. Rev. Lett. 42, p. 1375; Martinson, C.W.B. and Flodström, S.A.: 1979, Solid State Commun. 30, p. 671.

(53) Martinson, C.W.B., Flodström, S.A., Rundgren, J. and Westrin, P.: 1979, Surface Science 89, p. 102.

(54) Yu, H.L., Muñoz, M.C. and Soria, F.: 1980, Surface Sci. 94, p. L104.

(55) Payling, R. and Ramsey, J.A.: 1980, J. Phys. C13, p. 505.

(56) Gartland, P.O.: 1978, Surface Sci. 62, p. 183. Hofmann, P., Wyrobisch, W. and Bradshaw, A.M.: 1978, Surface Sci. 80, p. 344.

(57) For a convenient listing of bond lengths, see J.C. Slater, *Symmetry and Energy Bands in Crystals* (Dover, New York, 1972), pp. 308-346.

(58) Norman, D., Brennan, S., Jaeger, R. and Stöhr, J., Surf. Sci. Lett. (to be published).

(59) Johansson, L.I. and Stöhr, J.: 1979, Phys. Rev. Lett. 43, p. 1882.

(60) Jona, F. and Marcus, P.M.: 1980, J. Phys. C13, p. L477.

(61) Andersson, S. and Pendry, J.B.: 1975, Solid State Comm. 16, p. 563.

(62) Redhead, R.A.: 1964, Can. J. Physics 42, p. 886; Lichtman, D. and Kirst, T.R.: 1966, Physics Letters 20, p. 7; Riwan, R., Guillot, C. and Paigne, J.: 1975, Surf. Sci 47, p. 183; Dawson, P.H.: 1977, Phys. Rev. B15, p. 5522; Yu, Ming L.: 1979, Phys. Rev. B19, p. 5995.

(63) Madey, T.E., Czyzewski, J.J. and Yates, J.T., Jr.: 1975, Surface Sci. 49, p. 465.

NEUTRON SCATTERING APPLIED TO SURFACES

R.K. Thomas

Physical Chemistry Laboratory, Oxford.

ABSTRACT

The theory of neutron scattering and those of its features suitable for the study of surfaces are outlined.

Examples of the application of neutron diffraction to the study of the structures and phase diagrams of physisorbed layers are described. In some cases, cross interference between adsorbent and adsorbate scattering leads to information about the structure and configuration of the adsorbed molecule as well as longer range structural properties of the adsorbent.

Vibrational spectroscopy of physisorbed and chemisorbed species, using neutrons, is compared with other spectroscopic methods. At lower energies, transitions associated with rotational tunnelling have been observed for methane on graphite and hydrogen adsorbed into graphite-alkali metal intercalates. These spectra and how they lead to information on surface-adsorbate forces are discussed.

NEUTRON SCATTERING

There are several ways of using neutron scattering to study the properties of condensed phases. These lectures will outline the different types of scattering obtained from adsorbed systems, emphasizing the features that distinguish neutron scattering from other techniques.

P. Day (ed.), Emission and Scattering Techniques, 251–292.

The basic neutron scattering experiment is sketched in Figure 1. Neutrons from a reactor or the target of a particle accelerator are moderated to a temperature where their average velocity corresponds to a wavelength in the range 0.5 - 5 Å. The most probable wavelength of neutrons with an equilibrium velocity distribution is 1.1 Å at 300 K and 4.4 Å at 20 K (liquid hydrogen moderator). From the moderator is extracted a collimated beam, from which a monochromatic beam is selected, either by diffraction from a crystal or by a series of rotating choppers. The beam falls on the sample where it is scattered through a range of angles and where some of the neutrons may gain or lose energy. The scattered neutrons are detected by gas ionization counters, triggered by the recoil particles from suitable nuclear reactions. The two most widely used are

$$^{3}He + n \rightarrow {}^{3}H + p$$

and $$^{10}B + n \rightarrow {}^{7}Li + \alpha$$

where the boron is in the form of BF_3 enriched with ^{10}B. There may be one moveable detector or an array of detectors at different scattering angles, 2θ. The energies of the scattered neutrons are determined either by using diffraction to measure their wavelength ($E = h^2/2m_n\lambda^2$) or by time of flight techniques. In the latter, the incoming beam is divided into pulses and the times of arrival of neutrons at the detector are analysed. Since the initial velocity of the neutrons and the geometry of the

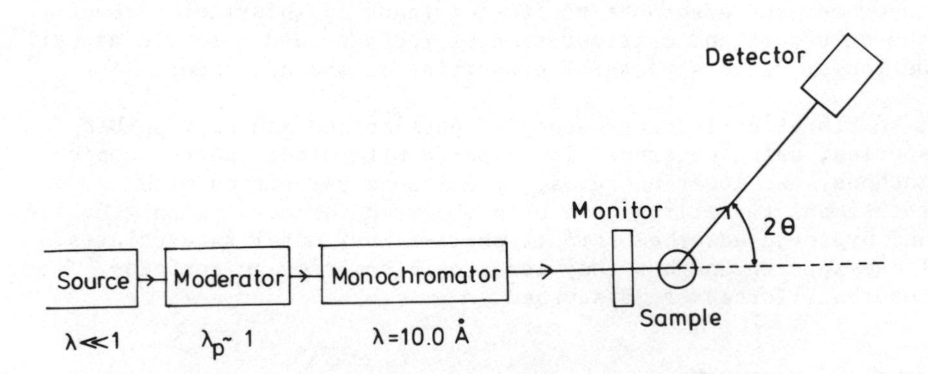

Figure 1. Schematic arrangement of a simple neutron scattering experiment.

experiment are known, the final velocities, and hence the energy transfer to the sample, can be calculated. Just in front of the sample, the incoming beam passes through a poor neutron detector which acts as a monitor of the incident flux, to which the spectrum can be normalized. What is then measured in the experiment is the partial differential scattering cross section, $d^2\sigma/d\Omega dE'$, the scattering $d^2\sigma$ into the solid angle between Ω and $\Omega + d\Omega$ with final energies between E' and E' + dE'. There are many variations of the basic experiment which depend on the range of Ω and E-E' of interest and also on what flux and resolution are needed. Some of these have been described by Stirling (1) and full specifications of actual instruments are given in the handbook of the Institut Laue-Langevin, Grenoble (2) where a large proportion of experiments are now done.

The wavelength of thermal neutrons is comparable with inter-atomic distances in condensed matter and neutrons scattered from a sample will be subject to interference effects, detected by the variation of scattering with angle or wavelength, and which will depend on the arrangement of atoms in the sample. In such a diffraction experiment no energy analysis is usually done; the measurement is of the differential scattering cross section, $d\sigma/d\Omega$, as in X-ray diffraction. In special cases analysis of the energies of the scattered neutrons may be used to separate the true elastic scattering (E = E').

Energies of thermal neutrons are comparable with the energies of most molecular motions. Thus in the scattering process a neutron may exchange an amount of energy comparable with its own energy. The energy exchanged with the sample is therefore easy to measure. The range of energy transfer that can be covered is limited at the high end by the shortest incident wavelength that can be made reasonably monochromatic, and at the low end by the resolution. With different spectrometers, optimized in quite different ways, the accessible range is at present about 4000 cm^{-1} down to about 10^{-4} cm^{-1}. This spans the energy ranges covered by infrared, Raman, electron energy loss and microwave spectroscopies, as well as the upper reaches of magnetic resonance. The experiment sketched in Figure 1 therefore not only gives information on the spatial properties of the system but also on a wide range of its temporal (frequency) properties. At the molecular level this is not the case for any spectroscopic technique using electro-magnetic radiation; the wavelength is too long to be directly sensitive to molecular dimensions. X-rays do probe spatial properties at the molecular level but their energy is too high for small exchanges of energy with molecular motion to be adequately resolved. A disadvantage in having so much information at once is that experiments may be difficult to set up correctly or to interpret. However, a simplifying feature, particularly useful to chemists, is the presence of an incoherent component in the scattering.

The amplitude and phase of a neutron scattered from nuclei at equivalent positions in a lattice may vary with the relative orientation of the spins of neutron and nucleus, so that there is not necessarily any interference between the neutrons scattered by equivalent nuclei. For example, the scattering from protons is quite different when neutron and proton spin are parallel from when they are opposed. Only coherent scattering would result if all the protons were to scatter the radiation identically, as does happen approximately for X-ray scattering from hydrogen atoms. However, if the protons scatter according to the orientation of their spins and if those spins are randomly oriented, an incoherent component, which is approximately isotropic, is introduced into the scattering. Only the coherent scattering contains information about the structure of the crystal and it consists of sharp and relatively intense peaks at angles satisfying the Bragg condition. It could be separated from the incoherent scattering by analysis of the spins of the neutrons scattered from a polarized beam (the neutron spin wavefunction is changed in incoherent scattering) but is usually obtained by subtracting a more or less flat incoherent background. The incoherent scattering is measured by placing the detectors at angles where there are no Bragg peaks. Since it contains no structural information, the incoherent scattering is of little interest except in inelastic scattering experiments. However, its inelastic component depends on the average motion of only those individual nuclei in the sample with reasonably large incoherent cross section. Thus protons have the largest incoherent cross section of all nuclei and the incoherent inelastic spectrum consists of the frequencies of <u>all</u> the motions of a hydrogen atom averaged over all the hydrogens in the system. On the other hand, when hydrogen is replaced by deuterium, the coherent scattering may be dominant and an excitation is only observed if it satisfies conservation of energy and conservation of momentum (Bragg condition).

The actual form of the interaction of a neutron and nucleus is not known, but it is of such short range compared with the wavelength of the neutron that the scattered wave is spherically symmetrical. The scattering amplitude from a single nucleus can be written as

$$b \exp\{i(\underline{k} - \underline{k}').\underline{r}\} \tag{1}$$

where b is an empirical parameter called the scattering length, which characterizes the phase and amplitude of the scattering. For most nuclei b is positive, though for some it is negative and it may occasionally be complex. $\underline{k}$ is the wavevector of the incoming neutron, related to its momentum by $\underline{k} = \underline{p}/\hbar$ and to its wavelength by $|\underline{k}| = 2\pi/\lambda$. $\underline{r}$ is the position of the nucleus relative to an arbitrary origin. It is convenient to abbreviate $(\underline{k} - \underline{k}')$ to $\underline{Q}$, $\underline{Q}$ being referred to as the momentum transfer. (The actual momentum transferred between neutron and sample is $\underline{Q}\hbar$.) $\underline{Q}$ is a more fundamental quantity than the scattering angle. The relation between the two is shown in Figure 2.

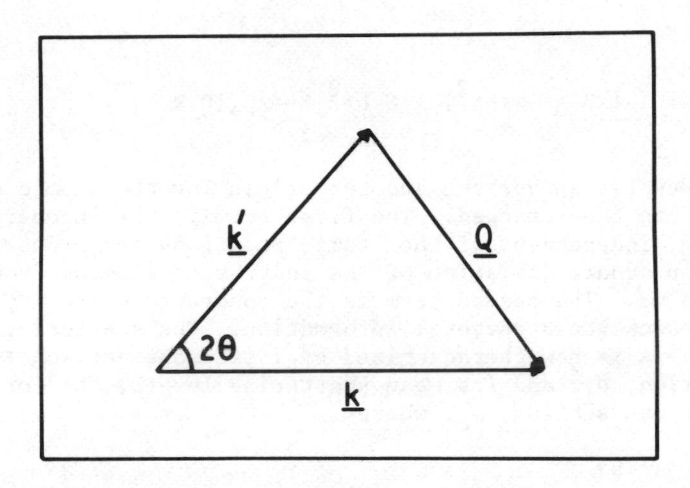

Figure 2. Definition of the momentum transfer, $\underline{Q}$. $\underline{k}$ and $\underline{k}'$ are the wavevectors of the incident and scattered neutron and 2θ is the scattering angle. For elastic scattering, $|\underline{Q}| = 4\pi\sin\theta/\lambda$.

The scattering amplitude from an assembly of atoms is

$$\sum_j b_j \exp(i\underline{Q}.\underline{r}_j) \tag{2}$$

If the nuclei in the system are assumed to be stationary, they exchange no energy with the neutrons and the differential scattering cross section is just the squared modulus of the scattering amplitude,

$$\frac{d\sigma}{d\Omega} = \left| \sum_j b_j \exp(i\underline{Q}.\underline{r}_j) \right|^2 \tag{3}$$

which may be separated into two terms:

$$\frac{d\sigma}{d\Omega} = \sum_j b_j^{\,2} + \sum_{i \neq j} b_j b_i \exp\{i\underline{Q}.(\underline{r}_i - \underline{r}_j)\} \tag{4}$$

As already mentioned, for neutron scattering there is not necessarily any correlation between the scattering of equivalent atoms in the lattice; b_j is independent of b_i and we may average over each separately to give

$$\frac{d\sigma}{d\Omega} = N\langle b^2\rangle + N \sum_{i\neq j} \langle b\rangle^2 \exp\{i\underline{Q}.(\underline{r}_i - \underline{r}_j)\} \qquad (5)$$

or, finally,

$$\frac{d\sigma}{d\Omega} = N\{\langle b^2\rangle - \langle b\rangle^2\} + N\langle b\rangle^2 \left|\sum_j \exp(i\underline{Q}.\underline{R}_j)\right|^2 \qquad (6)$$

where <> denotes an average and the origin for the atomic coordinates has been changed. The first term is the incoherent scattering, independent of the atomic positions and proportional to the mean square deviation of the scattering lengths from their average value. The second term is the coherent scattering, which contains the structural information. The scattering of a given nucleus is now characterized by its incoherent scattering cross section, σ_i, and its mean scattering length, <b>, or coherent cross section, σ_c, where

$$\sigma_c = 4\pi\langle b\rangle^2$$

and $\sigma_i = 4\pi(\langle b^2\rangle - \langle b\rangle^2)$

In terms of scattering cross sections,

$$\frac{d\sigma}{d\Omega} = \frac{N}{4\pi}\{\sigma_i + \sigma_c\left|\sum_j \exp(i\underline{Q}.\underline{R}_j)\right|^2\} \qquad (7)$$

In any form of spectroscopy the intensity of a feature in the spectrum depends on the matrix element of the appropriate operator between initial and final states, for example, the dipole moment operator in microwave and infrared spectroscopy. For inelastic neutron scattering, the corresponding operator is the scattering amplitude. The partial differential cross section is

$$\frac{d^2\sigma}{d\Omega dE'} = \frac{k'}{k} \sum_i \sum_f P_i \sum_j \sum_k \langle i|b_k \exp(i\underline{Q}.\underline{r}_k)|f\rangle$$

$$\times \langle f|b_j \exp(i\underline{Q}.\underline{r}_j)|i\rangle\ \delta(E - E' - \hbar\omega) \qquad (8)$$

where P_i is the probability of the initial state being occupied, the factor k'/k allows for the neutron flux being proportional to the neutron velocity, and the δ function imposes conservation of energy. Just as the differential scattering cross section splits into coherent and incoherent components so does the partial cross section. We write it in terms of the so-called scattering laws, $S(\underline{Q},\omega)$:

$$\frac{d^2\sigma}{d\Omega dE'} = \frac{k'}{k} \frac{N}{4\pi\hbar} \{\sigma_i S_i(\underline{Q},\omega) + \sigma_c S_c(\underline{Q},\omega)\} \tag{9}$$

The coherent scattering law, $S_c(\underline{Q},\omega)$, contains information about both the frequencies of motion and the structure of the system;

$$S_c(\underline{Q},\omega) = \frac{1}{N} \sum_i \sum_f P_i \sum_j \sum_k \langle i|\exp(-i\underline{Q}.\underline{r}_k)|f\rangle \times \langle f|\exp(i\underline{Q}.\underline{r}_j)|i\rangle \delta(E-E'-\hbar\omega) \tag{10}$$

The incoherent scattering law contains no interference terms and therefore depends only on the average motion of the single atom j rather than on correlations between the motions of different atoms, j and k;

$$S_i(\underline{Q},\omega) = \frac{1}{N} \sum_i \sum_f P_i \sum_j \langle f|\exp(i\underline{Q}.\underline{r}_j)|i\rangle^2 \times \delta(E-E'-h\omega) \tag{11}$$

In forms of spectroscopy with electromagnetic radiation, the electric field is effectively constant on the molecular scale, so that only its time dependence need be considered when evaluating the matrix element. However, even in incoherent scattering, the spatial variation of the wavefunction must be included, although this is not always easy to do. The corollary is that the neutron scattering spectrum contains information about the spatial properties of the motion. This is now discussed for incoherent scattering from vibrational motions.

The incoherent scattering law for excitation of a vibrational mode of an atom in a crystal can be expressed as a series, provided that $\underline{Q}.\underline{u}_o \ll 1$, where $\underline{u}_o$ is the amplitude of vibration, $\underline{u} = \underline{u}_o \cos(\omega_o t)$. The first two terms of the series are

$$S_i(\underline{Q},\omega) = h \exp\{-\langle(\underline{Q}.\underline{u}_o)^2\rangle\}\delta(\hbar\omega) + h(\underline{Q}.\underline{u}_o)^2\{P_o\delta(\hbar\omega - \hbar\omega_o) + P_1\delta(\hbar\omega + \hbar\omega_o)\} \tag{12}$$

where P_o and P_1 are the probabilities of ground and excited states being occupied. The first term is the elastic scattering and the second term represents energy transfer respectively to and from one quantum of vibration of frequency ω_o. Energy gain of the neutrons is weaker than energy loss by the Boltzmann factor, just as in Stokes and anti-Stokes lines in the Raman spectrum. Since the final spectrum depends on $\sigma_i S_i(\underline{Q},\omega)$, the intensity is proportional to the incoherent scattering cross section of the atom in motion. Other important features are that the intensity depends on the relative orientation of $\underline{Q}$ and $\underline{u}_o$, being a maximum when they are parallel, and that it increases with u_o^2 (Q and θ constant) and with Q^2 (u_o and θ constant). For a more complicated system with several modes of vibration, the requirements for a vibrational excitation to appear in the spectrum are that a nucleus with a non-zero incoherent cross section is involved in the motion and that $\underline{Q}.\underline{u}_o \neq 0$. There are therefore no selection rules corresponding with those in optical spectroscopy. Also, the neutron spectrum contains information not just about the frequency but about the amplitude of a vibration. At best this may be used as additional data for the calculation of the force field (3). At worst, it may be a qualitative aid to assignment of modes, either because of their relative intensity or because of the effects of isotopic substitution on the intensities. For example, substitution of deuterium for hydrogen will change both the frequency and the intensity of a mode involving that hydrogen atom.

Transfer of energy between a neutron and proton is not restricted to quantized motions such as vibration. For example, the neutron may exchange energy with a proton undergoing translational or rotational diffusion as part of a water molecule in liquid water. Diffusive motions are associated with a continuum of low energy levels and exchange of energy with this continuum appears as a broad envelope of scattering around the elastic peak, called quasielastic scattering. A quite different approach has to be used to derive the appropriate scattering laws, described fully in reference (4). Although there is not room to describe this approach here, quasielastic scattering is an excellent technique for probing diffusion at the molecular level, and has been used to study such motion on surfaces (4).

The range of wavelength and energy of neutron scattering is obviously well suited to the study of surfaces. What is less clear is whether or not the scattering from the surface layer can be distinguished from the scattering from the bulk. Compared with electrons and X-rays, neutrons are only very weakly scattered by matter. For example, it takes a thickness of about 2mm of H_2O (10^7 molecular layers) to extinguish half the beam, although H_2O has an unusually high cross section. This is a disadvantage when looking at gases adsorbed on solids in that it requires adsorbents with high ratios of surface area to mass, that is powders or partially oriented powders. The surfaces of such materials tend to be inhomogeneous. On the other hand, these are the kind of materials that are not easily studied by other techniques. There are some advantages of the weak scattering; experiments may be done at normal pressures of gas, containers may be used that are suitable for vacuum experiments and extreme conditions (low and high temperatures and pressures), and adsorbents that are optically opaque present no particular problem. The observation of surface properties with neutrons then depends on the success of choosing a strongly scattering adsorbate and a weakly scattering adsorbent. For diffraction experiments, σ_c is the important parameter and for incoherent scattering, σ_i. In practice, the latter simply requires the presence of hydrogen in the adsorbate but not in the adsorbent. Some examples of cross sections for various adsorbents and adsorbates are given in Table 1. With the further possibility restricting the range of Q or ω covered in an experiment, it should be possible to study systems with areas as low as 1 m^2g^{-1}.

Table 1. Scattering cross sections of adsorbents and adsorbates for an adsorbent area of 50 m^2g^{-1} and adsorbate areas given in the Table.

Substance	Area/$Å^2$	Coherent Cross Section x 10^5/ m^2g^{-1}	Incoherent Cross Section x 10^5/ m^2g^{-1}
Graphitized carbon black	-	2.8	0
Silica	-	1.1	0
Alumina	-	0.9	0
Rutile	-	0.7	0.2
Platinum	-	0.4	0.02
Nickel	-	1.4	0.5
Hydrogen(H)	16	0.016(D)	0.5(H)
Methane	16	0.08(CD_4)	2.0(CH_4)
Benzene	35	0.09(C_6D_6)	1.4(C_6H_6)

Given that the signal from the adsorbate is sufficient, a neutron experiment may give valuable information on its structure and dynamics. The types of information that may be derived are summarized in Table 2, together with the type of scattering and the necessary range of energy and momentum transfer.

Table 2. Neutron scattering experiments on surfaces

Type of Scattering	Range of ω (cm^{-1}) or Q ($Å^{-1}$)	Comments	Information derived
Coherent (no energy analysis)	0.5 < Q 15	Diffraction	Atomic arrangement in adsorbed layer, distance of layer from surface, incidence of defects on surface (5)
" "	Q < 0.1	Small angle scattering	Overall particle sizes and shapes (6)
" "	15 < Q < 100	Higher energy neutrons required from hot or pulsed source	Intramolecular structure of adsorbed species
Coherent inelastic scattering	0.5 < Q < 15 0 < ω < 1000		Cooperative motions within adsorbed layer (7)
Incoherent inelastic scattering	0.5 < ω < 4000		Vibration spectroscopy of chemi-and physisorbed species containing hydrogen (4)
" "	0.01 < ω < 200		Rotational spectroscopy of adsorbed molecules (4)
Incoherent quasielastic scattering	0.01 < ω < 5		Diffusion ($D > 5x10^{-11} m^2 s^{-1}$) of chemi- and physisorbed species. Rotational diffusion of adsorbed species (4).

Most of the experiments so far done have been neutron diffraction or incoherent inelastic scattering. These are now considered separately.

NEUTRON DIFFRACTION FROM ADSORBED LAYERS

All the surface experiments so far done have only been with physisorbed layers, although experiments have been done on intercalated species, either physi- or chemisorbed (8). This is partly because it is experimentally easier to do diffraction experiments on homogeneous surfaces, which, combined with the requirement of a fairly large surface area (see Table 1), has so far discouraged attempts to study chemisorbed species. It is also because there is at present considerable interest in phase transitions in two dimensional systems (9), physisorbed layers on homogeneous smooth surfaces often approximating sufficiently closely to free standing two dimensional layers. Most experiments have used graphite as adsorbent, either in the form of graphitized carbon black, or as recompressed exfoliated graphite. In both cases the surface is thought to be homogeneous, consisting only of the basal planes of graphite. Experiments have also been done using lamellar halides (10), magnesium oxide (11), spinel (11), and rutile (12).

The structure of a physisorbed layer is determined by the balance between surface-molecule and molecule-molecule forces. Since these are comparable, there is a variety of possible types of structure. At a temperature low enough to immobilize the molecules and where there are not enough molecules to cover the surface completely, these structures may be classified as follows:

(i) a two dimensional layer spread evenly over the whole surface with negligible interaction between the adsorbed molecules;

(ii) a close packed two dimensional layer with a lattice out of registry with the underlying surface lattice;

(iii) a nearly close packed two dimensional layer in registry with the surface lattice;

(iv) local regions of multilayer formation based on a two dimensional lattice;

(v) three dimensional crystallites nucleated on inhomogeneous regions of the surface.

For all five types of structure except (i) there will be areas of the surface that are bare so that in (iii), for example, the adsorbate may either form a single large patch or a number of relatively small two dimensional clusters. A particular adsorbate may have more than one stable phase of type (ii) or (iii) especially if it consists of anisotropic molecules. The equilibria between all types of phase will depend on the temperature and pressure, and may also be sensitive to the quality of the surface, which is likely to affect factors like the mean size of two dimensional clusters. At higher coverages (above a monolayer) the structure of the adsorbate may be a combination of the types of structure listed above. Finally at higher temperatures, where the adsorbate molecules are mobile, there will be equivalent fluid phases.

Neutron diffraction can distinguish the different types of phase quite easily. It is particularly successful at distinguishing changes of phase between structure of types (ii) and (iii), something which is much more difficult with LEED. This is a consequence of the weaker scattering which only results in single scattering from the adsorbed layer. The formulae for diffraction from the layer are therefore relatively simple and are now summarized.

The scattering amplitude from an adsorbed system may be divided into two parts, one from the adsorbent and one from the adsorbate (13):

$$F' \sum_p \exp(i\underline{Q}\,\underline{r}_p) + F'_A \sum_q \exp(i\underline{Q}\,\underline{r}_q) \qquad (13)$$

where p and q refer to adsorbent and adsorbate unit cells respectively. F' and F'_A depend on the contents of the unit cells. For single atoms they are merely b and b_A. As before, the differential cross section depends on the squared modulus of the scattering amplitude. After removal of the incoherent component of the scattering following equations (2) to (5), we are left with three terms describing the diffraction from the system:

$$|F|^2 \sum_p \sum_{p'} \exp\{i\,\underline{Q}(\underline{r}_{p'} - \underline{r}_p)\} \quad \text{.... adsorbent} \qquad (14a)$$

$$|F_A|^2 \sum_q \sum_{q'} \exp\{i\,\underline{Q}(\underline{r}_{q'} - \underline{r}_q)\} \quad \text{.... adsorbate} \qquad (14b)$$

$$FF_A^* \sum_p \sum_q \exp\{i\underline{Q}(\underline{r}_p - \underline{r}_q)\}$$

$$+ F^*F_A \sum_p \sum_q \exp\{i\underline{Q}(\underline{r}_q - \underline{r}_p)\} \quad \text{... interference} \qquad (14c)$$

F and F_A are now structure factors for the contents of the respective unit cells. (14a) is the diffraction from the adsorbent on its own and (14b) from the adsorbed layer alone. Just as always in a neutron or X-ray diffraction experiment, the sum in each term determines the position of the Bragg peaks, whereas the intensities are determined by $|F|^2$ and hence the unit cell structure. The adsorbate has a periodic structure only in the plane of the surface. No Bragg condition exists for the direction normal to the surface. The overall conditions for a diffraction peak are therefore not quite as restrictive as for a three dimensional crystal. After making allowance for the random orientation of the planes of the adsorbed layer, appropriate to a powdered adsorbent, the profile of any diffraction peak from the adsorbate is as shown in Figure 3 (14). The strict Bragg condition applies to the low Q edge of the peak. The intensity at each point in the profile also depends on $|F_A|^2$ and, since this is a function of Q, the ideal profile of Figure 3 may be substantially modified by the arrangement of atoms in the unit cell. The peak will also be broadened, especially at its sharp edge, if the two dimensional order extends over only a short range ($\lesssim$300 Å). Although a single diffraction peak of this kind may contain a fair amount of information about the layer, it says nothing about the arrangement of the adsorbed layer normal to the surface nor about its lateral position relative to the surface.

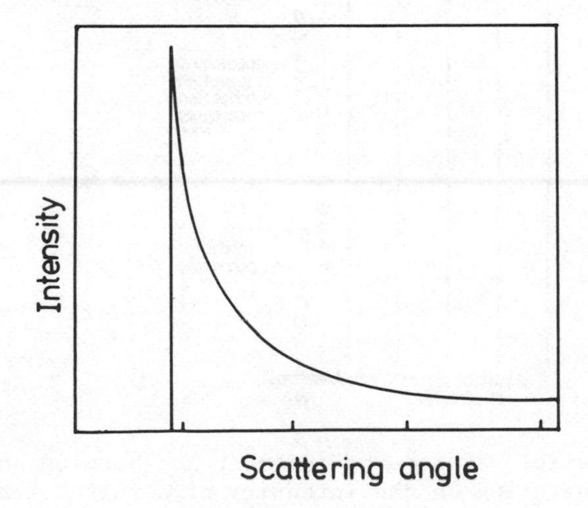

Figure 3. The profile of a diffraction peak from an assembly of randomly oriented two dimensional atomic lattices each with long range order.

There is, however, information about the structure normal to the surface in term (14c), the interference term. A schematic diagram of this effect is shown in Figure 4 for a reflection from the planes of the adsorbent parallel to the surface plane. The phase of the scattering from each plane is represented by the cosine wave at the top of the diagram. If a layer of say deuterium atoms is adsorbed at exactly the same distance from the surface as the spacing of the lattice planes of the adsorbent, it acts as an extra reflecting layer, and the intensity of the adsorbent reflection is enhanced. If the layer is adsorbed at half the lattice spacing, the peak decreases in intensity. With hydrogen as adsorbate the effect is reversed. Moreover, since powdered adsorbents are imperfect materials, the number of planes contributing to an adsorbent reflection may be quite small, and the interference effect may be relatively large (up to 5% of the (002) reflection from a graphitized carbon black). Once F is obtained from the pattern of the adsorbent alone, F_A may be calculated from the difference pattern between adsorbent + adsorbate and adsorbent (term (14c)), and hence the distance from the surface and the configuration of the adsorbed species determined accurately.

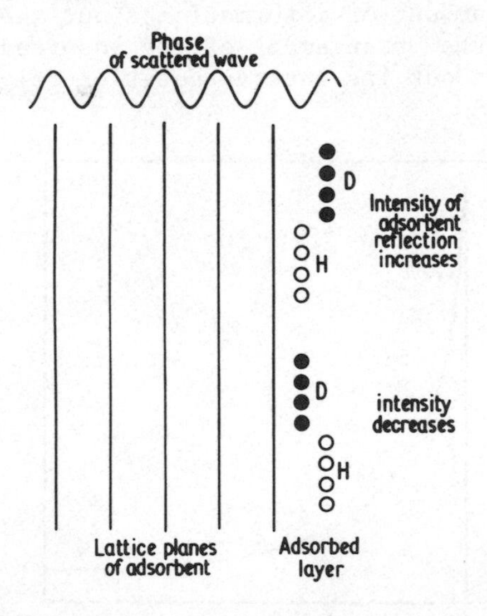

Figure 4. The possible effects of interference between adsorbate and adsorbent scattering on the intensity of a relfection from the adsorbent.

The difference in the shape of the reflection from the adsorbate alone readily distinguishes genuinely two dimensional layers from small three dimensional crystallites. In Figure 5 is shown the low temperature diffraction pattern from ND_3 on

Graphon (a graphitized carbon black of specific area 86 m^2g^{-1}) after subtraction of the graphite background. Two symmetrical narrow peaks are observed which index as reflections of bulk ammonia. At higher temperatures (>200K), ammonia adsorbs well onto graphite but the diffraction pattern shows that, at low temperatures, it nucleates into crystallites at inhomogeneities on the surface. The melting properties of these crystallites are, however, quite different from bulk ammonia (15), and depend strongly on the amount of ammonia present. In Figure 6 are shown difference patterns from CD_4 on Vulcan III (another graphitized carbon black of area 71 m^2g^{-1}). There are two peaks, the one at lower angles coming from the adsorbate alone and having the typical asymmetry of a reflection from a two dimensional layer. The higher angle peak is an interference with the (002) reflection from the basal planes of graphite. The asymmetric peak shows that methane forms a two dimensional layer at low temperatures. It is thought that a structure intermediate between those of CD_4 and ND_3 on graphite is formed by $C(CD_3)_4$ on rutile at low temperatures (12).

The surfaces of powdered adsorbents are imperfect. The average range over which perfect two dimensional order persists may be calculated from the exact shape of the reflection from the layer (13). It is determined primarily by the adsorbent and is typically 70-100 Å for graphitized carbon blacks (13) but larger for reconstituted exfoliated graphites such as Papyex (16), Grafoil (17), or XYZ graphite (17). The rather short range of the two dimensional order often means that higher order reflections are damped out. Nevertheless, it may still be possible to deduce the structure of the layer. For example, the position of the peak from CD_4 at low coverages is exactly as expected from a $\sqrt{3} \times \sqrt{3}$ structure in registry with the surface (Figure 7(a)), a structure that has also been observed for some other small molecules on graphite. The persistence of the peak to low coverages (0.36 monolayers) shows that the methane-methane forces are sufficiently strong to bind the molecules into two dimensional clusters with the $\sqrt{3} \times \sqrt{3}$ structure. However, in this structure the methane molecules are held further apart by the surface (4.26 Å) than their gas phase diameter (~3.92 Å). It is no surprise that, at a sufficiently high coverage (>0.9), a more compressed layer is formed (Figure 7(b)), it being more favourable to squeeze extra molecules into the layer next to the surface rather than adsorb them into the second layer. The change of structure is shown by the shift of the peak in Figure 6 to higher angles. Vora et al. have shown that the phase changes quite abruptly with coverage (18).

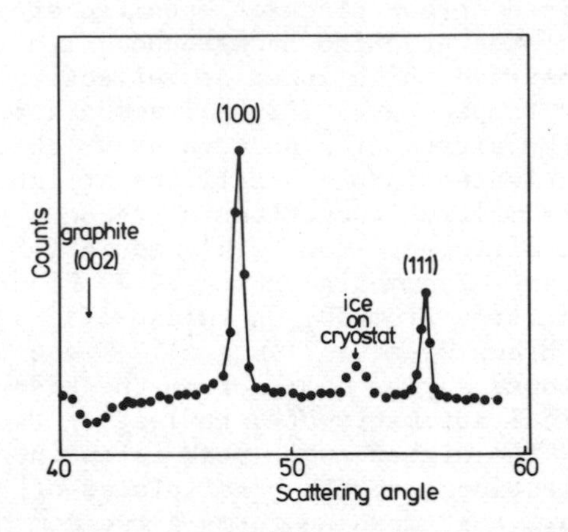

Figure 5. Diffraction from ND_3 on Graphon after subtraction of the background. Temperature = 140 K and coverage (as though for a uniform monolayer) = 1.0. λ = 2.4 Å.

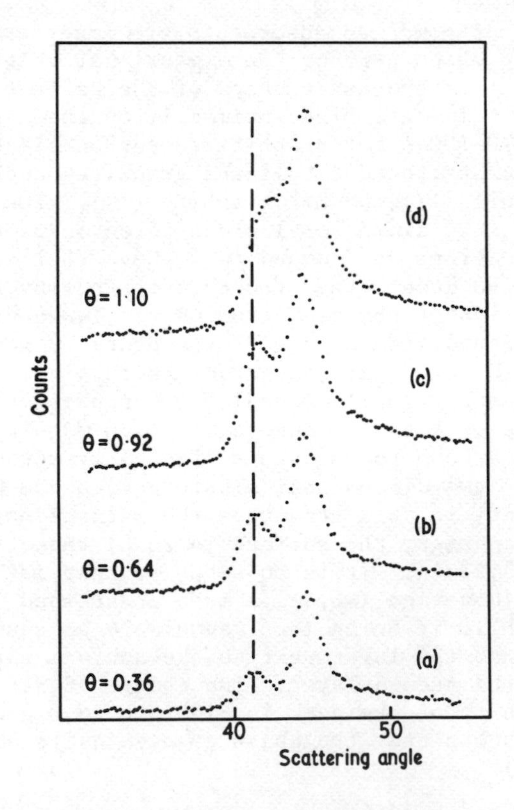

Figure 6. Diffraction from CD_4 on Vulcan III at different coverages (θ) after subtraction of the background. T = 20 K, λ = 2.52 Å.

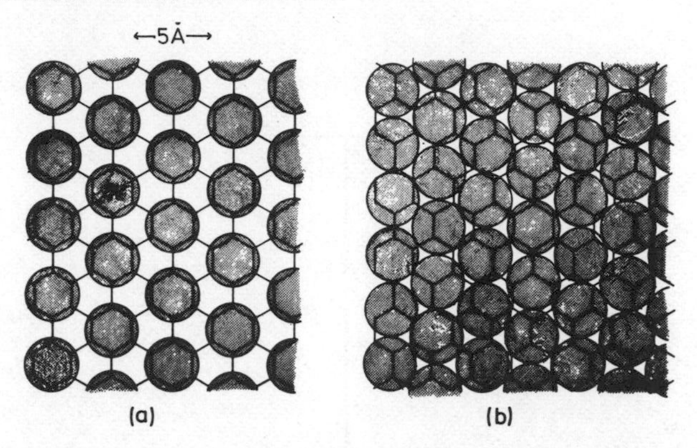

Figure 7(a) The structure of methane on graphite at coverages below 0.9 monolayers (√3 x √3). (b) The structure when θ > 0.9.

The second peak in Figure 6 results from the interference effect. Bomchil et al (13) have made a quantitative treatment of the diffraction from graphite alone, and of the interference effect, and deduced that methane adopts a tripod configuration on the surface with a carbon (graphite) - carbon (methane) distance of 3.35 Å. There are a number of different models that have been used to describe the structure of graphitized carbon blacks (19) and an interesting by product of the quantitative treatment of the interference effect is that it distinguishes between these models. The basic reason is that the pattern from the adsorbent alone depends on F^2 whereas the interference depends on F and contains the phase of F as well as its amplitude. The value obtained for the absorbent-absorbate distance seems, on the other hand, to be relatively independent of the model chosen to describe the adsorbent.

Apart from giving the structure of the two dimensional solid, diffraction may also be used to study the process of melting of the layer. This is illustrated, again for CD_4 on graphite, in Figure 8. The diffraction peak of the adsorbate broadens at first slowly with increasing temperature andthen quite suddenly. The patterns in Figure 8 are for two different coverages and it can be seen that the melting temperature depends on the coverage. At 55K, the layer has melted when the coverage is 0.64 but not at a coverage of 0.92. That a broad diffraction peak from the fluid remains at a similar angle to the sharper peak of the solid, indicates that methane remains clustered on melting. This has also been established by incoherent quasielastic scattering (20).

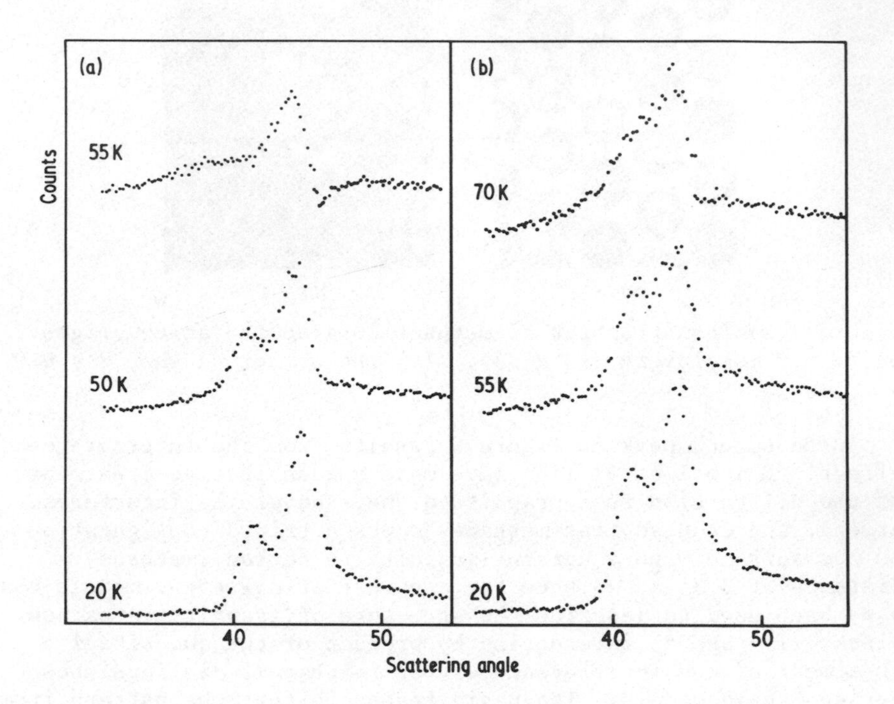

Figure 8. Diffraction patterns from methane (CD_4) on Vulcan at two different coverages (left hand side, 0.64, and right hand side, 0.92 monolayers) and different temperatures. The background has been subtracted. $\lambda = 2.52$ Å.

The structure of two dimensional solid methane is determined by the balance between the in-plane molecule-molecule forces and the variation of the potential energy of an isolated molecule as it moves across the surface. For less symmetrical molecules there is another factor, the variation of the molecule-surface potential energy with the orientation of the molecule. This is now illustrated for benzene, ethane, and oxygen, each on the surface of graphite.

An isolated benzene molecule would be expected to lie flat on the surface of graphite because this maximizes the dispersion interactions between the separate atoms and the surface. However, in solid benzene the molecules prefer a configuration where they are more nearly perpendicular to one another (21), and in the vapour phase the benzene dimer has a T shaped structure (22). Thus, we might expect benzene to lie flat on the surface at low coverages, but for the layer to adopt a more complicated structure satisfying the competing molecule-molecule forces at higher coverage. Diffraction patterns from benzene on Vulcan III are shown in Figure 9. At low coverage there are two peaks from the adsorbed layer, as well as an interference effect at ~ 20°, which index as (10) and (20) reflections of a two dimensional hexagonal lattice. On grounds of the large area per molecule in this lattice, the larger interference effect, and a full structure factor calculation (solid line in the Figure), Bomchil et al (23) have shown that benzene lies flat, or nearly so, on the surface. In forcing the benzene molecule to lie flat on the surface, the molecule-molecule interactions have been weakened to a point where they cannot prevent the layer going into registry with the surface with a ($\sqrt{7}$ x $\sqrt{7}$) structure (24,25). The benzene molecules remain flat until the monolayer iscomplete. No new two dimensional phase is formed when more benzene is added. All that happens is that the additional benzene nucleates to form three dimensional crystallites. The extra diffraction peaks from the crystallites can be seen in the top part of Figure 9. This behaviour is similar to that of ND_3 on the bare surface.

In the ethane-graphite system the competition between surface-molecule and molecule-molecule forces leads to different results. At lower coverages ethane forms a close packed layer in an oblique lattice with the molecules on their sides (Figure 10). This structure could be determined with good precision from the intensities of the reflections, of which there are eight from ethane on Papyex (a partially oriented graphite of area 20 m^2 g^{-1}) (26). The oblique lattice is sufficiently different from the surface that there is no chance of it distorting to form a registered layer. However, at higher coverages, the ethane molecules stand with their C-C axis perpendicular to the surface in a $\sqrt{3}$ x $\sqrt{3}$ registered layer with the same structure as a layer of methane (Figure 7(a))(27).

For small anisotropic molecules the balance between the two types of lateral force and the anisotropy of the molecule-surface potential is more delicate. Sometimes such molecules form layers in registry with the surface of graphite (H_2 and N_2) and sometimes they do not (H_2, N_2, O_2 and NO). Oxygen is particularly interesting since it has three stable solid phases at coverages below a monolayer, one of which is antiferro-magnetically ordered and none of which is in registry with the surface (28). At low coverages on graphite, oxygen forms a close packed hexagonal structure (δ-phase) with a shorter lattice parameter (3.37 - 3.47 Å) than the $\sqrt{3}$ x $\sqrt{3}$ registered structure (4.26 Å). This is too small for the oxygen molecules to rotate freely or even to lie with their molecular axes parallel to the surface. It is probable also that their molecular axes are not exactly perpendicular to the surface, but are tilted away

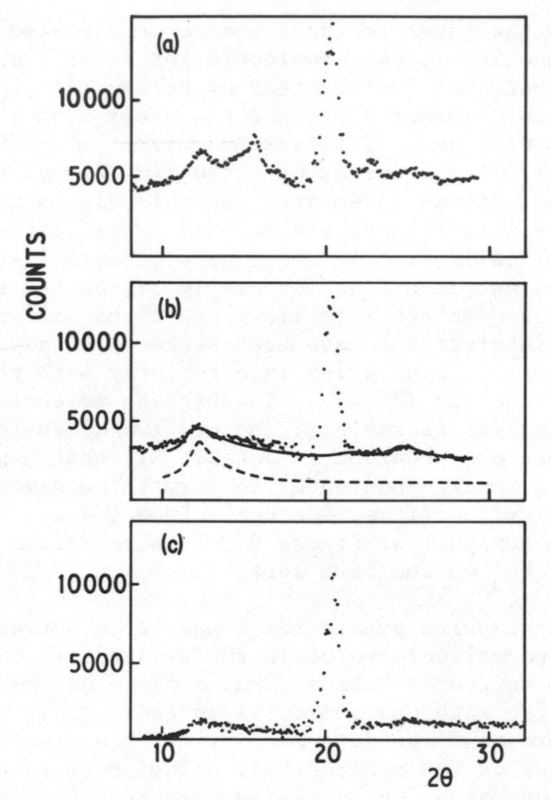

Figure 9. Neutron diffraction from benzene adsorbed on graphitized carbon black (Vulcan III, Å = 71 m^2g^{-1}) after subtraction of background. The area available per molecule is (a) 30 $Å^2$ (b) 47 $Å^2$ (c) 95 $Å^2$, the temperature is 100K. The continuous line in (b) is the pattern calculated using equation (14b) for benzene molecules lying flat and the dashed line for molecules freely rotating about all three axes. The extra peaks in (a) are from three dimensional crystallites of benzene.

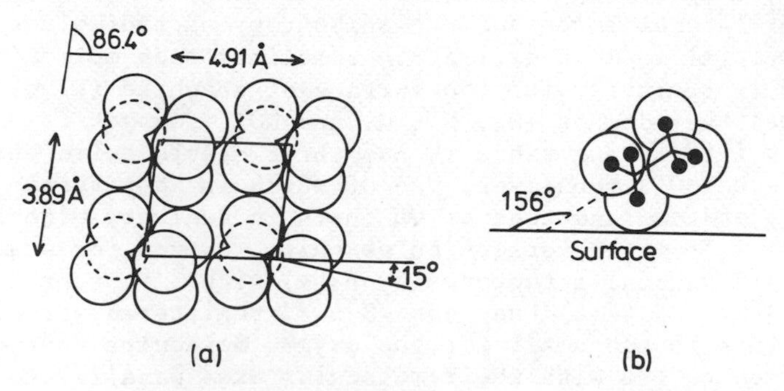

Figure 10. Structure of ethane on graphite at coverages well below monolayer. (a) View from above the surface (b) View parallel to the surface.

from the perpendicular direction by up to 40°. When the monolayer of δ-phase is complete, additional oxygen may be accommodated by closer packing into a distorted triangular (oblique) lattice with the molecular axes now exactly perpendicular to the surface. This is called the α-phase because its structure is similar to the basal planes of the α-phase of bulk solid oxygen. In bulk oxygen there is antiferromagnetic ordering of the magnetic moments of the molecules in the basal planes. Nielsen and McTague (28) have found a magnetic Bragg peak (from scattering of the neturon by the magnetic moment rather than the nuclei of the molecule) for the α-phase of adsorbed oxygen. The position of the peak corresponds to a magnetic unit cell double the structural unit cell (Figure 11) showing that the layer is antiferromagnetically ordered. The magnetic moment in oxygen ($^3\Sigma_g^-$) is perpendicular to the molecular axis. electrons on adjacent molecules may interact attractively through virtual ionic $O_2^+ - O_2^-$ states if the spins of the two molecules are opposed. Thus opposite magnetic moments are attracted leading to the distortion of the hexagonal structure to that of the α-phase shown in Figure 11. When the α-phase is heated it changes, at 11.9K (the equivalent transition is at 23K in bulk oxygen), into the β-phase, a hexagonally closed packed phase with the molecules exactly perpendicular to the surface. Although β and δ phase are similar, there is a definite range of temperature and coverage where they coexist as distinct phases.

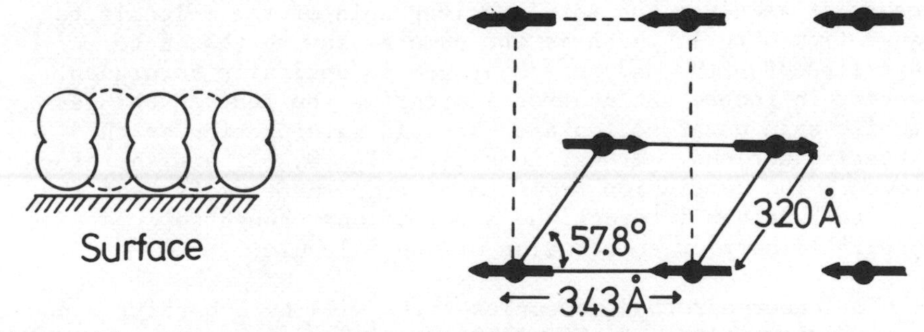

Figure 11. The structural and magnetic unit cells of the α-phase of oxygen adsorbed on Grafoil (recompressed exfoliated graphite of area ~20 m^2g^{-1}). The molecules are perpendicular to the surface but their magnetic moments are in the plane of the surface.

The examples given above should illustrate the point that, although only a relatively small number of systems can be studied by neutron diffraction, the amount of structural and thermodynamic information that can be derived from each system is large. In comparison with LEED, this is simply because the weaker scattering of neutrons can be described quantitatively and analytically. It should be noted, however, that comparable experiments could and are being done with X-ray diffraction (29,30).

INCOHERENT INELASTIC NEUTRON SCATTERING

At least as important as knowing the structure of adsorbed species and adsorbed layers is direct information on the forces binding them to the surface and on the interactions within the layer. Where the molecule is only bound to the surface by physical rather than chemical interactions, such forces may merely hinder reorientation of the molecule and so can be studied from the rotational spectrum. For physisorption and chemisorption, the forces will determine the vibrational frequencies and amplitudes of motion of the adsorbate, both of which can be obtained from the vibrational part of the incoherent scattering spectrum. Here rotational motion is considered first.

The only molecules known to rotate completely freely on a surface are H_2 and D_2 on graphite (31). At low temperatures, neutrons are able to excite the $J = 0 \rightarrow 1$ transition of H_2 at 15.0 meV (121 cm^{-1}), which is close to the value that would be obtained for H_2 in the gas phase, if the transition were optically allowed. It is forbidden in infrared or Raman spectroscopy because it requires the total nuclear spin of the molecule to change from 0 to 1, which is the same as saying that interconversion of ortho and para hydrogen is optically forbidden. However, in incoherent neutron scattering the neutron changes both its spin wavefunction and the spin wavefunction of the scattering nucleus. Transitions with $\Delta I = 0, \pm 1$ are therefore allowed. The conclusion from the neutron spectrum is that the graphite hydrogen interaction is not strong enough to create an appreciable barrier to rotation of the molecule.

For other adsorbed molecules there will be a barrier opposing their reorientation and this will complicate the pattern of rotational energy levels. The simplest case is for rotation of a species about a single axis. Although no such case has yet been observed for an adsorbed molecule, it has been observed several times in molecular crystals (4). In the molecule hexadiyne, CH_3-C≡C-C≡C-CH_3, for example, there is no internal barrier to rotation of a methyl group but, in the crystal, the methyl groups are in a potential of threefold symmetry generated by their interaction with the surrounding molecules. The simplest form that this potential could take is

$$V = V_o/2(1 - \cos 3\theta) \tag{15}$$

where θ is the angle of rotation and V_o the height of the barrier (Figure 12b). The qualitative pattern of the variation of energy levels with barrier height is established quite easily. When V_o is zero (left-hand side of Figure 12(a)) the energy levels are those of a free rotor; $E = BK^2$ where K is the angular momentum along the molecular axis and B is the inertial constant ($B = h^2/8\pi^2 cI$). K can be positive or negative, corresponding to clockwise or anticlockwise rotation, so that each level, except K = 0, is doubly degenerate. When V_o is large (right-hand side of Figure 12(a)) the methyl group oscillates about its equilibrium position with energy levels given by $E = (v+\frac{1}{2})h\nu$ where v is the vibrational quantum number and ν the frequency of the oscillation. If we label the protons, there are three equivalent equilibrium orientations (Figure 12(b)) about which the group can oscillate. Thus for each vibrational energy level there are three equivalent "pocket" states. To maintain the correct degeneracy of the levels as V_o changes, the correlation between the two extremes must be as drawn in Figure 12(a). At intermediate values of V_o the pattern corresponds to a series of vibrational levels but with each one split into two, one doubly degenerate (E) and one non-degenerate (A). When the pair of levels forming the ground vibrational state lies below the top of the barrier, the splitting is known as the tunnelling splitting because it is associated with reorientation of the methyl group by tunnelling through the barrier (32). For rotation about a single axis and for a periodic cosine potential such as (15), the Schrödinger equation can be solved to give the energy levels as a function of V_o/B. Thus if B is known, a measurement of either the tunnelling splitting or the torsional frequency, ν, gives the height of the barrier to rotation.

Just as in the case of H_2, only certain values of the total nuclear spin of the three protons of the CH_3 group may be associated with the different rotational levels. The total spin may be 3/2 or 1/2; doubly degenerate E levels have spin $\frac{1}{2}$ and A levels have spin 3/2. Neutrons may therefore cause transitions between the two levels split by tunnelling by flipping the spin of one of the protons. This is shown in the incoherent scattering spectrum of hexadiyne in Figure 13. As well as a strong peak due to elastic scattering, side peaks are observed at $\pm$ 1 μeV (0.008 cm^{-1}) corresponding to excitation or de-excitation of the tunnelling splitting. The barrier to rotation of a methyl group in the hexadiyne crystal is found from this splitting to be large (430 cm^{-1}) corresponding to the right hand extreme of Figure 12(a) (33).

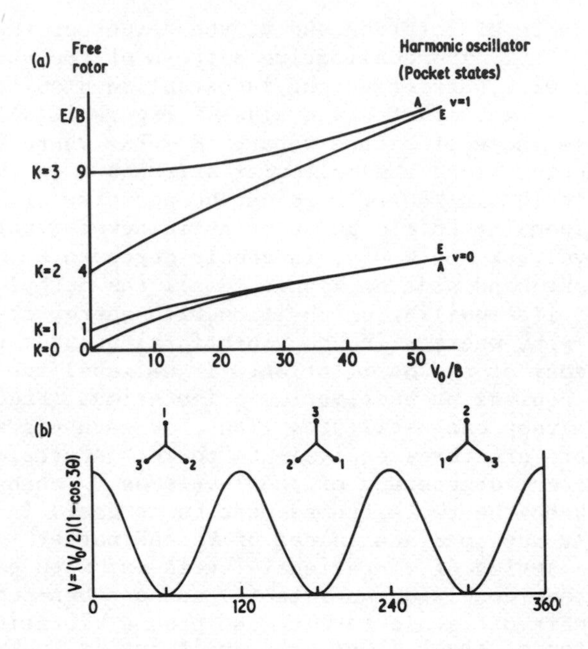

Figure 12.(a) The energy levels of a methyl group as a function of V_0/B for the potential given by equation (15) and shown in (b).

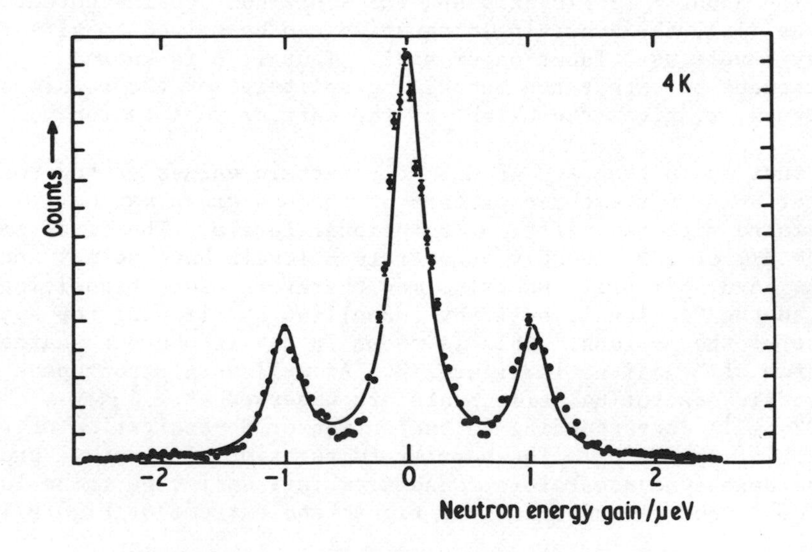

Figure 13. The incoherent neutron scattering spectrum of hexadiyne at 4K. The transitions at $\pm$ 1 μeV (0.008 cm^{-1}) are between the A and E levels of the vibrational ground state.

The motion of an adsorbed molecule is less easy to analyse because it may rotate about more than one axis. The qualitative pattern of the energy levels may, however, be constructed in a similar way to that used for Figure 12. For methane, for example, the free rotor energy levels are given by $E = BJ(J+1)$ where J is the quantum number of the total angular momentum. The degeneracy of each J level is $(2J+1)^2$ (34). When the hindering potential is large the methane molecule will oscillate about its equilibrium orientation and its energy levels will be approximately $(v+\frac{1}{2})h\nu$. On the surface of graphite the equilibrium position of an adsorbed methane molecule is known from diffraction experiments to be as shown in Figure 14. By interchanging the labels of the protons twelve equivalent configurations can be generated, of which three are shown in Figure 14. There are therefore twelve pocket states and each vibrational level has a degeneracy of twelve. We may then draw the correlation diagram shown in Figure 15(a). Since the ground vibrational state correlates with three rotational levels it must be split into at least three levels. The J=1 state of the free rotor itself consists of three threefold degenerate T levels and may be further split depending on the symmetry of the field. In Figure 15(b) are shown the different splitting patterns for a tetrahedral potential and a trigonal potential, the latter being what is expected for methane on graphite (35).

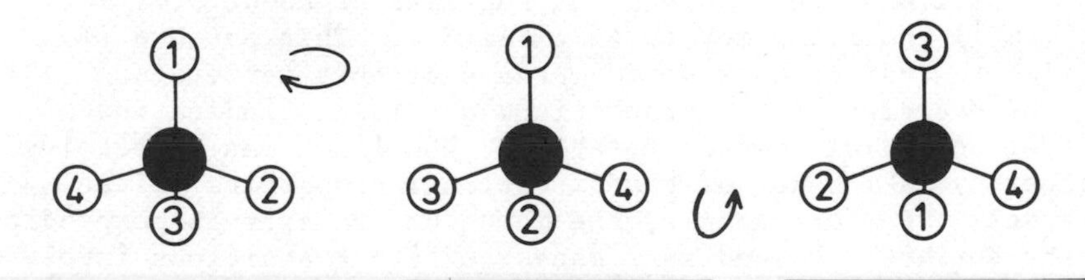

Figure 14. Three of the twelve possible equilibrium orientations of methane adsorbed on graphite.

Once again, only certain values of the total spin of the protons may be associated with the rotational levels. These are 2 for the A state (J=0), 1 for the T states (J=1), and 0 for the E state (J=2). The selection rule for incoherent scattering ($\Delta I = 0, \pm 1$) allows transitions between all the levels in Figure 15 except A ↔ E. We therefore expect a five line spectrum for methane on graphite.

Spectra of methane on graphite at coverages where it is in the registered ($\sqrt{3} \times \sqrt{3}$) phase are shown in Figures 16, 17 and 18. In Figure 16 the adsorbent is graphitized carbon black and only

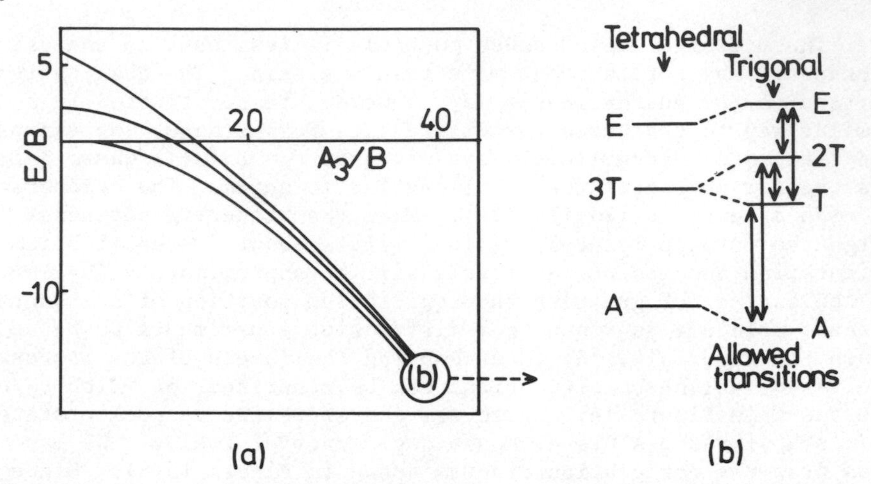

Figure 15. (a) Schematic diagram of the rotational energy levels of methane as a function of the barrier between the twelve equivalent orientations. (b) The tunnelling splitting of the ground vibrational state in fields of tetrahedral and trigonal symmetry.

two transitions are resolved, at energies of about 0.11 meV (0.9 cm^{-1}) and 0.055 meV (0.45 cm^{-1})(36). This pattern is similar to that of bulk solid methane at this temperature (37) but the energies of the transitions are lower, indicating a stiffer potential towards rotation. The lines can be resolved further by making use of the directional properties of the neutrons. If, for example, the momentum transfer is perpendicular to the surface, the neutrons cannot excite transitions involving only displacements of the protons parallel to the surface. Thus, depending on the details of the associated motion, a given transition may appear with different intensity for Q parallel or perpendicular to the surface. This polarization of the tunnelling spectrum is clearly seen in Figure 17 for methane on a partially oriented graphite. The conditions are otherwise the same as for Figure 16 but each line in Figure 16 is now split into two, about 18 μeV (0.14 cm^{-1}) apart. Of the four lines, only two appear in the Q perpendicular spectrum while three appear, though not with equal intensity, in the Q parallel spectrum. The assignment of the four transitions is as shown in the inset of Figure 18 and requires there to be a fifth line at 18 μeV. This was resolved in a separate experiment at much higher resolution and is shown in Figure 18. The observation of the frequencies alone would leave an ambiguity in the assignment over the order of the T levels. Fortunately, this may be resolved by a consideration of the intensities. The A $\leftrightarrow$ 2T transition is expected to be polarized parallel to the surface and to be much more intense than any other transitions in the Q parallel spectrum (38).

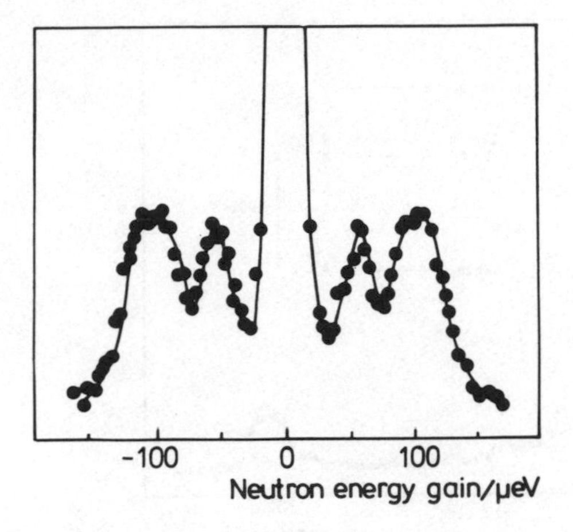

Figure 16. The rotational tunnelling spectrum of CH_4 adsorbed on Vulcan at a coverage of 0.7 and a temperature of 6K.

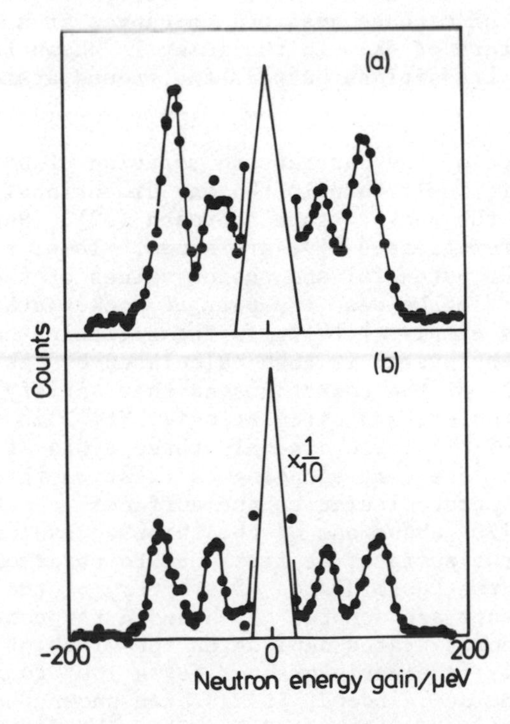

Figure 17. The rotational tunnelling spectra of CH_4 adsorbed on Papyex at a coverage of 0.7 and a temperature of 4K. (a) Q parallel to the surface. (b) Q perpendicular to the surface.

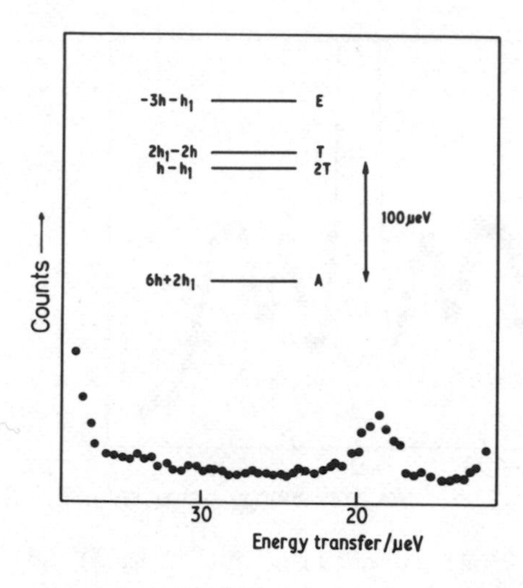

Figure 18. The transition between the T levels in the ground vibrational state of methane adsorbed on Papyex at a coverage of 0.7 and a temperature of 4K. In the inset is shown the assignment of the transitions between the ground state levels.

The derivation of the barriers to rotation of adsorbed methane is more difficult than in the one dimensional case but can be done using the pocket state approach (39). Suppose the pocket state wavefunctions, $|\phi_i\rangle$, are known. Then, for a chosen shape of the potential and chosen values of the barrier heights, the splitting between any pair of pocket states can be calculated. It is simply $\langle\phi_i|V|\phi_j\rangle$. The splitting between the actual levels of the system is then calculated by taking combinations, $|\psi_i\rangle$, of the pocket states that satisfy the symmetry of the problem, and calculating $\langle\psi_i|V|\psi_j\rangle$ in terms of the various $\langle\phi_i|V|\phi_j\rangle$. There are only three types of matrix element $\langle\phi_i|V|\phi_j\rangle$; one corresponding to rotation through 120° about the CH axis perpendicular to the surface; another to rotation through 120° about one of the three equivalent CH axes pointing towards the surface; a final one to rotation through 180° about one of the twofold axes of symmetry of the molecule. These matrix elements are denoted h_1, h and H respectively. The overlap between pocket states depends on the width of the barrier between them. This is clearly greater for a 180° rotation than for a 120° rotation and, indeed, it has been shown that H may be neglected in comparison with h_1 and h (40). The final values of the energy levels in terms of h and h_1 are given in Figure 18 (38).

The pocket state wavefunctions are determined by a variational treatment. An empirical form, containing an adjustable parameter, β, is chosen for the wavefunction. For a given potential, the value of β that minimizes the energy is found and then used to calculate the matrix elements, h and h_1, and hence the tunnelling splittings. For a chosen shape of the potential, the values obtained for the barrier heights are those that reproduce most accurately the observed splittings. The shape used for the potential was one which has the same threefold form as equation (15), both for rotation about the CH axis perpendicular to the surface and for rotation about each of the equivalent CH axes. The barrier height for rotation about the perpendicular axis is then found to be 205 cm^{-1} while that for rotation about one of the three equivalent axes is 175 cm^{-1}. Since the graphite surface is known to be fairly "smooth" (41), this result shows that molecule-molecule interactions are at least as important as molecule-surface interactions in determining the reorientation of adsorbed methane. It is possible to explore the balance of these interactions further by following the tunnelling spectrum of CH_4 as a function of coverage, and in mixtures with other small molecules or atoms, Kr, Ar, Xe and N_2, for example. Some experiments of this kind have been done but are not described here (36).

The requirements for observing a tunnelling spectrum can be seen from Figure 12 to be that V/B is not too large, that is, the moment of inertia of the molecule must be small and the barrier to rotation must not be too large. However, the smaller the molecule, the stiffer the potentials that can be investigated in this way. The H_2 molecule has an inertial constant, B, twelve times that of methane. It should therefore be possible to measure its rotational spectrum in potentials with barrier heights at least an order of magnitude greater than those found for adsorbed methane. The heights of such barrier would then be comparable with the changes in energy attending chemisorption and would be valuable measures of the more specific interactions in chemisorbed systems. Experiments of this kind have recently been done on hydrogen adsorbed into graphite alkali metal intercalation compounds (42).

When graphite and an alkali metal are heated together, the alkali metal may intercalate the graphite layers to form a series of approximately stoichiometric compounds with a layer of alkali metal atoms occupying either every interlayer space (Stage I compound), or every other space (Stage II compound), and so on (43). There is a considerable amount of empty space in these compounds. For example, in a Stage I compound of stoichiometry C_8M, the area of the layer available per atom M is 21 $Å^2$, and in the Stage II compound, $C_{24}M$, it is 31 $Å^2$. The value of the separation between the layers indicates that the alkali metal atoms are similar in size to their ions, in which case they themselves only occupy areas varying from 5.5 $Å^2$ for potassium to 8.8 $Å^2$ for caesium. In $C_{24}K$ the free area is thus about 26 $Å^2$, large enough to accommodate one methane molecule (~ 15 $Å^2$ see Figure 7) or two hydrogen molecules. At low temperatures (< 150K), the $C_{24}M$ compounds are indeed found to adsorb methane and hydrogen with stoichiometries one and two respectively, as well as other small molecules (44). Although the low temperatures required for adsorption and the low isosteric heat of adsorption (10 kJ mol^{-1} for H_2 into $C_{24}K$) suggest only weak interactions between adsorbate and adsorbent, the dissociative chemisorption of H_2 that occurs at higher temperatures in both C_8K and $C_{24}K$ suggests more specific interactions. The reactions of hydrogen with C_8K and $C_{24}K$ are summarized in Figure 19, which also shows their structures (44,45,46).

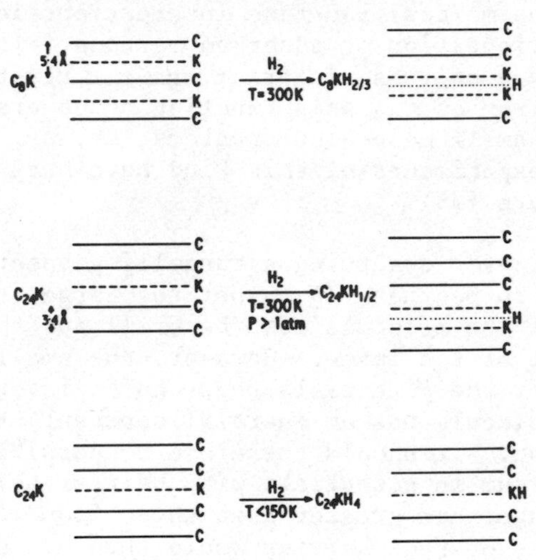

Figure 19. The reaction of potassium graphites with hydrogen.

The spectra of hydrogen in varying amounts in the compound $C_{24}Rb$ are shown in Figure 20 (43). At low coverages there is a single transition at an energy of 1315 μeV (10.6 cm^{-1}). At higher coverages, new lines appear at 600 and 1230 μeV while the original line remains constant in both intensity and frequency. This shows that there are two different sites for the adsorption of hydrogen and that one of them is very much preferred, although the possibility that hydrogen itself affects the site distribution cannot be excluded.

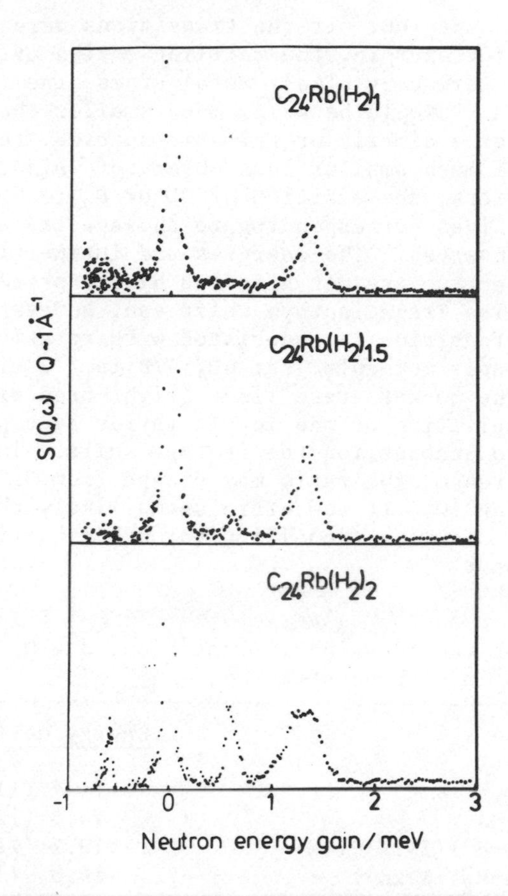

Figure 20. Incoherent inelastic spectra (neutron energy gain only) of hydrogen adsorbed into $C_{24}Rb$ at 30 K.

The energy of the observed transitions is so different from those normally observed for rotation of the hydrogen molecule in condensed phases (Table 3) as to suggest either that it is not a rotation or that it is a rotational transition of an associated species such as the dimer. However, the nature of the motion is easily ascertained by comparing spectra of the isotopic species H_2 and HD and by taking spectra of mixtures of isotopic species, H_2/HD and H_2/D_2. The spectrum of HD in $C_{24}Rb$, shown in Figure 21, is similar in pattern to that of Figure 20 but the energies of the lines, at 420, 345 and 140 μeV, are at about a quarter of the energies of the corresponding lines for H_2. There is also only a tiny trace of the pure H_2 spectrum in Figure 21, showing both that the spectra are not from dissociated molecules and that there is negligible dissociation followed by recombination to give a mixture of isotopic species. If the transitions were associated with a vibration, for example, the rattling motion of an H_2 molecule in a cage formed by alkali metal atoms, the ratio of the energies for H_2 and HD would be $\sqrt{3/2}$, much smaller than observed. For free rotation of a dimeric or trimeric species, the ratio would be 3/2, still much smaller than observed. Also, if the spectra were of dimers, the addition of HD or D_2 to H_2 in $C_{24}Rb$ would produce new lines corresponding to the spectra of (H_2D_2) or (H_2HD). None is observed. The energies and intensities of the lines in the H_2 spectrum are not affected by the presence of either HD or D_2. The large isotope shift can, however, be explained if the transition is associated with rotational tunnelling of a single molecule. For HD, V/B is 3/2 times greater than for H_2. In the pocket state limit (right-hand side of Figure 12(a)) the splitting of the levels varies as exp(-V/B), more than enough to account for the isotope shift. Thus, as the barrier height increases the ratio may change from 3/2 (free rotor) to as much as 10. It therefore seems likely that the spectra are of H_2 molecules tunnelling through a barrier of intermediate height.

Table 3. Energy of the rotational transition, $J = 0 \rightarrow 1$, of hydrogen molecules in various states.

State	Energy/meV(cm^{-1})
Gas	14.7 (119)
Liquid (30K) (47)	14.8 (120)
Solid (a) f.c.c.(o-H_2)($T<2K$) (48)	13.7 (111)
(b) h.c.p.(o-H_2)($T>2K$)	14.0 (113)
Adsorbed on graphite (31)	15.0 (121)
Adsorbed on alumina (31)	6.6 (53)
In alkali metal graphites (42)	~1 (8)

A quantitative analysis of the reorientational potential of H_2 in the intercalates has not yet been made. However, it is possible to estimate both V and B. A large value of V/B can arise from either a large value of V or a small value of B. A small value of B could be caused by expansion of the H_2 molecule, which is possible if the molecule has accepted electrons from the lattice. For a given shape of the potential, the isotope shift more or less fixes V/B while the absolute value of the splitting depends much more on B alone. The exact symmetry of the potential is not known but experiments on oriented samples suggest that the rotation is approximately isotropic, indicating a potential of high symmetry. Smith (49) has calculated the energy levels of a linear molecule in octahedral and tetrahedral fields. Fitting the observed splittings to his tables indicates that B is close to its value

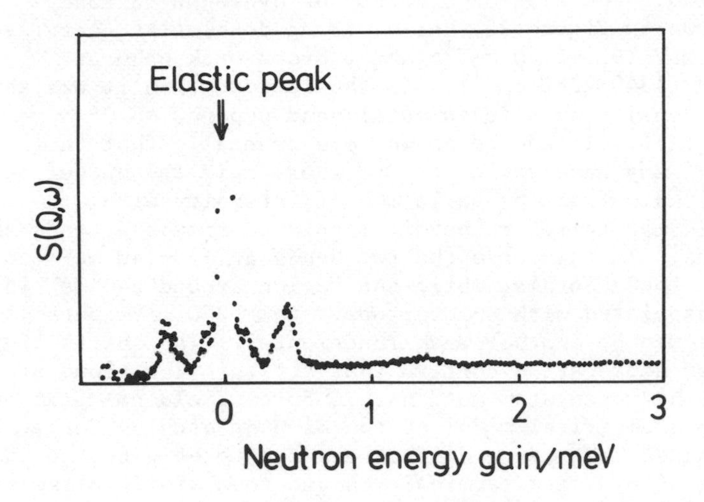

Figure 21. Incoherent inelastic spectrum of HD in $C_{24}Rb$ at 30 K.

for the free molecule, while the barrier height is in the range 2000-4000 cm^{-1}, depending on the particular choice of symmetry of the field and assignment of the levels (43). Arguments based on the volume available in the intercalate per hydrogen molecule also suggest that the size of the molecule changes little on adsorption. The height of the barrier (25-50 kJ mol^{-1}) shows that there is an unusually strong interaction of the hydrogen with the lattice.

Although rotational tunnelling spectra give detailed information on potential energy surfaces in special cases, there is no doubt that vibrational spectroscopy is more generally useful. However, there are also more competing experimental techniques for studying vibrations; infrared, Raman, and electron energy loss spectroscopy. Some examples of the application of neutron scattering to the vibrations of physi- and chemisorbed systems are now described, which illustrate the key features of the technique discussed in the introduction.

In comparison with other techniques the most valuable feature of incoherent neutron scattering is its ability to highlight the motion of hydrogen atoms. It is therefore a suitable technique for studying hydrogen chemisorbed on metals (see Table 1), although it is still necessary to use high surface area samples to have enough hydrogen in the neutron beam. An example is hydrogen on Raney nickel (50). This is a material of high specific area made by reaction of a nickel/aluminium alloy with KOH. It contains about 4% aluminium and is widely used as a catalyst for hydrogenation. The neutron spectrum of hydrogen on Raney nickel, shown in Figure 22, has a strong doublet at energies of 120-140 meV (970-1140 cm^{-1}) and a broad weak band at 240-280 meV (1940-2280 cm^{-1}). In the introduction it was shown that the intensity of a fundamental band depends on Q^2 (equation (12)). It can be shown more generally that the intensity of any band varies as Q^{2n} where n is the number of quanta excited. From the variation of intensity with Q^2, fundamentals may therefore be distinguished from overtones and combinations. In this case the two bands at 120-140 meV can be shown to be fundamentals, while the region around 240 meV is probably associated with an overtone. Near 280 meV there is thought also to be another weak fundamental. The shifts of frequency on deuteration confirm that all the vibrations are exclusively hydrogen atom motions. A force field calculation with a simple empirical model of the Ni-H stretching force constants shows that large differences are to be expected in the frequencies of hydrogen terminally bound to a single nickel atom and of multiply bound hydrogen. For hydrogen bound to three nickel atoms in a C_{3v} configuration, two stretching frequencies of symmetry A and E, are expected in the region 120-140 meV, whereas a terminally bound hydrogen would have a frequency at about 275 meV. The dominant contribution to the spectrum is therefore from multiply bound hydrogen, although it is not possible to distinguish C_{3v} and C_{4v} sites. A small fraction of hydrogen on Raney nickel is terminally bound, accounting for the weak band at 275 meV.

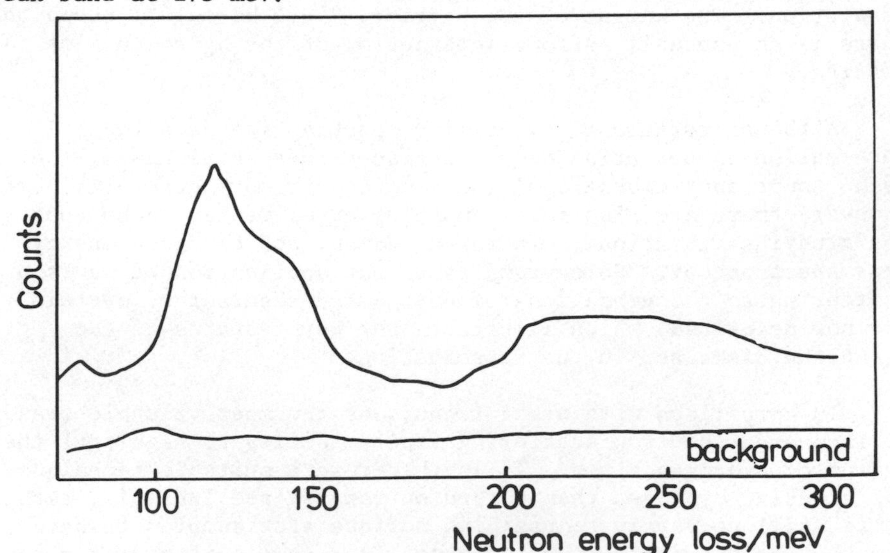

Figure 22. Incoherent neutron inelastic spectrum of hydrogen adsorbed on Raney nickel. The nickel background has been subtracted.

The interpretation of the spectrum of hydrogen on Raney nickel is similar in style to what is done in more conventional forms of vibrational spectroscopy. This is also true of the next example where neutron scattering is used just to identify the chemical species present on the surface. Renouprez et al. (51) have used the sensitivity of neutrons to hydrogen to show what happens to water when adsorbed on Raney nickel. When water is first adsorbed the spectra obtained are identical to those of hydrogen alone on Raney nickel (Figure 22), showing that water is dissociatively adsorbed. There are, however, two possibilities:

$$H_2O \rightarrow H_{ads} + OH_{ads} \qquad (I)$$

$$H_2O \rightarrow 2H_{ads} + O_{ads} \qquad (II)$$

Any OH_{ads} would also give strong bands in the neutron spectrum but none is observed, showing that II is the process that occurs. Magnetic evidence suggests that the remaining oxygen atom is adsorbed by the aluminium present in the catalyst. At higher coverages (> 4mg/g of H_2O), new bands appear at 65 and 105 meV (525 and 845 cm^{-1}), which are identified as H_2O bonded to the surface by its oxygen atom.

It should not be thought that neutrons only observe motions that are purely hydrogen atom vibrations. Any vibration which requires accompanying motion of a hydrogen atom will appear in the spectrum. Thus, when hydrogen is chemisorbed on a metal, the vibrations of the metal atoms at the surface appear in the spectrum since the hydrogen atom must follow the metal atom motion. For both Raney nickel and platinum black (52) this effect has been observed. Mahanty et al. (53) have made a detailed analysis of the intensity changes to be expected in the vibrational density of states of the metal when hydrogen is adsorbed on the surface, and have shown that the changes may lead to information about the local configuration of metal atoms about the adsorbed hydrogen. The effect on the metal density of states is largest when the frequency of the hydrogen stretching mode falls below the high frequency edge of the band of surface modes. In principle, both enhancement and depletion of the density of states may occur. However, the quality of the neutron spectra are not yet of a standard for a quantitative interpretation of this effect.

The feature that most distinguishes neutron scattering from other techniques of vibrational spectroscopy is that both the frequency and intensity of a mode are determined by its mechanical properties. The value of this is well illustrated by the spectrum of benzene adsorbed on Raney nickel (54). Low energy electron diffraction and electron energy loss spectroscopy indicate that benzene lies flat on Ni(111) and (100) faces but a complete assignment of the vibrational modes could not be made because only a small number of bands are observed in the electron energy loss spectra. In contrast, the incoherent neutron spectrum of benzene in Raney nickel is a rich one (Figure 23(b)). Comparison with the spectrum of solid benzene (Figure 23(a)) shows that many of the internal modes of adsorbed benzene appear in the neutron spectrum, although some of them are changed in frequency. The qualitative similarity of the spectrum of adsorbed benzene to that of (C_6H_6) $Cr(CO)_3$ suggests that benzene is lying flat on the surface and that a first approximation to the valence force field would be the force field of $(C_6H_6)Cr(CO)_3$. Jobic et al. (54) have calculated all the frequencies and intensities of adsorbed benzene using the force field of $(C_6H_6)Cr(CO)_3$ with changes in one or two of the most important force constants, the Ni-C and C-C stretching force constants, for example. The calculated spectrum (Figure 23(c)) agrees well with the observed spectrum provided that a small contribution (15%) is included from adsorbed hydrogen atoms (Figure 22). It is also consistent with the few bands observed in the electron energy loss spectrum. The final conclusions are

(a) benzene is definitely lying flat on the surface because there is no splitting of degenerate vibrations, which would be expected in a less symmetrical environment,

(b) The C-C bond in benzene is weakened by about 20% from its value in the free molecule,

and (c) the Ni-C force constant is twice as strong as in $(C_6H_6)Cr(CO)_3$ and also stronger than for benzene on the clean (111) or (100) faces of nickel.

A final feature of incoherent neutron scattering is that it can observe all the modes in a crystal, not just those excitations with a long wavelength (wavevector, $\underline{k}$, close to zero), to which optical spectroscopy is limited. Neutron scattering may therefore be used to look at cooperative motions within a layer and hence to probe the lateral forces between the adsorbed molecules. In incoherent scattering all the motions are observed simultaneously, irrespective of their wavevector. The resulting spectrum therefore resembles the density of vibrational states of the layer. Since the lateral forces in a chemisorbed layer are usually insignificant compared with the forces normal to the surface, we here consider an example of incoherent scattering from a physisorbed layer, methane adsorbed on graphite (5).

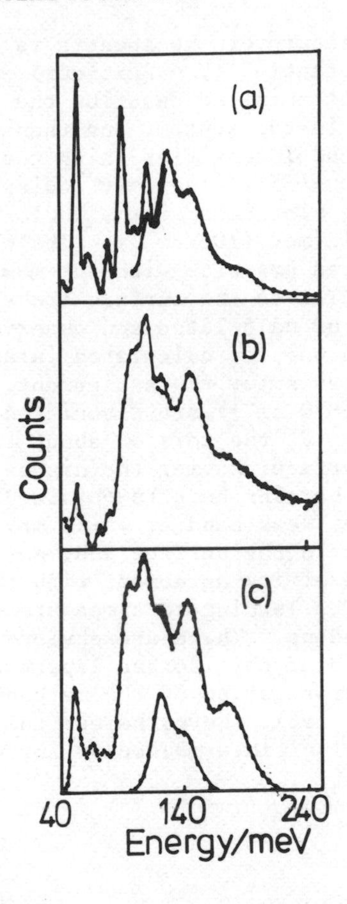

Figure 23. Incoherent neutron inelastic spectrum of (a) solid benzene (b) benzene on Raney nickel after subtraction of the nickel background. (c) Calculated spectrum for benzene on Raney nickel including the contribution of 15% adsorbed hydrogen atoms.

Because of the relatively large contribution of the methane-methane forces to the energy of the system, all of the vibrational modes of the lattice, except the vertical motion of the molecule against the surface, are expected to show strong dispersion. Since all the modes occur in a limited frequency region there will therefore be considerable overlap and it will be difficult to make a definite assignment. The problem is further complicated by anharmonicity. For example, if the potential energy for torsion of the methane molecule about an axis perpendicular to the surface is of the form of equation (15), the torsional energy levels are not those of the harmonic oscillator. The procedure that has been used to interpret the spectra is to use a semi-empirical Buckingham potential ($V = A\exp(-Cr) - B/r^6$ where A, B and C are empirical constants) to describe the interaction between pairs of atoms in the system, and then to calculate the vibration frequencies and intensities using the harmonic oscillator approximation (55). The only undispersed motion is predicted to be T_Z, the vibration of the whole molecule against the surface, at about 12 meV (100 cm^{-1}). The observed spectra of methane on an oriented graphite with the momentum transfer perpendicular and parallel to the surface are shown in Figure 24. Although the agreement of calculated and observed frequencies is poor, it is possible to use the calculated intensities for the two orientations as a basis for the assignment. The strongest features at 8-9 meV (65-70 cm^{-1}) are associated with the three torsions, R_x, R_y and R_z, R_z the torsion about the axis perpendicular to the surface, having the highest frequency. This is consistent with the barrier heights obtained from the tunnelling spectra. The weak band at 12-13 meV appears to be polarized perpendicular to the surface and, partly because of this and partly because of its agreement with the calculated value, is assigned to T_z, leading to a measure of the strength of the methane-surface binding. There are obvious difficulties in interpreting the spectra of physisorbed layers and indeed a slightly different interpretation of the methane spectra has been given by Maki and Klein (56). Nevertheless they remain the most direct test of any model of intermolecular forces at interfaces.

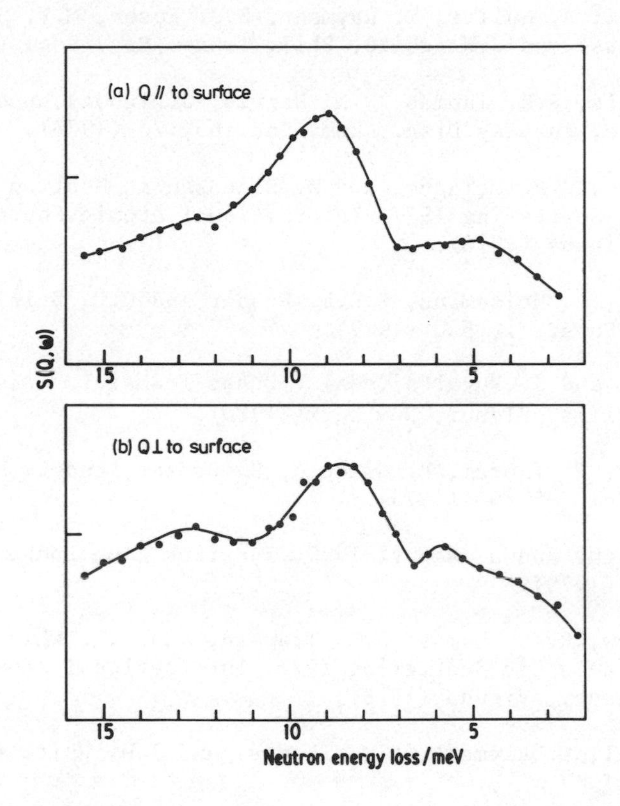

Figure 24 Incoherent inelastic spectra of methane on Papyex with (a) Q parallel to the surface, (b) Q perpendicular to the surface. Coverage is approximately 0.7 monolayers.

REFERENCES

(1) G.C. Stirling in Chemical Applications of Thermal Neutron Scattering, ed. B.T.M. Willis. Clarendon Press, Oxford (1973).

(2) Institut Laue-Langevin: Neutron Beam Facilities Available for Users, ILL. Grenoble (1980).

(3) M.W. Thomas and R.E. Ghosh, Mol. Phys. 29, 1489 (1975).

(4) R.K. Thomas in Chemical Society Specialist Reports on Molecular Spectroscopy, 6, 232 (1979).

(5) G. Bomchil, A. Hüller, T. Rayment, S.J. Roser, M.V. Smalley, R.K. Thomas, and J.W. White, Phil. Trans. Roy. Soc. (in press).

(6) D.J. Cebula, R.K. Thomas, N.M. Harris, J. Tabony, and J.W. White, Faraday Disc. Chem. Soc. 65, 76 (1978).

(7) M. Nielsen, J.P. McTague, and W. Ellensen in Neutron Inelastic Scattering 1977, International Atomic Energy Agency, Vienna (1978).

(8) C. Riekel, A. Heidemann, B.E.F. Fender and G.C. Stirling, J. Chem. Phys. 71, 530 (1979).

(9) J.C. Dash and J. Ruvalds (eds) : Phase Transitions in Surface Films, Plenum, New York (1980).

(10) Y. Larher, P. Thorel, B. Gilquin, B. Croset, and C. Marti, Surface Sci. 85, 94 (1979).

(11) Annex to the Annual Report 1979, Institut Laue-Langevin, Grenoble (1979).

(12) I. Marlow, R.K. Thomas, T.D. Trewern, and J.W. White, in Neutron Inelastic Scattering 1977, International Atomic Energy Agency, Vienna (1978).

(13) G. Bomchil, T. Rayment, R.K. Thomas, and J.W. White (to be published).

(14) B.E. Warren, Phys. Rev. 59, 693 (1941).

(15) P.H. Gamlen, R.K. Thomas, T.D. Trewern, G. Bomchil, N. Harris, M. Leslie, J. Tabony, and J.W. White, J. Chem. Soc. Faraday I, 75, 1542 (1975).

(16) C. Marti and P. Thorel, J. Physique, 38, C-4 26 (1977).

(17) M. Nielsen, p. 132 in reference (9).

(18) P. Vora, S.K. Sinha, and R.K. Crawford, Phys. Rev. Lett. 43, 704 (1979).

(19) S. Ergun, Carbon, 14, 139 (1976).

(20) J.P. Coulomb, M. Bienfait, P. Thorel, Phys. Rev. Lett. 42, 733 (1979).

(21) G.E. Bacon, N.A. Curry, and S.A. Wilson, Proc. Roy. Soc. A278, 98 (1964).

(22) K.J. Janda, J.C. Hemminger, J.S. Winn, S.E. Novick, S.J. Harris, and W. Klemperer, J. Chem. Phys. 63, 1419 (1975).

(23) P. Meehan, T. Rayment, R.K. Thomas, G. Bomchil, and J.W. White, J. Chem. Soc. Faraday I, 76, 2011 (1980).

(24) P. Meehan, T. Rayment, and R.K. Thomas, p.187 in reference (11).

(25) M. Monkenbusch and R. Stockmeyer, Ber. Buns. Phys. Chem., 84, 808 (1980).

(26) J.P. Coulomb, J.P. Biberian, J. Suzanne, A. Thomy, G.J. Trott, H. Taub, H.R. Danner and F.Y. Hansen, Phys. Rev. Lett. 43, 1878 (1979).

(27) J. Regnier : Thèse de 3 me Cycle : University of Nancy (1976).

(28) M. Nielsen and J.P. McTague, Phys. Rev. B19, 3096 (1979).

(29) P.W. Stephens, P. Heiney, R.J. Birgeneau, and P.M. Horn, Phys. Rev. Lett. 43, 47 (1979).

(30) T. Ceva and C. Marti, J. Physique, 39, L-221 (1978).

(31) L.F. Silvera and M. Nielsen, Phys. Rev. Lett. 37, 1275 (1976).

(32) C.H. Townes and A.L. Schawlow : Microwave Spectroscopy, Dover, New York (1975).

(33) M. Batley, R.K. Thomas, A. Heidemann, A.H. Overs, and J.W. White, Mol. Phys. 34, 1771 (1977).

(34) G. Herzberg : Infrared and Raman Spectra, van Nostrand, New York (1955).

(35) T. Nagamiya, Prog. Theor. Phys. 6, 702 (1951).

(36) M.W. Newbery, T. Rayment, M.V. Smalley, R.K. Thomas and J.W. White, Chem. Phys. Lett. 59, 461 (1978).

(37) W. Press and A. Kollmar, Sol. State Comm. 17, 405 (1975).

(38) A. Hüller, T. Rayment, M.V. Smalley, R.K. Thomas and J.W. White (to be published).

(39) A. Hüller and D.M. Kroll, J. Chem. Phys. 63, 4495 (1975).

(40) A. Hüller and J. Raich, J. Chem. Phys. 71, 3851 (1979).

(41) W.A. Steele : The Interaction of Gases with Solid Surfaces, Pergamon, Oxford (1974).

(42) J.P. Beaufils, T. Crowley, T. Rayment, R.K. Thomas, and J.W. White (to be published).

(43) G.R. Hennig, Prog. Inorg. Chem. 1, 125 (1959).

(44) K. Watanabe, T. Kondow, M. Soma, T. Onishi, and K. Tamaru, Proc. Roy. Soc. A333, 51 (1973).

(45) D. Saehr and A. Herold, Bull. Soc. Chim. Fr. 3130 (1965).

(46) M. Colin and A. Herold, Bull. Soc. Chim. Fr. 1598 (1971).

(47) P.A. Egelstaff, B.C. Haywood, and F.J. Webb, Proc. Phys. Soc. 90, 681 (1967).

(48) H. Stein, H. Stiller, and R. Stockmeyer, J. Chem. Phys. 57, 1726 (1973).

(49) D.F. Smith, J. Chem. Phys. 68, 3222 (1978).

(50) A.J. Renouprez, P. Fouilloux, G. Goudurier, D. Tocchetti, and R. Stockmeyer, J. Chem. Soc. Faraday I, 73, 1 (1977).

(51) A.J. Renouprez, P. Fouilloux, J.P. Candy, and J. Tomkinson, Surface Sci. 83, 285 (1979).

(52) J. Howard, T.C. Waddington, and C.J. Wright, J. Chem. Phys. 64, 3897 (1976).

(53) J. Mahanty, D.D. Richardson, and N.H. March, J. Phys. C. 9, 3421 (1976).

(54) H. Jobic, J. Tomkinson, J.P. Candy, P. Fouilloux, and A.J. Renouprez, Surface Sci. 95, 496 (1980).

(55) T. Rayment, S.J. Roser, and R.K. Thomas (to be published).

(56) K. Maki and M.L. Klein, Phys. Rev. (in press).

(e,2e) SPECTROSCOPY

A. Giardini-Guidoni*, R. Fantoni*

Comitato Nazionale Energia Nucleare, Centro di Frascati,
C.P. 65, 00044 Frascati, Rome, Italy

R. Camilloni, G. Stefani

C.N.R., Laboratorio di Metodologie Avanzate Inorganiche,
Montelibretti, Rome, Italy

ABSTRACT

Basic (e,2e) experiments carried out by our group are reviewed with the aim of outlining the main objectives and the results obtained. First of all the reaction mechanism has been investigated in order to find a working theory successful in interpreting experimental data. The basic theory whose validity has been investigated is the eikonal-averaged, distorted-wave impulse approximation. In this framework the (e,2e) amplitude factorizes into parts depending on the target structure and on the two-body interaction. Through absolute value measurements and test of the energy and momentum dependence of the cross section performed on He, Ne and Xe, it has been ascertained that the approximation works well in a large kinematic range going not under 400 eV for the incident energy and not under 80° for the angle between the outgoing electrons. Dependence of the validity limits on the binding energy and symmetry of the state investigated is evident. Secondly some results on molecules are discussed in terms of information obtainable on the dynamics of the electronic structure.

1. INTRODUCTION

Among the atomic scattering processes, those by electron impact have played an outstanding role in the atomic physics field. The most recent example in which problems of fundamental physics

P. Day (ed.), Emission and Scattering Techniques, 293–317.

are involved are the (e,2e) electron impact ionization processes. In these reactions the kinematics is fully determined by carefully measuring energy and momenta of the three electrons involved in the process and by detecting in coincidence the two final electrons. The first coincidence measurement was reported by Erhardt [1] in 1969 for He. The aim of this experiment was to investigate the reaction mechanism at low energies. The second experimental paper on (e, 2e) [2], appeared also in 1969, was concerning the use of (e,2e) reactions as a fine probe of the single particle character of electrons bound in atoms, molecules and in solids, in particular as a test of reliability of theoretical wave functions by comparing measured electron momentum distributions with calculated ones. Following this preliminary experiment, our group, in 1972, published the first momentum profile measured for a resolved single-particle state [3]. This was the 1s state of solid carbon (Fig. 1) and the experiment clearly demonstrated the

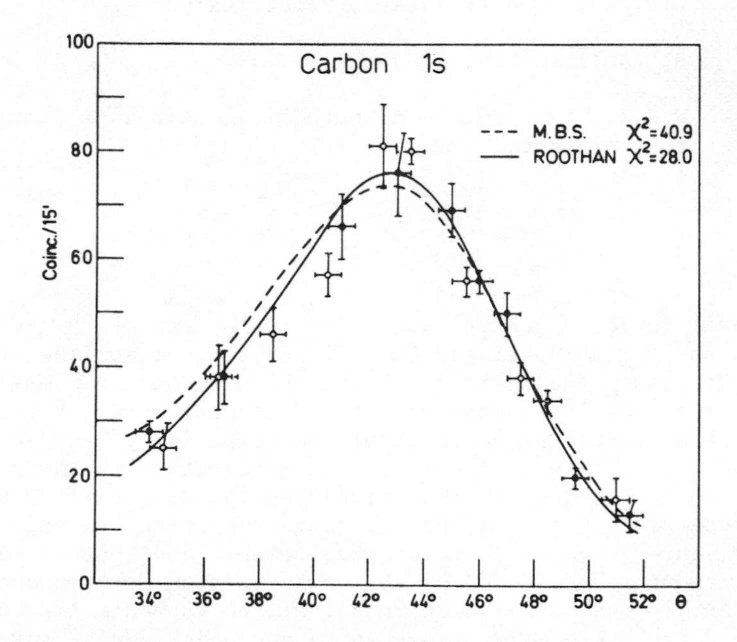

Figure 1. Angular distribution for the 1s state of carcarbon. The data belong to two different series of measurements on two 250-Å-thick carbon films. The dashed and solid curves are the predictions obtained with MBS and Roothaan wave functions, respectively. Incident electron energy 9.3 keV.

sensitivity of the momentum profile to the shape of the orbital wave function. In 1973 the experiment of Weigold, et al.[4] on argon resolved for the first time the valence states. They found strong electron-electron correlation effects in the 3s state and demonstrated the power of the technique in extracting structure information. Since these first studies, (e, 2e) experiments have been performed on the valence electrons of atoms and molecules with energy resolutions always improving [5,6].

The relevance of the (e,2e) reactions as a tool to test different reaction theories and to extract structure information about atoms molecules and their ions is nowdays amply demonstrated. In Figure 2 a sketch of the variety of information obtainable from the (e,2e) process is shown.

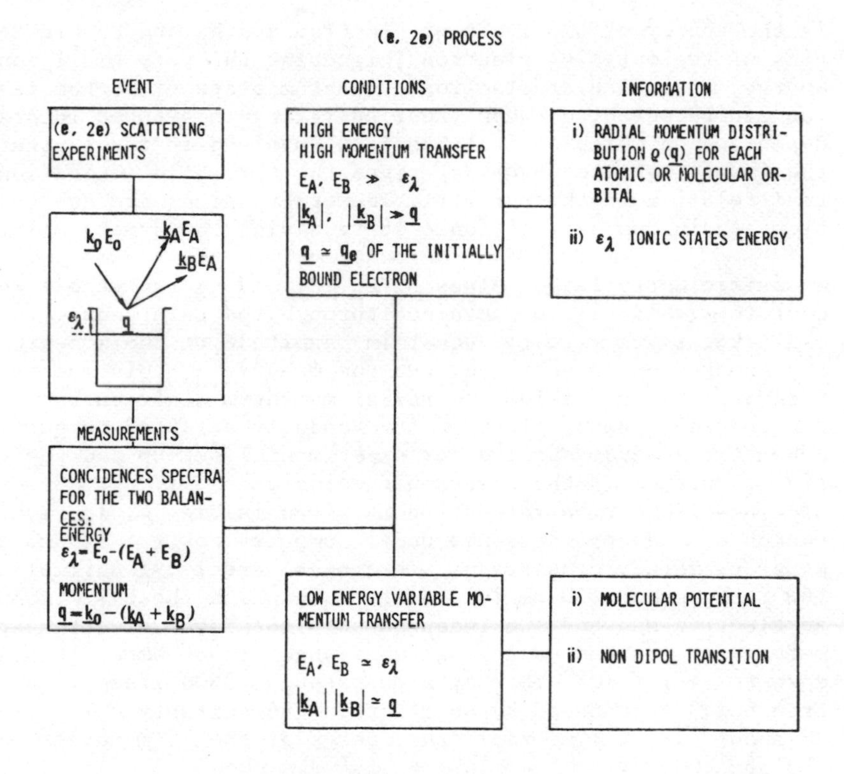

Fig. 2

2. EXPERIMENTAL

Figure 3 shows a schematic of the kinematics of the (e,2e) reaction. A beam of electrons of energy E_o and momentum $\underline{k}_o$ impinges on a gaseous or solid target, which is assumed to be at rest. To identify the two electrons coming from the knock-out process a fast timing coincidence technique is used, which discriminates against other events that yield electrons having the same $\underline{k}_A$ and $\underline{k}_B$ momentum.

From the energy conservation relation $\varepsilon_\lambda^f = E_o - E_A - E_B$ where E_o

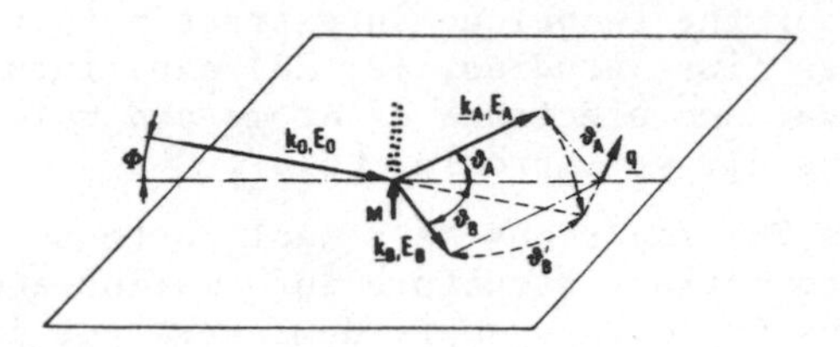

Figure 3. Kinetmatic of the (e,2e) process.

is the energy of the incident electron and E_A and E_B are the energies of the outgoing electrons, ignoring the very small ion recoil energy, a particular electronic quantum state ε_λ^f of the residual ion can be selected. When these energies are grouped into sets, depending on which orbital has been involved in the ionization, the lowest value of each set gives the threshold ionization potential relative to that orbital. The other values are due to ionization leading to a final ionic state having the same symmetry.

At sufficiently large values of E_o, E_A and E_B the recoil momentum $\underline{q}$ of the residual ion, obtained through the relationship $\underline{q} = \underline{k}_o + -\underline{k}_A - \underline{k}_B$, is very rearly equal in magnitude to the momentum of the knocked out electron $\underline{q}_e$ and the measure of this momentum at fixed E_A, E_B, E_o yields the radial momentum distribution $\rho(\underline{q})$ of the initially bound electron for randomly oriented targets. A schematic diagram showing our experimental set-up and the essential structure of the electronic coincidence circuit [3] capable of 4 nsec FWHM time resolution is given in Fig. 4. In Figure 5 a sketch of the apparatus now under computer control is shown. The appartus mainly consists of a stainless steel cylindrical chamber (60 cm high and 130 cm in diameter) in which the basic components, an electron gun and two independently rotatable electron spectrometers are mounted on the bottom flange. The chamber is pumped down to $\simeq 1 \times 10^{-7}$ Torr by a turbo pump (3500 1/sec). The electron beam is produced by an electron gun suitably collimated $\Delta\theta$ about $\pm 1 \times 10^{-2}$ rad), the energy of the beam ranges between 150 and 4000 eV.

The azimuth angle ϕ of the gun can be varied with respect to the detection plane of the electrostatic detectors from $- 10°$ up to $+ 30°$ with a precision of $\pm 1°$. The two twin electron spectrometers, which detect the two emerging electrons at different θ's angles, are formed by hemispherical electrostatic selectors having a retarding field at the entrance and a channeltron as electron detector at the exit. The gaseus beam is obtained by allowing the gas to effuse through a Bendix multichannel array whose thickness is 0.25 mm.

This apparatus allows to measure also the absolute value of the (e,2e) cross section when the e-e elastic cross section is known

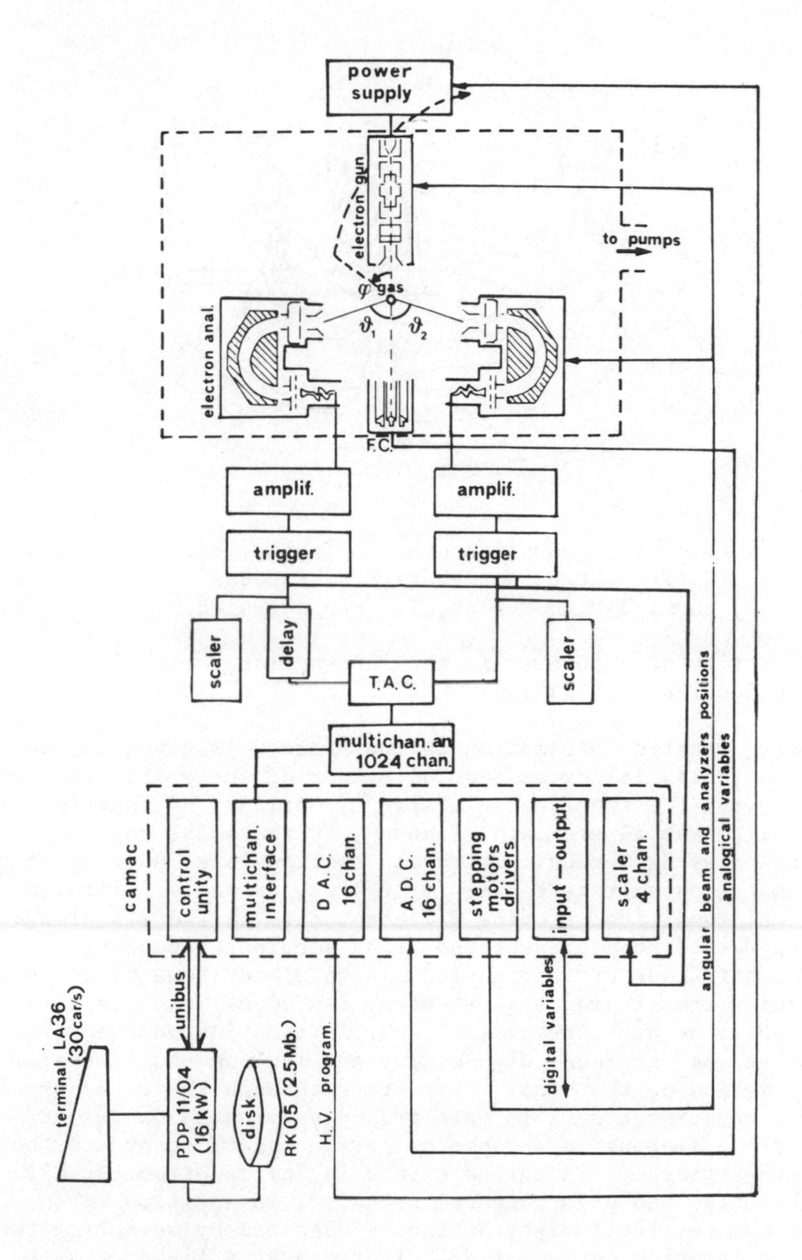

Figure 4. Schematic of the experimental apparatus and computer control (ϕ variation is out-of-plane)

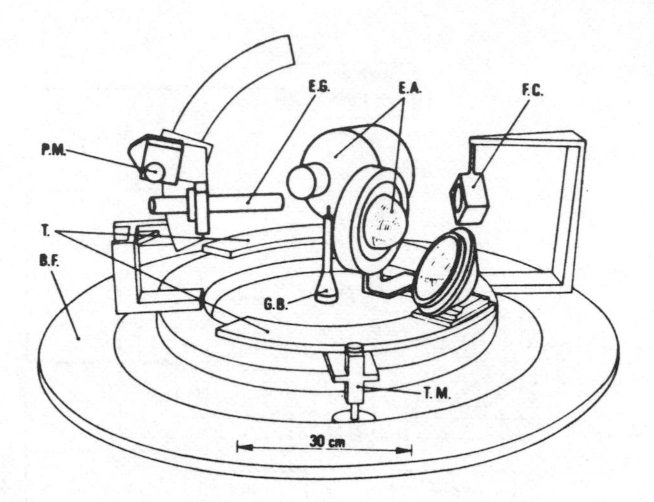

Figure 5. Experimental apparatus: E.G. electron gun; E.A. electron analyzer; F.C. Faraday cup; G.B. gaseus beam; T.M. movement; T. turntables; B.F. bottom flange.

and density of molecules in the interaction region is derived from the molecular flux [18].

A very detailed information on the process is given by the five fold differential cross section measuring the probability that in each event the two final electron directions and energies lie in the differential elements ϑ_A and ϑ_B at the solid angle Ω_A and Ω_B with energy E_A (or E_B). The cross section is a function of the kinematic parameters E_o, E_A (or E_B) ϑ_A, ϑ_B and Φ. Different kinematical conditions provide different cuts in the six dimensional space needed to represent the cross section. Depending on the cut, different kinds of information are obtained. Some times collision dynamics are studied at low energy ($\leq$ 200 eV) and small momentum [8] while at high energy ($\gtrsim$ 1 keV) information on the target structure can be obtained. High energy and low momentum transfer gives information on the final ionic state through the ε_λ separation energy measurements; ε_λ is selected by varying E_O or $E_A + E_B$ so that spectra analogous to the photoelectron spectroscopy are observed. This spectroscopy is called dipole (e,2e) spectroscopy [9]. At high energy and high momentum transfer, in addition to the energy spectrum, q-recoil distribution is obtained by measuring the angular correlation between final electrons at a fixed ε_λ value. Each q-profile is closely related to the symmetry of the initial state and is peculiar to this so called binary (e,2e) spectroscopy.

Typical spectra coming from different spectroscopies (binary (e,2e) dipole (e,2e), E.S.C.A.) are compared in Fig. 6 for NO, showing the identification of the valence states of this molecule. Figure

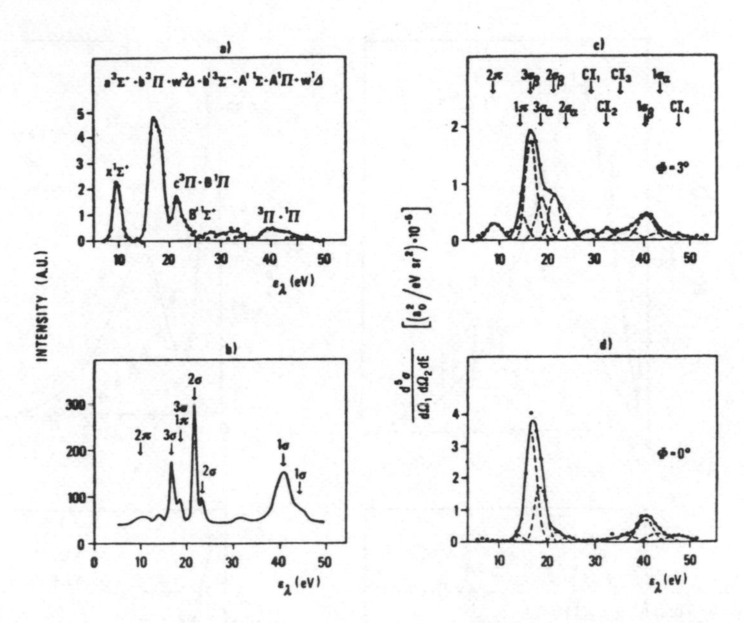

Figure 6. Comparison among NO valence shell energy spec trac measured with different spectroscopies (a) dipole (e,2e); (b) E.S.C.A.; (c),(d) binary (e,2e)

7 shows the momentum distribution measured for the previously identified states of NO. From this example it can be seen that two types of electronic properties of the target are obtained in binary (e,2e) spectroscopy:

i) on the final ionic state trhough the ε_λ^f separation energy measurements (I.P.'s. for ground and excited states). It has to be noted that the (e,2e) energy spectrum can put in evidence a larger number of states than P.E.S. because the (e,2e) process does not obey the dipole selection rules.

ii) On the initial neutral state through the q-distribution, i.e. electron momentum distribution (e.m.d.), obtained by measuring the angular correlation between the final electrons at a fixed ε_λ value. The e.m.d., related to the symmetry of the initial state [7], is usually described within the H.F. single particle approximation. For a state belonging to a σ bonding symmetry the small momentum components are strongly favoured so that its e.m.d. presents a maximum for q values near zero. This value roughly corresponds to $\theta_A = \theta_B \simeq 44°$ and $\phi = 0$. σ antibonding and π orbitals present a maximum at q values different from zero; from the maximum q value they are uniquely identified.

Due to the non orthogonality of the neutral and ionic wave func-

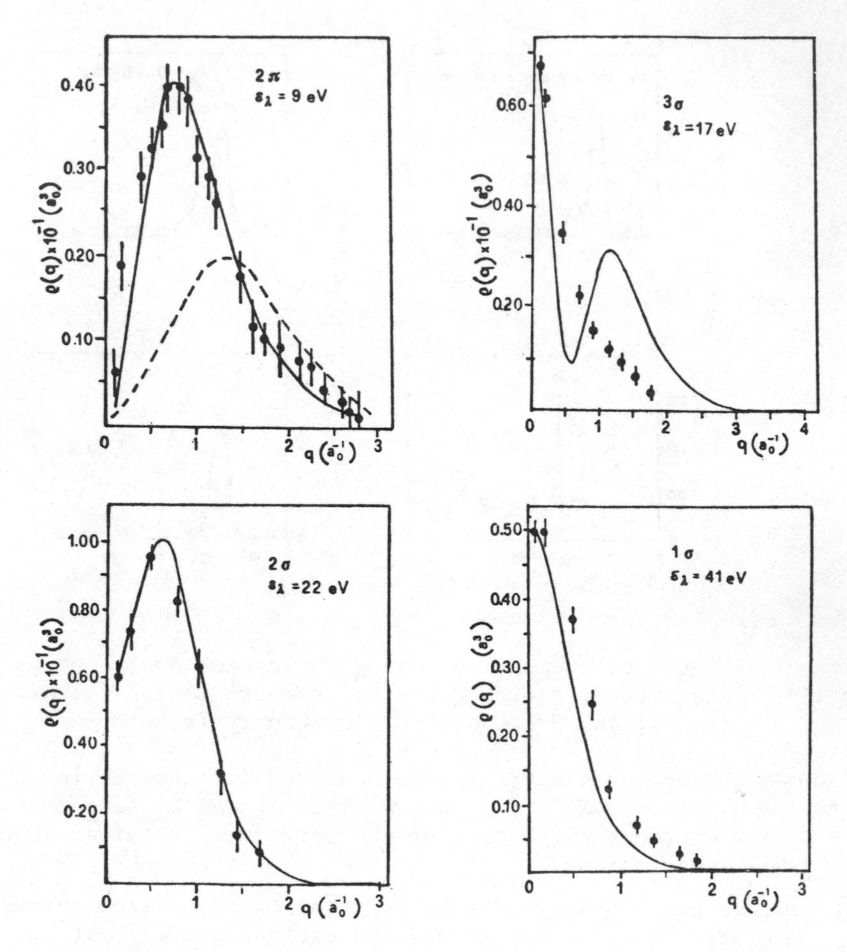

Figure 7. Angular distribution measured by binary (e,2e) spectroscopy for the valence shell of NO molecule. Full curves are the space averaged momentum distribution from Kouba et al. wave functions.

tions more than one ε_λ^f level can show the same q-distribution. Such a series of ε_λ^f states is usually called a satellite transition and labelled C.I. It can be accounted for these transitions by representing the final open shell ionic state wave function within the C.I. approximation which considers a suitable mixing of H.F. configurations. In this framework the cross section expected for a single particle ionization is divided with given probabilities (spectroscopic factors [10]) over all the ε_λ^f peaks. Spectroscopic factors measured in (e, 2e) spectra verify the sum rule and are different from those of photoelectron spectra which originate from a different interaction.

3. THEORETICAL BACKGROUND

The cross section expression fully clarify the type of information obtainable by an (e, 2e) process. As in the case of many theoretical problems involving scattering, the (e,2e) reaction must be reduced to a problem involving a few degrees of freedom. By considering the reaction where the residual ion is left in its ground state, the simplest model incorporating the important features of the reaction is a quasi-three body problem, in which two electrons interact with each other through the Coulomb potential and with the ion through optical model potentials where unobserved channels are treated by polarization and absorption terms.

It was shown by McCarthy and Weigold [10], with approximations amounting to closure over target states and weak coupling between channels in both two and three body systems, that is in the quasi--three body model with implied antisymmetry, the (e,2e) amplitude can be represented by:

$$T(\underline{k}_A, \underline{k}_B) = < \chi_A^{(-)} (\underline{k}_A) \chi_B^{(-)} (\underline{k}_B) \tag{1}$$

$$(f|v + v \frac{1}{E^{(-)} - k_1 - k_2 - V_1 - V_2 - v} v|g> \chi_o^{(+)} (\underline{k}_o)>$$

$\chi_A^{(-)}$ and $\chi_B^{(-)}$ are distorted waves computed in the optical model potentials V_1 and V_2. $\chi_o^{(+)}$ is a distorted wave computed in the entrance channel optical model potential. The ground state $|g>$ of the target and the final state $(f|$ of the ion are functions of the many body co-ordinates of the molecule and of the ion although these co-ordinates are assumed not to affect the potentials V_1 and V_2 (quasi-three-body approximation). This expression is already an approximation of an amplitude assuming neither factorization of the many body target nor of the scattering wave function [11]. The cross section at low energies has been calculated for the simple systems H, He and H_2 by using the unfactorized scattering amplitude in the Born approximation [11],[8] where higher order terms in v are dropped.

For kineamtical conditions where the residual ion can be treated as a spectator (binary (e,2e)) the Impulse Approximation accurately describes the process [12]. In this approach the electronic and ionic coordinates can be separated and the (e,2e) amplitude can be factorized [10]. This is called the factoriezed Distorted Wave Impulse Approximation (D.W.I.A.). The cross sections can be more easily computed if the optical potentials V_1, V_2 can be approximated by an average constant potential $V = \overline{V} + i\,\overline{W}$. This approximation, the averaged eikonal D.W.I.A., amounts to employing plane waves for the distorted waves in the quasi-three body scattering

Eq. (1). Effective vectors $\underline{\kappa}_I$ (I = 0, A, B), of which the magnitude is determined from the phenomenological potential $\overline{V}$ through the relationship $\underline{\kappa}_I^2 = (2m/\hbar^2)(E_I + \overline{V})$, the directions being those of $\underline{k}_I$, are used [13].

The imaginary potential $\overline{W}$ takes into account absorption effects produced by unobserved channels. It gives rise to an attenuation factor given by:

$$\gamma = \exp(-\overline{W}\overline{R}\ \Sigma_I k_I/E_I),\ I = (0,A,B),\ (0 \leq \gamma \leq 1)$$

$\overline{R}$ being the parameter which characterizes the atomic region in which distortion occurs. The effective potentials $\overline{V}$ and $\overline{W}$ cannot be predicted a priori from elastic scattering measurements. This is due to the fact that the relevant radial region is too broad to give a good approximation for χ_I. The parameter therefore must be extracted from the (e,2e) data. For a zero value of $\overline{V}$ and $\overline{W}$ the distorted waves becomes plane waves and the simple Plane Wave Impulse Approximation (P.W.I.A.), in which the distorting potentials are neglected, is obtained. Assuming the validity of all the approximations outlined so far, the explicit expression of the cross section derived by Eq. (1) is given by,

$$\frac{d^5\sigma}{d\Omega_A\ d\Omega_B\ dE} = \frac{4\ k_A\ k_B}{k_o} f_\lambda\ \gamma\ \rho(\underline{q}) \tag{2}$$

where f_λ is the antisymmetrized e-e Mott Scattering cross section calculated half-off the energy shell, in t-matrix or Born v-matrix approximation. The v-matrix approximation is:

$$f_v = \frac{1}{|\underline{\kappa}-\underline{\kappa}'|^4} + \frac{1}{|\underline{\kappa}+\underline{\kappa}'|^4} - \frac{1}{|\underline{\kappa}-\underline{\kappa}'|^2|\underline{\kappa}+\underline{\kappa}'|^2} \cos\left[\ell n \frac{|\underline{\kappa}+\underline{\kappa}'|^2}{|\underline{\kappa}-\underline{\kappa}'|^2}\right] \tag{3}$$

The t-matrix element is

$$f_t = C_o^2(\eta) \left\{\frac{1}{|\underline{\kappa}-\underline{\kappa}'|^4} + \frac{1}{|\underline{\kappa}+\underline{\kappa}'|^4} - \frac{1}{|\underline{\kappa}+\underline{\kappa}'|^2 \cdot |\underline{\kappa}-\underline{\kappa}'|^2} \cdot \cos\left[\eta\ \ell n \frac{|\underline{\kappa}+\underline{\kappa}'|^2}{|\underline{\kappa}-\underline{\kappa}'|^2}\right]\right\} \tag{4}$$

γ is the attenuation factor already mentioned and

$$C_o^2(\eta) = \frac{2\pi\eta}{e^{2\pi\eta^{-1}}} \; ; \; \eta = \frac{me}{2\hbar^2 \kappa} \; ; \; \underline{\kappa} = \frac{1}{2}(\underline{k}_o + q) ; \underline{\kappa}' = \frac{1}{2}(\underline{k}_A - \underline{k}_B)$$

In expression (2) $\rho(\underline{q})$ is simply:

$$\rho(\underline{q}) = |\langle \underline{\kappa}_A \, \underline{\kappa}_B \, |(f| \, g\rangle \, \underline{\kappa}_o\rangle|^2 \; ; \; \underline{q} = \underline{\kappa}_o - \underline{\kappa}_A + \underline{\kappa}_B \qquad (5)$$

The function $\rho(\underline{q})$ which carries the structure information about the target system needs to be further specified in the description of the initial neutral and final ionic states.

In case of gaseous free molecules the Born-Oppenheimer approximation can be used, separating rotational vibrational and electronics terms. Because the actual energy resolution in present (e,2e) experiments is insufficient to resolve rotational and vibrational structures $\rho(\underline{q})$ in expression [5] can be considered a function of the electronic contribution alone [10,14]. As a first approximation $\rho(\underline{q})$ may be interpreted in terms of the independent particle model as the spherically averaged momentum density for the Hartree-Fock orbital of initially bound electron. However, a satisfactory description of electronic properties of the target needs to include correlation effects. Among the ways to account for these effects, one is to write the ground and ionized state wave functions in a C.I. expansion of the H.F. basis set [10]. For closed shell systems the expansion may be limited to the final ionic state and the (e,2e) cross section then splits into the momentum density of the characteristic orbital ψ_j of the initially bound electron multiplied by a spectroscopic factor $S_j^{(f)}$, which is the probability that the many-body wave function $(f|$ contains the one-hole configuration with a hole in the orbital j of the ground state Hartree-Fock wave function. As previously mentioned, the spectroscopic factors obey the sum rule: $\Sigma_f \, S_j^{(f)} = 1$. In the case of open shell systems C.I. expansion of the ground $\langle g|$ state is required, so that more than one characteristic orbital can contribute to the momentum density of the ε_λ^f ionic state observed.

Another approach to the determination of the $(f|g\rangle$ overlap can be made in terms of Green's functions [15]. In this case the ionization energies ε_λ^f appear as the poles of the Green's function G and can be best calculated by solving Dyson equation [15], connecting it to the free Green's function $G°$ via the self energy Σ. Ionization energies ε_λ^f are calculated through the self energy, which is the effective energy dependent potential seen by an electron due to interaction with its surroundings. The relative intensities of the lines arising from ionization of electrons in orbital ψ_j (spectroscopic factors) are given by the residues of the poles of jth eigenvalue of G [15].

Koopmans approximation is recovered in this formalism by the first order approximation to Σ, which is zero in the case of a H.F. Green's function [16].

4. TEST OF THE APPROXIMATION USED IN DESCRIBING THE IONIZATION PROCESS

A series of tests has been performed in order to determine the range of validity of the approximations outlined in Sec. 3. Simple atoms or molecules with well known structures have been used as targets. At low energy and small momentum transfer the angular distribution of the lower energy ejected electron has been measured [2] for a fixed momentum change of the scattered higher energy electron. These experiments are suitable for investigating the interaction mechanism and show that in case of H, He, Ne and H_2 the Born approximation gives only a qualitatively correct description of the interaction [11]. From 200 eV on factorized and unfactorized cross sections have been calculated from expression (1) and tested on H and He in conditions of large momentum transfer [17,18]. Data are in qualitative agreement with both types of calculations and show that distorsion effects on the free electrons are more prominent than predicted. At higher incident energies ($\sim$ 400 eV) factorized cross sections in D.W.I.A. (P.W.I.A.) have been calculated and tests have been performed in three different geometries each of which is sensitive to different factors appearing in the cross section expression. For sake of clarity the geometries and details of information available are here briefly recalled.

i) Coplanar symmetric geometry: equal kinetic energies $E_A=E_B=E$ and equal scattering angles $\vartheta_A = \vartheta_B = \vartheta$ for the electrons emerging in the plane containing the incident beam. The factor f_λ and $\rho(\underline{q})$ both have a large variation over the range of ϑ for which the cross section is measurable. By suitable variation of ϑ, it is possible to scan $\underline{q}$ values parallel to $\underline{k}_o$. The $\rho(\underline{q})$ form factor is thus measured twice in the angular correlation spectrum, while the f_λ is continuously varying adding a further test of factorization [7].

ii) Non coplanar symmetric geometry: equal kinetic energies $E_A = E_B = E$ and equal scattering angles $\vartheta_A = \vartheta_B \sim 44°$, and Φ variable. The f_λ factor remains essentially contant as the variable azimuth Φ is changed from 0° corresponding to $q \simeq 0$ on, as to vary the recoil momentum q. Distortion effects are less evident in this geometry so that it is most suitable for studying form factors $\rho(\underline{q})$ [7].

iii) Symmetric ϑ and Φ variable geometry: equal kinetic energies $E_A = E_B = E$, $\vartheta_A = \vartheta_B = \vartheta, \Phi$ variable in such a way to keep contant $\rho(\underline{q})$. The f_λ factor variation can be observed [12, 13].

The magnitude of the momentum $\underline{q}$ of the initially bound electron in these geometries is

$$q = (k_o^2 + 4\, k_A^2 \cos^2\theta - 4\, k_A k_o \cos\theta \quad \cos\phi\ )^{\frac{1}{2}} \tag{6}$$

In what follows data obtained with the Frascati apparatus, in which these different conditions are allowed, are reported and discussed.

Absolute (e,2e) cross section measurements performed on the n = 1 shell of He [17, 18] the n = 2 shell of Ne and the n = 5 shell of Xe [19] have been shown to be in rather good agreement with the averaged eikonal D.W.I.A. calculations. Few significative examples of absolute measurements at various energies are reported in Fig.8 for He and Xe in the coplanar geometry, most suitable to test validity of approximations. They show that the agreement between theory and experiments improves with increasing energy. The limit of validity of D.W.I.A. is about 800 eV while the P.W.I.A. appears to hold at energies of the order at 2000 eV. Both the real and the imaginary part of the eikonal potential depend on a given orbital; $\overline{V}$ of the order of few eV has been obtained for He, Ne and Xe and non zero $\overline{W}$ only for more internal orbitals like 2s in Ne and 5s in Xe [19]. A fully distorted wave approximation, in which the optical potential was semiempirically derived [17] was also in agreement with absolute data. Calculations in Born approximation are however in agreement with experimental results only in the high energy limit ($E_0 \simeq 3600$ eV) where they are almost indistinguishable from t-matrix calculations [20].

q-distributions obtained in all the geometries previously mentioned are in satisfactory agreement with calculation in the H.F. model [21]. Discrepancies appear only at q-momenta larger than $\simeq 2a_o^{-1}$ where failure of the approximations in the description of the interaction is expected [13].

The (e,2e) energy spectra, being more sensitive than q distributions to relaxation and correlation effects, are in disagreement with Koopmans energies also for simple systems [5], and need to be interpreted in terms of C.I. or Green's function methods. As an example the Xe energy spectrum showing 5s satellite structure extending up to the first ε_λ values of the $4d^{-1}$ level is shown in Fig. 9. No comparison with Green's function calculations has yet been made. The cross section for the ejection of intermediate 4d electrons, measured in two different geometries is reported in Fig. 10. Although the simple factorized eikonal averaged D.W.I.A. should account for data taken at $E_0 = 3600$ eV, they show a very peculiar behaviour. The non coplanar symmetric results in the high q region are in good agreement with the calculated H.F. momentum distribution, while around q = 0 a sharp relative maximum instead of a minimum is observed. This maximum could be explained in terms of a very weak $5s^{-1}$ satellite transition whose strength is only

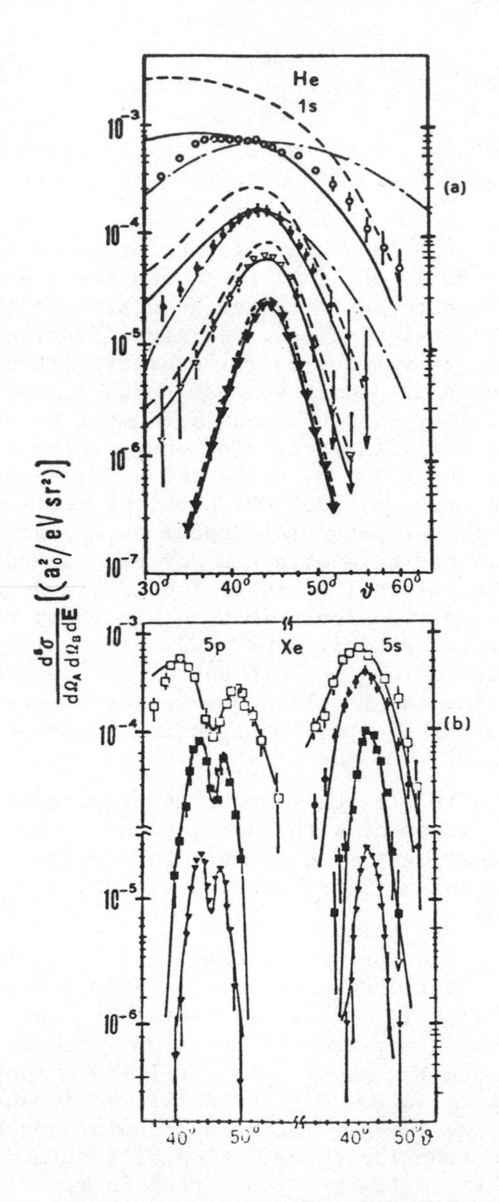

Figure 8. Coplanar symmetric (e,2e) cross section as a function of scattering angles taken at various incident energies: ○ 200 eV, □ 400 eV, ● 800 eV, ▽ 1600 eV, ■ 2600 eV, ▼ 3600 eV. (a) He 1s: --- factorized born approximation, —— eikonal averaged D.W.I.A. ($\overline{V}$ = 20 eV, $\overline{W}$ = 0), -·-·- C.P.B.E. calculation. (b) Xe 5p and 5s: —— eikonal averaged D.W.I.A. ($\overline{V}$ = 10 eV, $\overline{WR}$ = 7 Å eV)

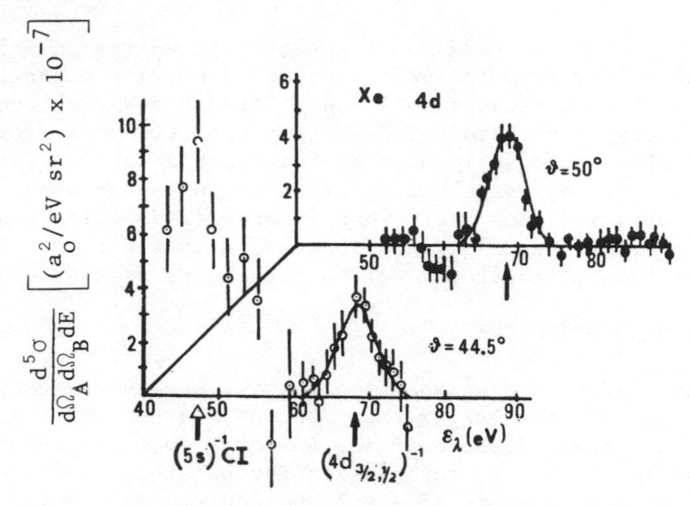

Figure 9. Xe 4d orbital (e,2e) cross section measured at E_0 = 3600 eV. Coplanar symmetric energy spectrum as a function of binding energy ε_λ taken at ○ ϑ = 44.5° and ● ϑ = 50°.

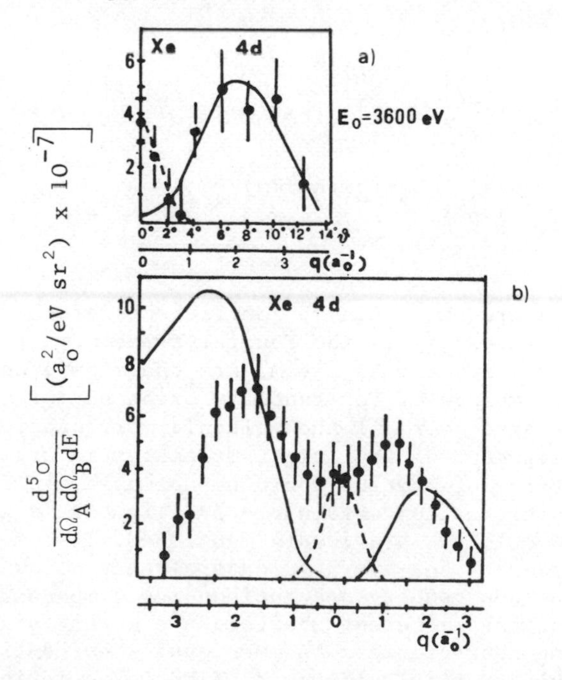

Figure 10. Xe 4d angular distributions measured at E_0 = = 3600 eV. (a) Non coplanar symmetric cross section measured at ε_λ=68 eV and ϑ=44.5°.(b) Coplanar symmetric cross section measured at ε_λ=68eV, Φ=0°. P.W.I.A. calculations: —— 4d orbital x.30, --- 5s orbital x.005.

.5% of the total $5s^{-1}$ and whose position in energy is only few tenths of eV far from the $4d^{-1}$ doublet. A similar situation has been found for the states $3\sigma^{-1}$ and $1\pi^{-1}$ of the N_2O molecule [22] and the energy resolution (2.6 eV) for the Xe 4d measurement is not sufficient to discriminate against such a satellite transition. The coplanar measurements are not consistent with the out of plane data and show that, when intermediate orbitals are involved in binary (e,2e) events, the validity limits of approximations for the interaction model have to be reassessed [21].

5. MOLECULAR STRUCTURE

The finding of a working reaction theory, namely the factorized P.W.I.A. at energies larger than 2000 eV and the D.W.I.A. eikonal averaged at intermediate energies (∿ 800 - 2000 eV), has allowed us to extract structure information for external shells of atoms and for valence orbitals of simple molecules. As pointed out in Sect.4 the best way to obtain such information is the use of energy and angular correlation spectra in various geometries. The explicit expression of the form factor $\rho(\underline{q})$ in H.F. approximations is:

$$\rho(\underline{q}) = S_j^{(f)} |\psi_j(\underline{q})|^2 = S_j^{(f)} \left\{ \sum_{1=p}^{N} C_{j,p} \; |\Phi_p(\underline{q})|^2 + \sum_{1=p}^{N} \sum_{p \neq s}^{N} C_{j,p} C_{j,s} e^{-i\underline{p}\cdot(\underline{R}_a - \underline{R}_b)} \times \Phi_p^{\star}(\underline{q}) \; \Phi_s(\underline{q}) \right\} \tag{7}$$

where $C_{j,k}$ (k = p,s) are the L.C.A.O. coefficients of the j molecular orbital, $\Phi_k(\underline{q})$ (k = p,s) is the Fourier transform of the atomic basis set and $\underline{R}_a$, $\underline{R}_b$ ($1 \leqslant a, b \leqslant M$) are the atomic co-ordinates of the M-atomic molecule. For randomly oriented molecules $\rho(\underline{q})$ has to be averaged over all the molecular orientation [22]. However spatial alignment of the target molecules in order to obtain better structure information would be valuable. In Figure 11 an example of the information obtainable by aligment is given for the orbitals of NO molecule previously mentioned. It comes out clearly that, although space-averaged data already allow to distinguish betwen the non bonding and antibonding components, momentum density maps of the different orbitals are strongly different, so that the bonding character can be more easily evidentiated in the aligned molecule. Partial alignment of NO molecule in the gaseous phase has already been achieved [23], so that this does not appear an impossible task.

For several molecules in the gas phase (e,2e) energy spectra and space averaged momentum densities have been measured in various

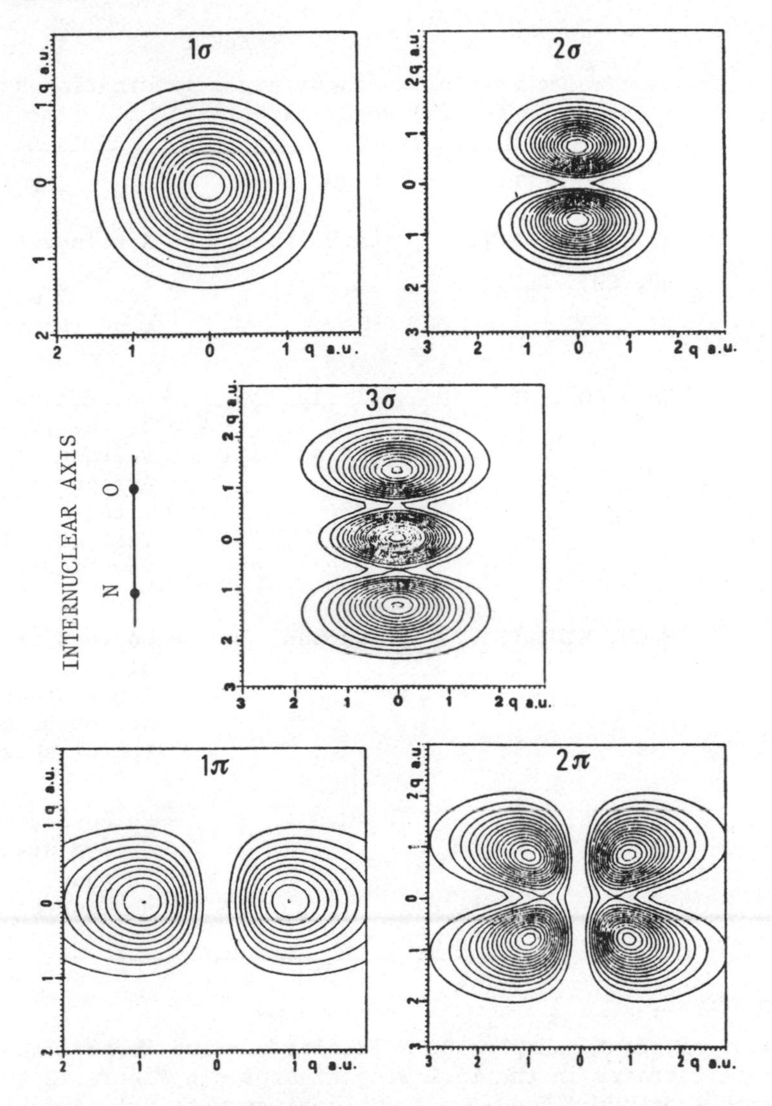

Figure 11. NO momentum density maps for valence shell orbitals of the space aligned molecule as calculated from Brion et al. basis set. The innermost is referred respectively to $\rho(\underline{q}) = .262\ a_o^3$ for the 1σ ,$\rho\ (\underline{q}) = .197\ a_o^3$ for the 2σ, $\rho(\underline{q}) = .114\ a_o^3$ for the 3σ, $\rho(\underline{q}) = .108\ a_o^3$ for the 1π, $\rho(\underline{q}) = .036\ a_o^3$ for the 2π. Continues curves join $\rho(\underline{q})$ isovalues and are equally spaced.

laboratories. In Table I some interesting molecules studied up to now by this tecnique are reviewed and the information obtained, which are complementary with respect to E.S.C.A. results, are

TABLE 1

Some molecules for which original information have been obtained by (e, 2e) spectroscopy

MOLECULE	GROUP	INFORMATION
H_2,O_2, N_2, HCl, HF, CH_4, C_2H_4	Australia	- Assignment of satellites - Test of molecular wave functions
NH_3, CO_2, SF_6, N_2O, CH_3F, CF_4	Italy	- Correct assignment of the innermost valence states - Assignment of satellites - Test of molecular wave functions
H_2CO, H_2O, PH_3	Canada	- Correct of assignment of the outermost valence states - Non bounding character of some orbitals
C_3H_6, C_2H_4	U.S.A.	- Evidences of unpredicted distortion in shape - Assignment of satellites

summarized.

Few significant examples of molecules studied in our laboratory are illustrated in the following figures. In Figure 12 the N_2O energy spectrum taken in non coplanar symmetrical condition is compared with a 2ph-TDA calculation [24]. It is shown that such an accurate model is able to predict features observed in the spectrum. In fact the calculation successfully reproduces the split of the pole strength among many lines. Significant in this respect is the peak near ε_λ = 20 eV where two distinct poles $3\sigma^{-1}$ and $1\pi^{-1}$ separated by 0.4 eV are predicted. Their convolution in the q-profile can be clearly recognized (Fig. 13). On the basis of the q-distribution the satellite at 40.6 eV must be assigned to the ejection of the 1σ electron. In Figure 13 cross sections of all the valence orbitals of N_2O are compared with values calculated with different basis sets [25] (S.T.O. and G.T.O.); these

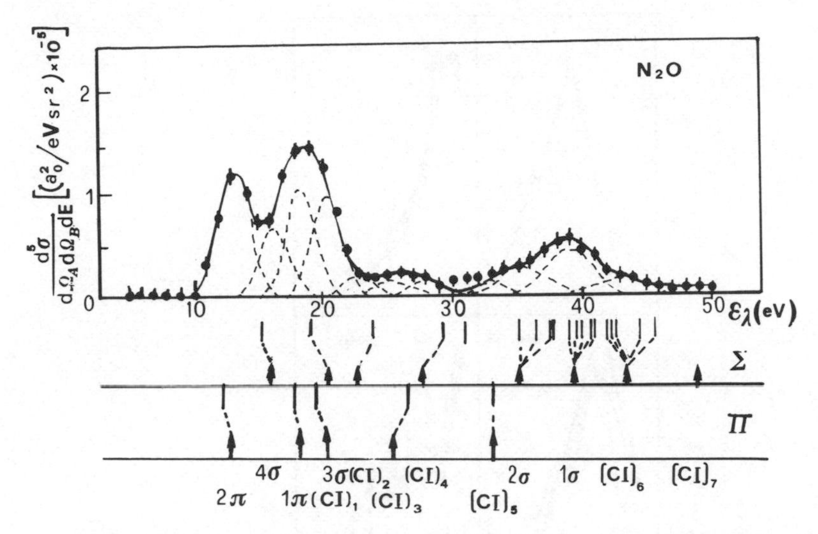

Figure 12. (e,2e) cross section of N_2O valence shell measured at E_0 = 2600 eV. The energy spectrum is taken at ϑ = 44.5°, Φ = 4° as a function of the energy balance ε_λ. Dashed curves are deconvoluted peaks folded into the gaussian resolution of the apparatus (ΔE_{FWHM} = 2.6 eV). Full curve is the best fit of data. Experimental ε_λ for the valence orbitals are marked by arrows. Calculated (2ph-TDA) ε_λ values are reported under the abscissa for Σ and Π symmetry states respectively..

basis sets for N_2O gave rather similar results.

The sensitivity of the electron momentum distribution measurements to distinguish among different basis set is shown in the case of SF_6 molecule [26] for the $2a_{1g}$ orbital (Fig. 14) and for the 3σ orbital of NO molecule (Fig. 15). In both the cases the simplest basis set, which does not give a good agreement for the total energy, appears to reproduce in a better way the contribution of each atomic orbital to the molecular wave function. This is not surprising because till now the only criterion which is followed in writing a good basis set through a self consistent procedure is the one of reaching, as fast as possible, the total energy minimum of the system. According to this idea basis sets are opened to include larger number of atomic orbitals (including also d and f wave functions) even when their contribution to the effective chemical bond should be very unlikely.

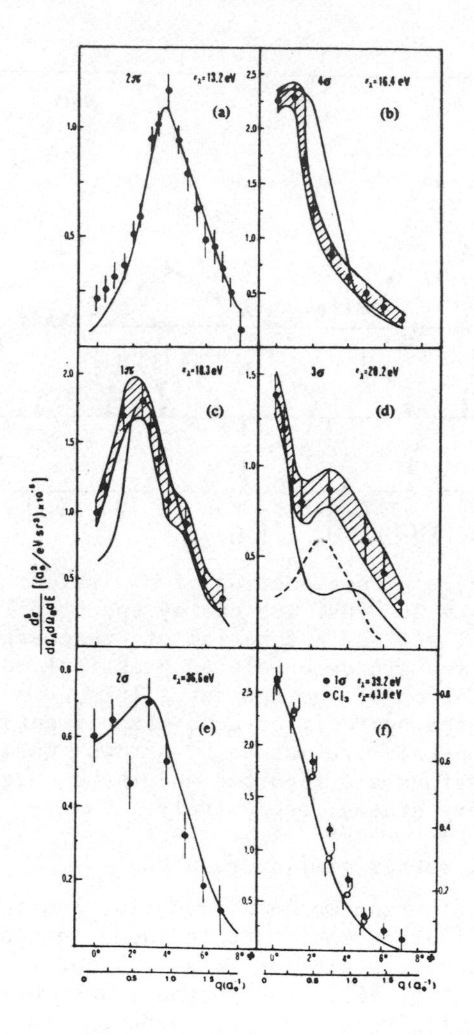

Figure 13. Absolute (e,2e) cross section of N_2O valence shell taken at E_o = 2600 eV. Angular distributions for the most intense peaks observed in the energy spectrum. A gray band envelops data obtained by a deconvolution procedure. Full curve is the cross section calculated for each orbital from single particle wavefunctions, multiplied respectively x.80 2π, x.60 1π, x.40 3σ, x.65 2σ, x.45 1σ to best fit data. Dashed line under data taken at $\varepsilon_\lambda \sim 20$ eV where $(CI)_1$ is convoluted with 3σ orbital is the calculated cross section of the 1π orbital x.20. $(CI)_6$ data (○) are referred to the right scale and compared with 1σ ●.

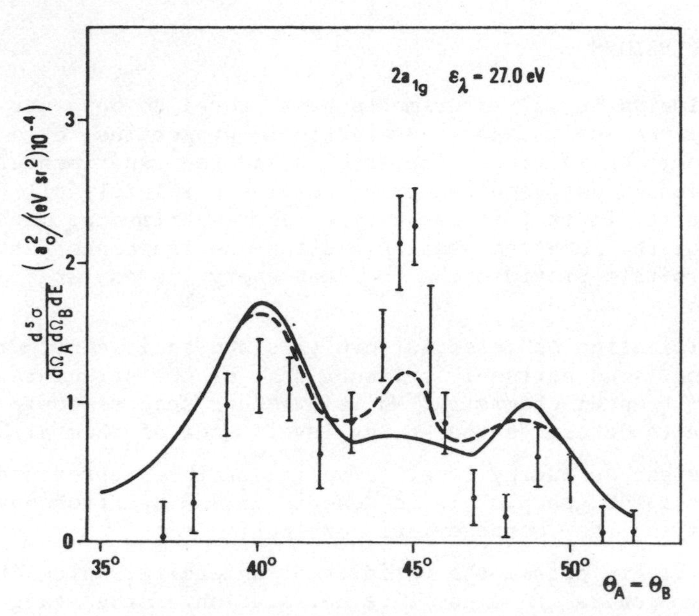

Figure 14. Absolute (e,2e) cross section for the SF_6 $2a_{1g}$ orbital in symmetric coplanar conditions. Filled circles experimental points. Comparison is made with cross sections calculated in P.W.I.A. by using $2a_{1g}$ wave function of Gianturco et al. with sulphur d-orbital excluded (- - -) and included (————)

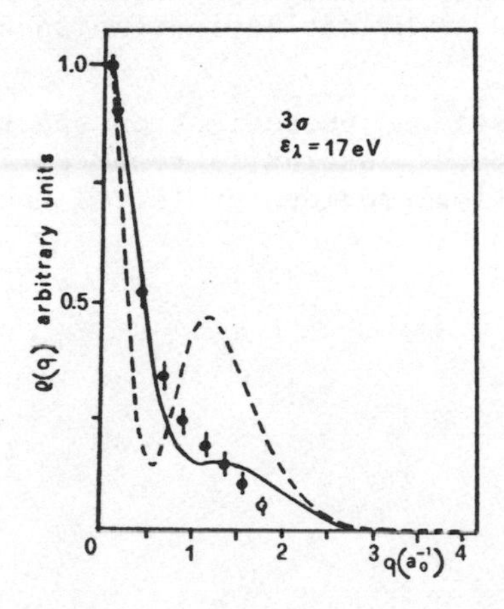

Figure 15. Angular distribution measured for the 3σ orbital of NO. Calculation from Kouba et al.(——) and Brion et al. (---) basis set are reported normalized to experimental data at q = 0

6. CONCLUSIONS

In conclusion (e,2e) experiments have proved to be a very useful tool for the analysis of the electronic properties, even when rather complex molecules are studied and the experimental energy resolution is not good enough to resolve completely all the measured states. In fact it is an excellent experimental method to determine the momentum profile and the spectroscopic factor of valence orbitals provided the incident energy is as large as 1500 eV.

This information on molecular orbitals and their configuration interaction is an extremely critical test of the structure calculations of quantum chemistry. While by other spectroscopies it is possible to determine the energy levels none of them allows to:

i) Assigne an energy level to an irreducible representation of the point group. This is done in (e,2e) spectroscopy by identifying its characteristic orbital;

ii) Verify in detail the validity of an orbital calculation, or more generally a complete calculation of the target-ion overlap. This is done in (e,2e) spectroscopy by comparing the details of experimental and theoretical profiles $\rho(\underline{q})$;

iii) Determine spectroscopic factors which are a critical test of configuration interaction. This is done in (e,2e) spectroscopy by comparing intensities for excitation of ion states belonging to the same group representation and, most important verified in all cases studied by satisfying the spectroscopic sum rules.

A final comparison between (e,2e) and photoelectron spectroscopy used as methods for structure determination is reported in Table 2 [19], and the momentum information provided by (e,2e) tecnique is evidenced.

TABLE 2

COMPARISON BETWEEN (,2) AND P.E.S. USED TO EXCTRACT MOLECULAR STRUCTURE INFORMATION

	(e, 2e)	PHOTOELECTRON SPECTROSCOPY
BINDING ENERGIES ε_λ	YES	YES
RELATIVE SENSITIVITY	VALENCE > CORE	VALENCE ONLY (U.P.S.) CORE > VALENCE (X.P.S.)
ENERGY RESOLUTION	~1 eV	~.001 eV (U.P.S.) ~. 5 eV (X.P.S.)
ASSIGNMENT OF STATES	UNAMBIGUOS FOR EACH ORBITAL	COMPARE ε_λ AND FROM ANGULARLY RESOLVED SPECTRA
INTERPRETATION OF SATELLITE STRUCTURES AND C.I. EFFECTS	UNAMBIGUOUS	EXAMINE STRUCTURES AT DIFFE- RENT ENERGIES AND COMPARE WITH C.I. CALCULATION
MOMENTUM INFORMATION PROVIDED	$\rho(\underline{q})$ SEPARATELY FOR EACH ELECTRONIC STATE	INDIRECT
TYPICAL MOMENTUM RESOLUTION (a. u.)	~.08 (DEPEND ON INCI- DENT ENERGY AND ANGULAR RESOLUTION)	-
GAS SAMPLES	NO RESTRICTION	NO RESTRICTION
SOLIDS	CORE LEVELS PLASMONS BY THIN FILMS	BASICALLY A SURFACE TECHNIQUE
BACKGROUND	VERY SMALL	SOME PROBLEMS IN DETECTION OF SATELLITE STRUCTURE
COUNTING EFFICIENCY	TYPICAL COUNTING PERIODS OF ONE DAY PER ANGULAR CORRELATION	RELATIVELY HIGH

REFERENCES

★ Univ. di Roma, Istituto di Chimica, Generale e Inorganica (also)

1. H. Ehrahrdt, M. Schulz, T. Tekaat, and K. Willmann, Phys.Rev. Lett. 22, 89 (1969).
2. U. Amaldi, A. Egidi, R. Marconero, and G. Pizzella, Rev. Sci. Instr. 40, 1001 (1969).
3. R. Camilloni, A. Giardini Guidoni, R. Tiribelli and G. Stefani Phys. Rev. Letts. 30, 475 (1973) and references therein.
4. E. Weigold, S.T. Hood, and P.J.O. Teubner, Phys. Rev.Letts. 30, 475 (1973).
5. S.T. Hood, A. Hamnett and C.E. Brion, J. Electron Spectr. 11, 205 (1977).
6. J.H. Moore, M.A. Coplan, T.L. Skillman Jr. and E.D. Brooks III Rev. Sci. Instr. 49, 463 (1978).
7. J.F. Williams, J. Phys. B11,2015 (1978)
8. H. Ehrhardt, K.H. Hesselbacher, K. Jung, M. Shulz and K. Will mann, J. Phys. B5, 2107 (1972).
   E.C. Beaty, K.H. Hesselbacher, S.P. Hong and J.H. Moore, J.Phys. B10, 611 (1977)
9. C.E. Brion and M. Van der Wiel, Radiat. Res. 64, 37 (1975); J.Phys. B9, 945 (1976). ESCA Applied to Free Molecules, North Holland, Amsterdam 1969, p. 74
10. I.E. McCarthy and E. Weigold, Phys. Reports, 27C, 275 (1976)
11. J.J. Smith, K.H. Winters and B.H. Bransden, J. Phys. B12, 1723 (1979), S. Geltman, J. Phys. B7, 1994 (1974) private communication; S. Geltman and M.B. Hidalgo, J. Phys. B7, 831 (1974); B.H. Bransden, J.J. Smith and K.H. Winters, J.Phys. B11, 3095 (1978); B.H. Bransden and J.P. Coleman, J. Phys. B5, 537 (1972) E. Weigold, XIth I.C.P.E.A.C., Kyoto 1979, Invited Paper
12. A. Giardini-Guidoni, R. Camilloni and G. Stefani, in "Coherence and Correlation in Atomic Collisions", H. Kleinpoppen and J.F. Williams (Eds), Plenum Press, New York, 1980; E. Weigold and I.E. McCarthy, Adv. At.Mol.Phys; 14, 127 (1978)
13. R. Camilloni, A. Giardini-Guidoni, I.E. McCarthy and G. Stefani, Phys. Rev. 17A, 1634 (1978); R. Camilloni, A. Giardini--Guidoni, I.E. McCarthy and G. Stefani, J.Phys. B13, 397 (1980)
14. R. Camilloni, G. Stefani, R. Fantoni, A. Giardini-Guidoni, J. Electr. Spectr. 17, 209 (1979)
15. B.T. Pickup and O. Goscinski, Molec. Phys. 26, 1013 (1973)
    O. Goscinski and P. Linder, J. Math. Phys. 11, 1313 (1970)
    G.R.J. Williams, J. Electr. Spectr. 15, 247 (1979)
16. L.S. Cederbaum, Theor. Chim. Acta, 31, 239 (1973)
    W. von Niessen, W.P.Kraemer and G.H.F. Diercksen, Chem. Phys. 41, 113 (1979)
17. E. Weigold, C. Noble, S.T. Hood and I. Fuss, J. Electr. Spectr. 15, 253 (1979)
    A.J. Dixon, I.E. McCarthy, C. Noble and E. Weigold, Phys. Rev. 17A, 597 (1978)

18. J. Fuss, I.E. McCarthy, C. Noble and E. Weigold, Phys. Rev. 17A, 604 (1978)
B. Van Wingerden, J.T. Kimman, M. Van Tilburg, E. Weigold, C.J. Joachain, B. Piraux and F.J. De Heer, J. Phys. B 12, L627 (1979)
B. Van Wingerden, J.T. Kimman, M. Van Tilburg and F.J. De Heer (in press)
19. G. Stefani, R. Camilloni and A. Giardini-Guidoni, J. Phys. B12, 2583 (1979)
A. Giardini-Guidoni, R. Fantoni, R. Tiribelli, R. Marconero, R. Camilloni and G. Stefani, Phys. Lett. 77A, 19 (1980)
20. E. Clementi and C. Roetti, Atomic data and Nuclear Data Tables, 14, 177 (1974)
R. Camilloni, A. Giardini-Guidoni, G. Missoni, G. Stefani, R. Tiribelli and D. Vinciguerra, "Momentum Wave Functions 1976" A.I?P. Conf. Proc. 35, 205 (1977) ed. D.W. Devins
21. A. Giardini-Guidoni, G. Missoni, R. Camilloni and G. Stefani in "Electron and Photon Interaction with Atoms" 149 H. Kleinpoppen and M.R.C. McDowell (Eds), Plenum Press N.Y. 1977
E. Weigold, Proc. 2nd Int. Conf. on Inner Shell Ionization Phenom. Freiburg 1976, pg. 367, W.Mehlhorn and R. Brenn Eds
A. Giardini-Guidoni, R. Fantoni, R. Marconero, R. Camilloni and G. Stefani, XIth I.C.P.E.A.C. book of Abstracts, 212 (1979)
22. R. Fantoni, A. Giardini-Guidoni, R. Tiribelli, R. Camilloni, and G. Stefani, Chem. Phys. Lett. 71, 335 (1980)
23. H. Twiss, S. Stolte and J. Reuss, Chem. Phys. 43, 351 (1979)
H. Twiss, S. Steolte and J. Reuss (in press)
24. W. Domcke, L.S. Cederbaum, J. Smirner, W. Von Niessen, C.E. Brion and K.H. Tan, Chem. Phys. 40, 171 (1979)
25. A.D. McLean and M. Yoshimine, IBM J. Res. Dev. pg. 1 (1967)
L.C. Snyder and H. Basch "Molecular Wave Functions and Properties" Wiley, New York (1972)
26. A. Giardini-Guidoni, R. Fantoni, R. Tiribelli, D. Vinciguerra R. Cammilloni and G. Stefani, J. Chem. Phys. 71, 3182 (1979) and references there in
27. J.E. Kouba and Y. Ohrn, Int. J. of Quantum Chem. 5, 539 (1971); H. Brion, C. Moser, M. Yamazaki, J. Chen. Phys. 30, 673 (1959)
28. C.E. Brion, E. Weigold, to be published

ION SCATTERING SPECTROSCOPY AND PARTICLE INDUCED X-RAY EMISSION.

Orazio Puglisi

Istituto Dipartimentale di Chimica e Chimica Industriale
Viale A.Doria 6-95125 Catania, Italy.

ABSTRACT.

Physical processes occurring during ion bombardment of solids are suitable for the analysis of thin layers providing information on the atomic masses present on the bombarded surface.This lecture deals with the wide angle scattering of light particles at MeV energy (R.B.S.) or at keV energy (I.S.S.). The obtained information comes from the first few microns in R.B.S. technique, while the sampling depth in I.S.S. is limited to the first one or two monolayers. The last topic of the lecture is PIXE technique which is based on the collection of beam generated photons and gives qualitative and quantitative informations about the sample constituents.

When an ion beam impinges on a solid surface many physical processes occur. Some of these processes are suitable for the analysis of thin layers providing informations on the atomic masses present on the bombarded surface. One of these processes is the wide angle scattering of the bombarding ions. The so called Rutheford Backscattering (RBS), already known from many years to nuclear physicists (I), has been only recently applied to other fields. The backscattering technique consists in analysing the energy spectrum of the particles elastically scattered at large angles from target atoms. The experimental apparatus required is sckecthed in Fig.I . The particle beam is usually produced by a Van de Graaff machine at an energy typically in the MeV region; light and medium particles like protons, helium, carbon and oxygen are used as projectiles. The beam is collimated by two or more slits a few meters apart and the beam spot is about Imm diameter; its intensity is of the order IO^{-9} Amps.

P. Day (ed.), Emission and Scattering Techniques, 319–333.

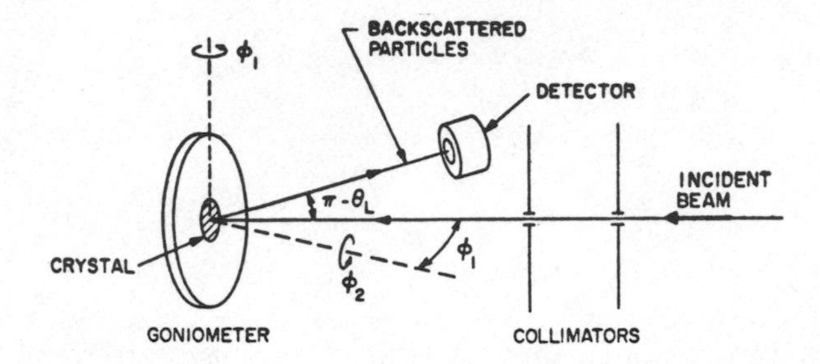

Fig.I-Experimental arrangement used for backscattering measurement.

The energetic particles scattered by target atoms at large angle (typical angle I60°) are energy detected by a solid state surface barrier detector. The amount of charge collected by the two electrodes is proportional to the energy lost by the particle in traversing the sensitive region of the detector; the signal is then handled by usual electronics and stored typically in a multichannel analyser which collects the energy spectrum of backscattered particles.

MASS AND DEPTH PERCEPTION.

Backscattering technique provides information on the depth distribution of elements present in a sample. To achieve this, the technique must give different signals for different atomic masses and different signals for the same atoms distributed at different

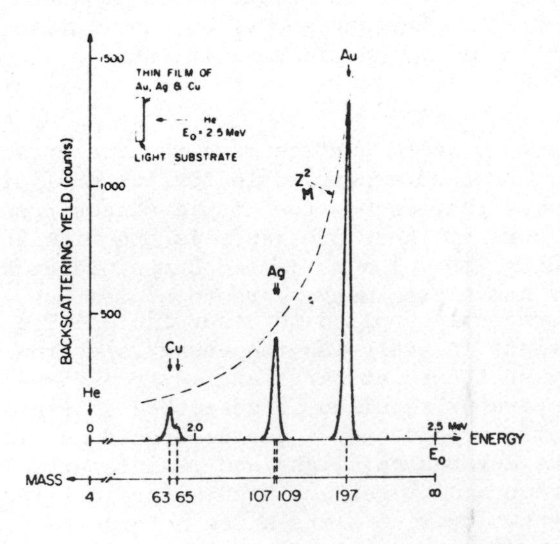

Fig.2-Backscattering spectrum of a very thin film (see text).

depths inside the sample. The first characteristic, namely mass perception is illustrated by the analysis of a carbon substrate onto which small and equiatomic amounts of different atoms (Cu,Ag and Au) have been evaporated (Fig.2). The energy spectrum, namely the number of helium particles versus detected energy, of 2.5 MeV helium ions scattered through 164° from the sample is shown. The spectrum shows three peaks corresponding to elastic recoil of the impinging helium from gold, silver and copper atoms respectively. The first observation is that we are able to distinguish the atomic masses in the backscattered energy scale. This is due to different recoil energies for collision between the projectile and different atomic masses. The recoil energy is given by

$$E_I = K^2 E_0 \tag{I}$$

where E_I is the backscattered energy and E_0 is the incident beam energy; K^2 is given by :

$$K^2 = \left(\frac{M_I \cos\theta + \sqrt{M_2^2 - M_I^2 \sin^2\theta}}{M_I + M_2} \right)^2 \tag{2}$$

where M is the mass (label I for projectile and label 2 for target) and θ is the scattering angle in the laboratory system. The relation is obtained simply assuming elastic collisions, like that between two billiard balls, and considering conservation of kinetic energy and momentum in the collision event. If the projectile mass is much lower than that of target atoms, the equation (2) becomes:

$$K^2 \simeq \left(\frac{M_I \cos\theta + M_2}{M_I + M_2} \right)^2 \tag{3}$$

When the projectile is reflected back exactly through 180°, then:

$$K^2 = \left(\frac{M_2 - M_I}{M_2 + M_I} \right)^2 \tag{4}$$

Then by a measurement of recoil energy we are able to identify the mass of the atoms present at the surface of the sample.
Other two features can be noted in the spectrum. First,the mass resolution is not extremely good. In particular we are able to distinguish between the two stable isotopes of copper but not between the two isotopes of silver. On going towards higher and higher masses the mass resolution becomes very bad.

The second feature is that the areas under the peaks, although coming from the same number of atoms, differ considerably one from the other. This is related to dependence of the differential Rutheford scattering cross section, in a given direction, on the atomic number of the atoms. In fact note the dependence of the peaks area on the square of atomic number. The differential Rutheford scattering cross section in a given direction is given by the following equation:

$$\frac{d\sigma}{d\Omega} = 1.3\ 10^{-27}\left(\frac{z_1 \cdot z_2}{E_0}\right)^2 \cdot \frac{1}{\sin^4 \theta/2} \quad (cm^2/sr), \text{with } E_0 \text{ in MeV.} \quad (5)$$

Note the dependence on the square of atomic numbers of both target and projectile atomic numbers. It is to note also the strong dependence on the scattering angle, so that the Rutheford scattering cross section is strongly forward peaking and only a small number of the incident particles is scattered at large angles from target atoms. The area under the peaks is proportional to the total number of atoms per unit area and to scattering cross section:

$$A = Q \cdot \Omega \cdot \frac{d\sigma}{d\Omega} \cdot N \quad (6)$$

where Q is the incident ion number per unit area, Ω the solid angle subtended by the detector and N the number of target atoms per unit area. Therefore backscattering technique gives qualitative and quantitative informations on the analysed sample. In conclusion we see that, by increasing the atomic number of the target atoms the sensitivity increases but the mass resolution decreases. Backscattering is then an useful technique to detect trace of heavy elements on a light substrate with a sensitivity limit of about 10^{12} atoms/cm^2.

In addition to the mechanism of elastic recoil, which forms the basis for mass analysis, there is a second mechanism of energy loss in a target. This effect is responsible of the major property of the technique, namely depth perception. Beam particles, in traversing the target lose energy due to electronic and nuclear interactions with the atoms of the sample. If they are backscattered from atoms located at a certain distance from the target surface, their detected energy will be lower than that corresponding to bacscattering from surface atoms. So that for thick samples we consider three energy loss events:

a) energy lost by projectile in its inwards path;
b) energy lost by projectile in the collision event;
c) energy lost by projectile in its outwards path.

The two additional energy loss events a) and c) result in the broadening of the backscattering peaks because, in addition to He^+ particles backscattered at sample surface, we detect also He^+ particles backscattered by bulk and rear surface. By increasing the sample's thickness, the FWHM of the spectrum increases. This energy broadening is correlated to the thickness of the sample by the simple relation:

$$E = S.\ t \qquad (7)$$

where t is the thickness and S is the so called backscattering energy loss factor.

BACKSCATTERING FACTOR |S|

$$\Delta E\ (KeV) = [S]\ \frac{KeV}{10^3 \mathring{A}}\ .\ \Delta x\ (10^3 \mathring{A})$$

Atom	No.	$1MeVHe^+$	$2MeVHe^+$	Atom	No.	$1MeVHe^+$	$2MeVHe^+$
B	5	-	50	Ni	28	126.2	118.8
C	6	-	48.3	Cu	29	110	108
N	7	-	20.6	Ge	32	66.5	58.5
O	8	-	26.7	Nb	41	131.6	106
Ne	11	16.8	16.5	Mo	42	144.8	118
Mg	12	37.3	34.0	Pd	46	143	120.2
Al	13	51	46.2	Ag	47	120.8	104.8
Si	14	54.5	46.0	In	49	83.1	75.4
S	16	42.8	38.0	Sn	50	82.8	71.5
Cl	17	44.8	35.6	Sb	51	74.9	65.2
K	19	20	17.1	Ta	73	130.5	114.2
Ca	20	32.7	27.4	W	74	145	129.8
Ti	22	93.5	78.9	Pt	78	162	149
V	23	113.1	96.6	Au	79	148	133.8
Cr	24	125.8	110.5	Pb	82	90.8	80.3
Mn	25	120.8	106.2	Bi	83	76.2	66.6
Fe	26	133.1	115.0	U	92	157	132
Co	27	131.1	117.8				

The S factor is depending on the stopping power of the material and is almost constant in the energy range usually adopted in the experiments. S values are given in the table at two typical energies. The linear relation between thickness and energy broadening holds within 5% for thickness of gold up to 5000 A°.
It is then possible to measure the depth of scattering event if is known the stopping power (dE/dx) of the investigated layer. The energy loss per unit lenght (dE/dx) is a convenient method of expressing the stopping power of a medium. However a more useful concept is that of energy loss based on the number of atoms/cm^2 that are traversed by the ion beam. Thus,if ρ is the density of target, M its atomic mass, then the stopping cross section ε is:

$$\varepsilon = M/N_o\ .\ (dE/dx)\ .\ 1/\rho \qquad (8)$$

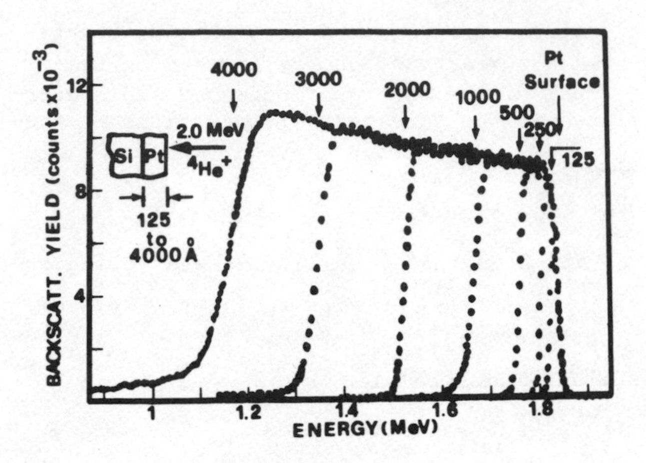

Fig.3-Backscattering of Pt films with different thickness.

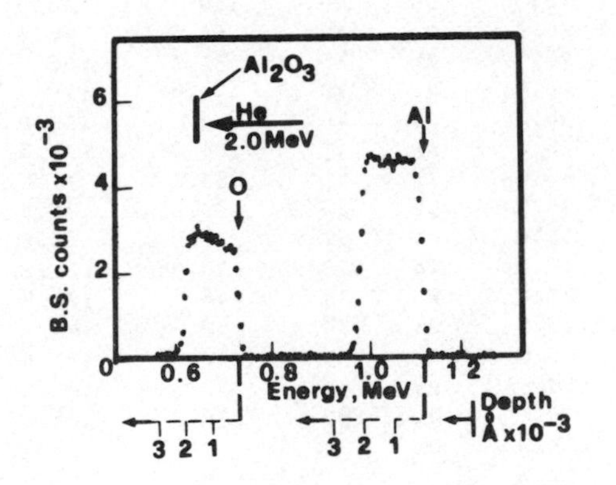

Fig.4-Backscattering spectrum of a self supporting aluminum oxide thin film.Each element generates its depth scale.

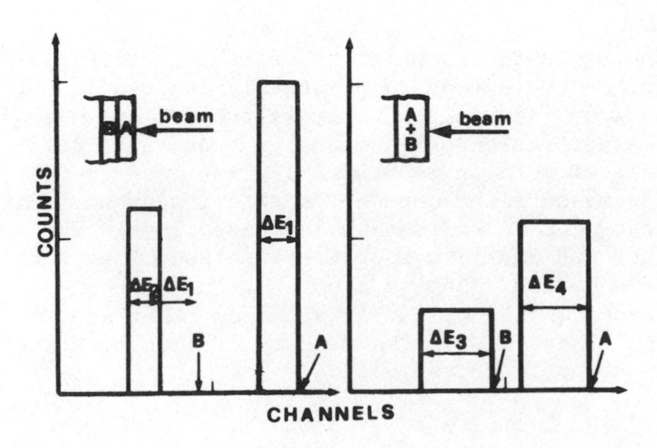

Fig.5-Backscattering spectra (schematic) of a thin film couple A-B ($M_A > M_B$) as evaporated (left) and after complete mixing (right).

where No is Avogadro's number. Stopping cross sections expressed in this way (for compilation of experimental data see Ref.2) are linearly combined for a compound material:

$$\varepsilon_{Al_2O_3} = 2\,\varepsilon_{Al} + 3\,\varepsilon_{O} \tag{9}$$

The linear relation between thickness and energy broadening (eq.7) is illustrated in Fig.3, which shows backscattering spectra of Pt films with different thickness. The signals at higher energy are coming from atoms located at surface of sample, when the signals at lower energy are generated by He particles backscattered from atoms lying on the rear surface of the specimen.
Fig. 4 shows a spectrum of a self supporting aluminum oxide thin film. The arrows indicate the energies of Helium particles backscattered from Al and O atoms, respectively, present at the surface; the signals at lower energy come from atoms present in the bulk. From the FWHM one can also compute the depth scale shown at the bottom of figure.
Fig. 5 shows a schematic diagram of a thin film couple analysed by backscattering. On the left is shown the spectrum of the mixture as evaporated. Note that the B signals are shifted towards lower energies than that corresponding to those of atoms lying at the surface. From this shift one can also compute the depth of A film which covers the B film. Obviously this shift is equal to the FWHM of the A signal . On the right is shown the spectrum after complete mixing of A and B films. Note that the FWHM are in this case larger than that before mixing; note also that now we detect B atoms present at target surface.
The analysis so far considered can be extended to determine depth profile.

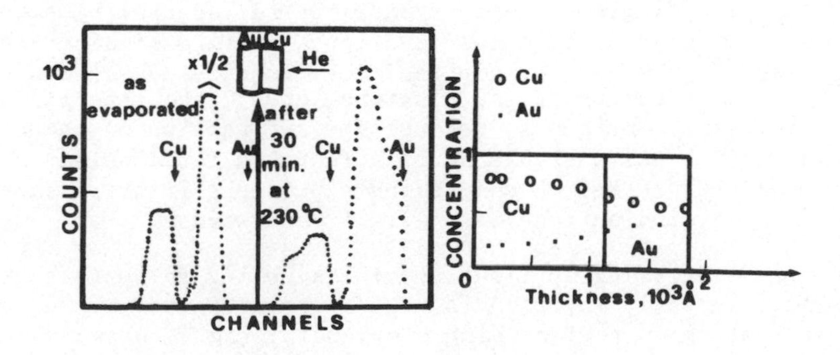

Fig.6- Backscattering analysis of a thin film couple of Cu (1200 Å) and Au (800 Å) ; see text.

Fig. 6 shows the spectrum of a thin film of copper on gold, as evaporated, on the left, and, on the middle, the spectrum of the same films annealed at a temperature which gives a not uniform concentration profile. Note that, because of interdiffusion, a part of the gold peak is shifted towards higher energy and that a part of copper peak is shifted toward lower energy (both Au and Cu atoms are now detected at surface). The concentration profile computed from this analysis is also shown (right part of Fig. 6).

The introductory aspect of this lecture does not allow to illustrate the analytical procedure employed for this calculation, so that only this qualitative approach will be given (details of the calculation procedure are reported in ref. 3).

SENSITIVITY TO CRYSTALLINE STRUCTURE.

In addition to mass and depth perception, backscattering technique is sensitive to the crystalline structure of the target. The more important characteristic of this effect is related to the decrease of two or more orders of magnitude of the backscattering yield, for beam incidence parallel to an atomic row of low Miller index (Fig.7, left). The decrease of yield occurs because the projectile particles that are steered along the channels (rows or planes of a single crystal), do not approach the crystal atoms closely enough to undergo wide angle elastic processes; the minimum in the backscattering yield, by tilting the sample, is observed for beam incidence parallel to atomic row of low Miller index (right part of Fig. 7). This is the so called "Channeling phenomenon".

Displaced atoms in a single crystal increase the Backscattering yield in the channeling direction. If randomly displaced atoms are located in a single perfect crystal the (de)channeling technique provides with accuracy number and depth profile of disordered atoms (Fig. 8). In the case of not randomly distributed defects (dislocation loops, stacking fault, clusters, etc.), the interpretation of the spectra is ambiguous, because the interaction of channeled projectiles with all these kinds of defects is not completely understood at the present. A channeling experiment in these cases will provide only indications of the lattice disorder.

An interesting application of channeling to surface studies has been recently reported (4) and is related to the study of surface relaxation in perfect single crystals. Fig. 9 shows the basic concepts of these experiments. When an ion beam at MeV energy penetrates into a crystal along a crystallographic direction of low Miller index, a "surface peak" appears in the backscattering spectrum. The area of surface peak as compared to random yield, gives a quantitative measure of the number of atoms per row "seen" by beam particles.

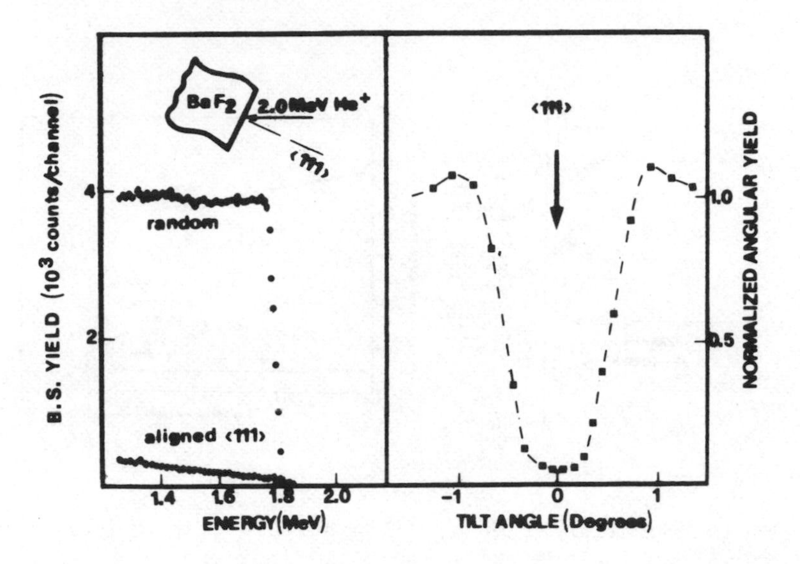

Fig.7-Backscattering spectra of BaF_2 single crystal in random and aligned <111> direction; on the right is shown the normalised backscattering yield versus tilt angle around <111> direction.

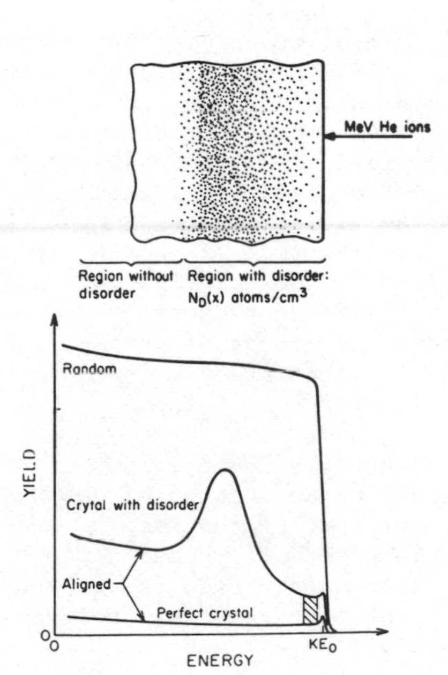

Fig.8-Disorder analysis. Backscattering spectra of a single crystal with displaced atoms randomly distributed.

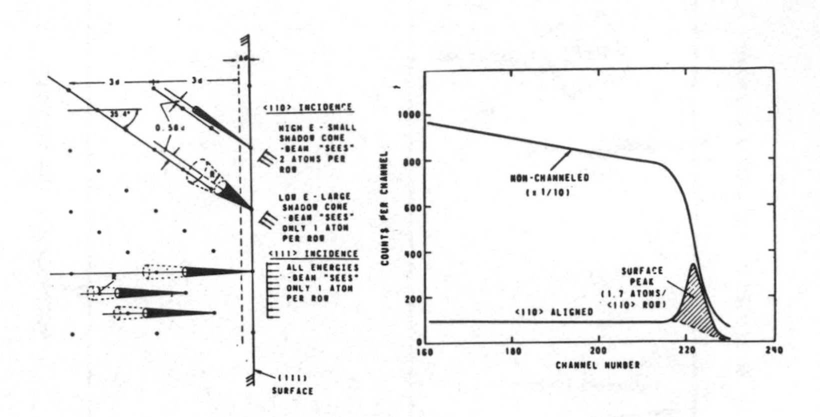

Fig. 9-Atomic configuration of <111> surface of a single crystal (Pt) and channeling analysis. For <111> incidence the ion beam "sees" always 1 atom/row. For <110> incidence the beam "sees" 1 atom/row at low energy and 2 atoms/row at high energy because the "shadow cone" depends on the ion energy. The area of surface peak reflects the number of atoms per row.

At normal incidence (<111> in Fig.9) the second atom of each row is perfectly screened because it is placed within the shadow cone of the first one; a surface relaxation does not produce any effect in this case. Indeed the beam "sees" only 1 atom/row.
At not normal incidence (<110> in Fig.9) the shadow cones of lower atomic rows are shifted and the beam sees 2 atoms/row.
By decreasing the energy beam the radius of shadow cone increases (deflection in a coulombic field) so that the surface peak changes from 2 atoms/row (high energy) to 1 atom/row (low energy). Therefore, a measure of surface peak at different energies and at different incidences allow to measure the amount of surface relaxation (Δd) in addition to the sign (i.e. outwards or inwards relaxation; see ref. 4).

In conclusion Rutheford backscattering provides information on the distribution of atomic masses as a function of depth, up to thickness of few microns, without sputter the sample, and on the crystalline disorder of a single crystal. Typical depth resolution is of the order of 100 A°. This technique requires in general only a poor vacuum (except surface peak measurements), because the information comes from regions below the surface.

By contrast, the low energy backscattering technique, the so called Ion Scattering Spectroscopy or Ion Surface Scattering (5,6) requires high vacuum because the information comes from the first

one or two monolayers. I.S.S. makes use of noble gas ion beam in the keV energy region. This energy is much lower than that used in RBS, so that the sampling depth is limited to the first atomic layer. Each interaction between a projectile and a target atom can be described by a binary collision. At high energy (RBS), this interaction can be described by the Coulomb potential, while at lower energy the electron screening has to be taken into account.
I.S.S. spectrum is represented as counts of backscattered beam particles versus detected energy, and the mass perception is based on the same arguments of RBS (see eq. 1-2-3-4).
The problems encountered in I.S.S. are similar to those of SIMS (Secondary Ion Mass Spectrometry). In particular, the main limitation depends on the charged fraction of the backscattered particles and to its variation with hydrocarbons contamination and, generally speaking, surface composition. Indeeed,at this low energy, the projectile can be scattered as neutral or as an ion. As in ISS we detect only ionised backscattered beam particles, then it is important to know the neutralisation yield of the surface. Therefore quantitative evaluations from ISS spectra require reference standards of very similar composition and structure for calibration.

In conclusion the main feature of ISS is to provide information on the atomic masses present at the first one or two monolayers.

PIXE

Particle Induced X ray Emission has been extensively used in these last few years (7). Appropriate ion beam from an accelerator bombards the sample situated in an evacuated scattering chamber. This chamber can be the same of RBS and the sample can be analysed by both PIXE and RBS in order to clarify the backscattering spectrum which sometimes does not have a good mass resolution.
The coulombic interaction between projectile and atomic electrons results also in the ionization of inner shell electrons. The vacancies so created can be filled by other electrons, and the process gives rise to the emission of photons. The energies of the emitted X photons are characteristic of the different elements, and then from a measurement of these energies one can ascertain the presence of elements in the target. The development of solid state detectors as Si(Li) or Ge(Li) or intrinsic Germanium has allowed a fast growth of this technique. The simultaneous detection of several and different X rays with a satisfactory energy resolution, makes this technique a fast multielemental analytical tool. X rays from elements with $z > 12$ can be detected.
Several characteristic lines are associated to each element as illustrated in Fig. 10a in which the energy spectrum obtained by 1.5 MeV protons bombardment of a thin target of Cu is shown. Two lines are characteristic of Cu atoms and they are associated to the transition of electrons from L shell (K_α) and M shell (K_β) to

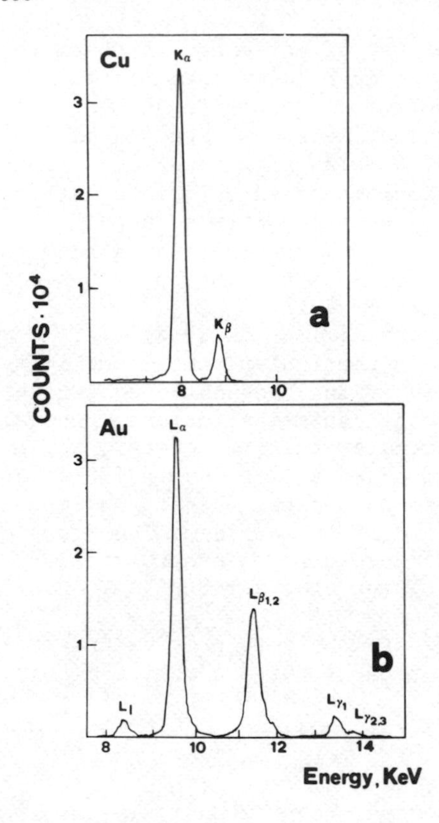

Fig.10-X rays spectra of Cu K_α and K_β lines (a) and Au L lines (b) induced by 1.5 MeV protons bombardment on thin films of Cu and Au.

the empty K shell. The energies and the ratio between K_α and K_β lines intensities are characteristic of the element. Fig. 10b shows the spectrum of a thin film of gold. The gold K lines cannot be detected by Si(Li) detector because of their high energy. Instead L lines appear in number of five and in an intensity ratio characteristic of gold and obviously at characteristic energies. The various elements will be then identified by number of lines, their energy and intensity. With a Si(Li) detector elements with z higher than 12 can be identified through their characteristic lines; this limitation is due to the thin Berillium window which protects the detector from scattered beam particles and absorbs with high efficiency soft X rays. Making use of a not Berillium protected crystal detector (Bragg detector), elements also with atomic number lower than 12 can be detected; it is to note, however, that with Bragg detector the resolution increases at expense of analysis time which is about ten times longer than that of solid state detectors.

The yield (Y) of an X ray line is related to the areal density of the element through the relation:

$$Y = \sigma_x^i . \omega_x . N_x . Q . \Delta\Omega . \varepsilon \tag{8}$$

where σ_x^i is the ionization cross section of the element which genetes the photons, ω_x its fluorescence yield, N_x the number of atoms per unit area, Q the total number of projectile particles, $\Delta\Omega$ the solid angle subtended by the detector and ε is the efficiency of detection system including also filters which depend on the line energy. This formula holds for thin films, namely for sample thickness lower than about 5 microns. For thick films corrections for autoabsorption of emitted radiation and slowing down of projectiles are necessary.

The most important feature of PIXE is the high ionization cross section which is of the order of several barns. In Fig.11 is shown the product of ionization cross section by fluorescence yield (namely X ray cross section) versus atomic number of elements at three different energies for proton bombardment.

An other important feature of PIXE is the very low level of Bremsstrahlung, namely continuous electromagnetic radiation emitted by slowing down charged particles. The most important contributions to Bremsstrahlung are: slowing down of primary projectiles and of the electrons emitted by the bombarded material (like electrons emitted during ionization and Auger electrons). The sum of these contributions constitutes the Bremsstrahlung of a PIXE experiment which is some order of magnitude lower than that of electron induced X ray emission. Both two features, namely very low noise and very high cross section, puts PIXE among the best multielemental microanalytical techniques. The amount of sample required is of the order of microgram (10^{-6} g.), revealing elements present at ppm level, so that the minimum detectable amount is of the order of 10^{-12} g.

In the case of thick samples PIXE gives information on the composition of the first few microns of sample. A better surface sensitivity can be achieved by tilting the sample and then decreasing the sampling depth and by decreasing the beam energy (again decreasing sampling depth) but at expense of sensitivity. In the best conditions, like for instance 300 keV He^+ at about grazing

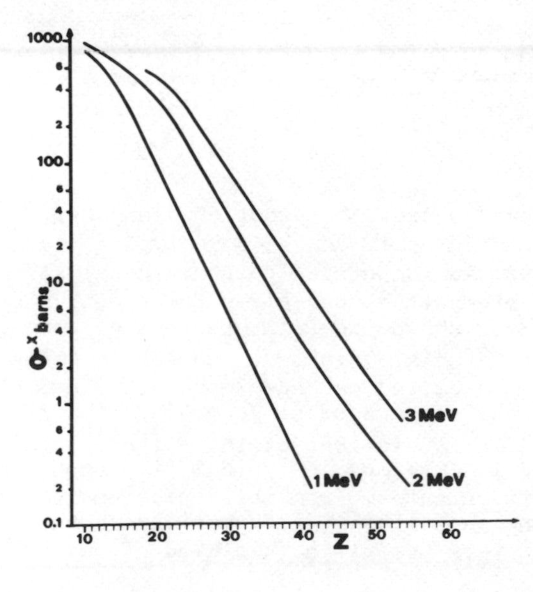

Fig.11-Cross sections for proton induced X ray Emission (average literature (2) data) versus atomic number at three proton energies.

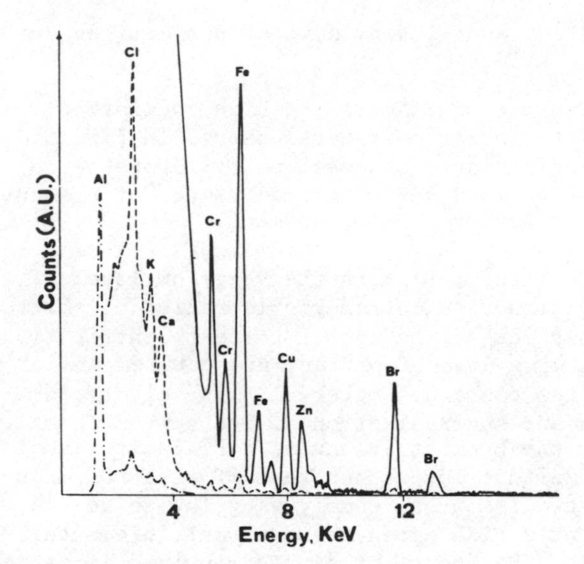

Fig.12-PIXE spectrum of an industrial poisoned catalyst (2.0MeV protons); ---- x 10; ——— x 200.

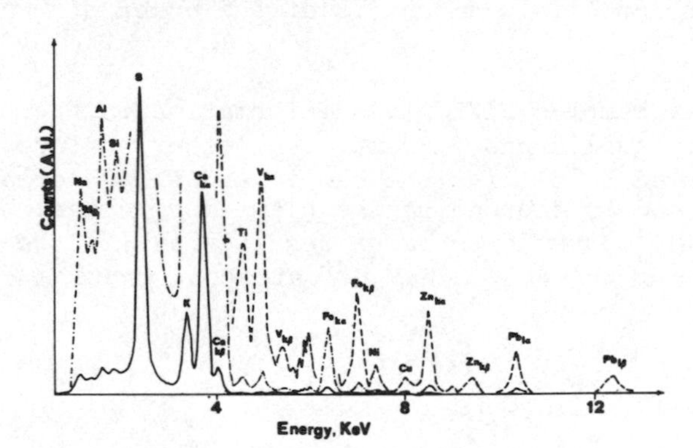

Fig.13-PIXE spectrum of a nuclepore filter with atmospheric particulate (2.0 MeV protons); —·—·— x 10; - - - - x 100.

angle, sampling depth of some hundred A° (about 500) and sufficient sensitivity are achieved. In figures 12 and 13 some typical examples of PIXE application are shown. An industrial poisoned catalyst (note the very low level of Bremsstrahlung) gives the spectrum shown in Fig. 12, while in Fig. 13 is shown the PIXE spectrum of atmospheric particulate collected in Catania. This last sample was collected on nuclepore filter in a so called cascade impactor with 5 stages, in order to split the collected aerosol according to particle size. The spectrum shown in Fig. 13 is that of the filter containing aerosol size between 0.7 and 1.2 microns. A typical time required to obtain this spectrum is two minutes. From this spectrum one can easily obtain qualitative and quantitative informations on the sample constituents, so PIXE is very useful in this kind of research.

REFERENCES

1) Rubin,S., Passel,T.O. and Bailey,L.E. : 1957, Anal. Chem.,29, p. 736. See also: Turkevich, A.L., Franzgrote,E.J. and Patterson, J.H. : 1967, Science, 158, p. 636 (in which is described the famous experiment which allowed the first direct elemental analysis of the moon surface from a Surveyor spacecraft.).

2) Ion Beam Handbook for Material Analysis (Mayer,J.W., Rimini,E., eds.),1977, Acad. Press, N.Y. .

3) Brice,D.K., : 1973, Thin Solids Films, 19, p.121.

4) Davies,J.A., Jackson, D.P., Mitchell, J.B., Norton,P.R., Tapping, R.L. : 1976, Nucl. Inst. Meth. , 132, p.609.

5) Heiland,W. and Taglauer, E. : 1976, Nucl. Inst. Meth., 132,p.535 and references therein.

6) Taglauer, E. and Heiland, W. : 1976, Appl. Phys., 9, p.261 and references therein.

7) See for example: Nucl. Inst. Meth. , Vol 142, which is devoted to PIXE technique and applications.

VALENCE-SHELL PHOTOELECTRON SPECTROSCOPY OF MOLECULES[1,2]

ASSIGNMENT OF RADICAL CATION STATES AND ITS APPLICATION TO ANALYZE AND TO OPTIMIZE GAS REACTIONS IN FLOW SYSTEMS

Hans Bock

Department of Chemistry, Johann Wolfgang Goethe University, D-6000 Frankfurt (M) 50, Niederurseler Hang, West Germany

A neutral molecule in its ground state $\Gamma(M)$ by taking up the ionization energies IE_1 or $IE_{n(>1)}$ will loose an electron and the corresponding radical cation will be generated in its ground state $\Gamma(M^{\cdot\oplus})$ or in its electronically excited states $X_1^j(M^{\cdot\oplus})$. These $M^{\cdot\oplus}$ states are characterized by a typical electron hole, which usually can be rationalized on the basis of simple models. The information obtainable from the vertically produced radical cations in their individual states will be illustrated by examples of small molecules such as $HAH^{\cdot\oplus}$ which adiabatically relax undergoing structural changes.

Photoelectron spectra allow to observe all 'Koopmans' radical cation states and, therefore, the recorded ionization band patterns represent a 'molecular fingerprint'. If PE spectra of various molecules do not overlap completely and/or display recognizable characteristic bands, gas phase reactions can be monitored visually: the PE fingerprints of the reactants give way to those of the products, depending on temperature and/or pressure: with only millimole quantities of compounds a reaction at .1 torr can be monitored between 300 K and 1500 K in 50 K intervals within a single day. This new real-time gas analysis in flow systems proves to be especially valuable in the detection and the characterization of intermediates with half-lives as short as 10^{-3} sec as well as for the optimization of heterogenously catalyzed reactions.

1. PHOTOELECTRON SPECTRA AND THEIR ASSIGNMENT TO RADICAL CATION STATES [2,3,4]

Radical cation states are well-suited for an introductory discourse of molecular states e.g. within the framework of a general chemistry lecture series: they can be observed by the straightfor-

P. Day (ed.), Emission and Scattering Techniques, 335–352.

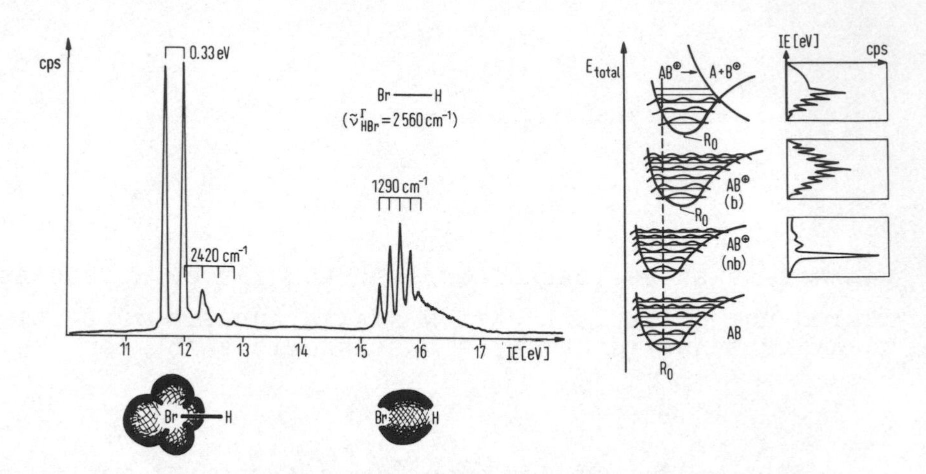

Figure 2. The helium(I) photoelectron spectrum of HBr displays 3 of the 4 ionizations expected for its 8 valence electrons. The radical cation states in the sequence $^2\Pi_{3/2}$ spin/orbit coupled with $^2\Pi_{1/2}$ and $^2\Sigma^+$ differ for instance in their stretching frequencies $\nu^{\oplus}_{H-Br}$, which are hardly changed at all ($^2\Pi$) or almost halved ($^2\Sigma^+$) relative to uncharged HBr. Together with the differing band shapes, this finding can be interpreted in terms of potential energy curves as follows: if an (almost) non-bonding electron is ionized, the equlibrium distance R_O remains (almost) constant, and owing to the most frequent $0 \rightarrow 0$ transition a needle-like band results. If, on the other hand, a bonding electron is removed, R_O increases, the force constant f_{HBr} decreases and also the stretching freqency $\nu^{\oplus}_{HBr}$; furthermore the most frequent transition is now shifted towards the centre of the band. In addition, the third $HBr^{\cdot\oplus}$ state is unstable towards dissociation yielding $HBr^{\cdot\oplus} \rightarrow H^{\cdot} + Br^{\oplus}$ at 15,85 eV and the vibrational fine structure vanishes there accordingly: in the potential diagram, the dissociative curve crosses the one of the binding state.

of a 'nonbonding' electron exerts no significant effect on the force constants ($f^{\oplus}_{HBr} \approx f_{HBr}$) as well as on the equilibrium distances ($R^{\oplus}_O \approx R_O$). In contrast, the third PE band assigned to σ_{HBr} ionization exhibits an approximately halved frequency $\nu^{\oplus}_{HBr}$ duê to the diminished force constant $f^{\oplus}_{HBr}$ and the increased equilibrium distance $R^{\oplus}_O$ caused by the expulsion of a 'bonding' electron. The characteristic properties of the individual states are further demonstrated by their differing reactivities: contrary to the 2 lower radical cation states, the third one decomposes at 15.85 eV according to $HBr^{\cdot\oplus} \rightarrow H^{\cdot} + Br^{\oplus}$ (Figure 2) as confirmed by independant mass spectroscopical data [2-4].

ward 'anti-black-box' technique of photoelectron spectroscopy [3], are characterized by a typical electron hole [2] , and can usually be rationalized based on simple MO models [2-4]. In general, the state of a molecule is defined by the difference between its energy and that of the preceding initial state or that of the subsequent final state, as well as by the respective charge distribution i.e. the molecular structure (Figure 1).

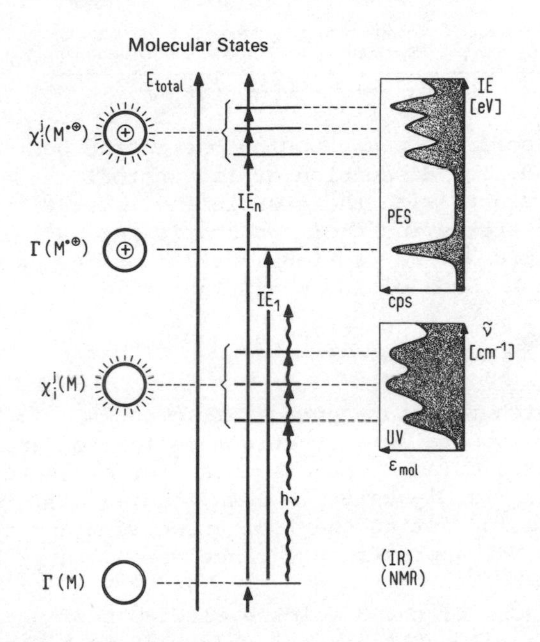

Figure 1. A neutral molecule in its ground state $\Gamma(M)$ can take up certain amounts of energy e.g. the absorption of UV radiation $h\nu$ will transfer it into one of its numerous electronically excited states $X_i^j(M)$. Larger energies may lead via loss of an electron, to the ground state of the corresponding radical cation $\Gamma(M^{\cdot\oplus})$ or one of its many electronically excited states $X_i^j(M^{\cdot\oplus})$ likewise observable in its photoelectron spectrum (PES).

The PES information about the radical cation states will be exemplified here by presenting two small molecules, HBr (Figure 2) and HSH (Figure 3), respectively. In the PE spectrum of HBr [2-4] the first 2 ionizations can be assigned to the bromine lone pairs perpendicular to the molecular axis, because the resulting radical cation states, $^2\Pi_{3/2}$ and $^2\Pi_{1/2}$, are coupled by the bromine spin/orbit coupling constant of .33 eV . The vibrational fine structure recognizable on the high-energy side of the second band exhibits a stretching frequency only slightly lower than that of the neutral HBr molecule (Figure 2): hence the electron hole arising on removal

The delicate interrelationship between energy content and charge distribution i.e. structure as simplified for e.g. H_2SiCl_2 by

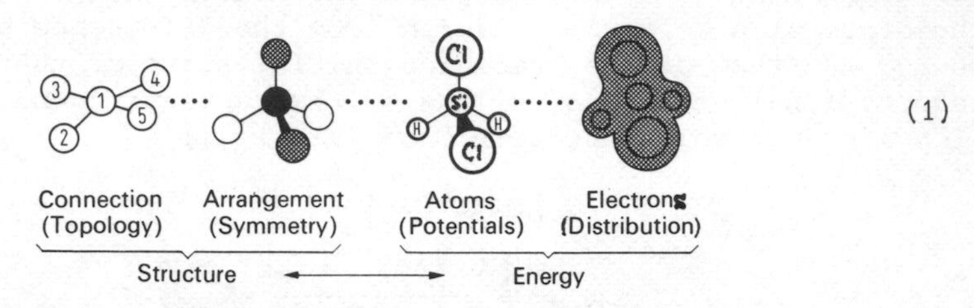

(1)

can be best elaborated upon choosing one of the bent triatomic molecules like HSH. The discussion of its photoelectron spectrum [5] advantageously starts with the 'vertically' i.e. within 10^{-16} sec generated $HSH^{\cdot\oplus}$ states and their adiabatic relaxation after $< 10^{-14}$ sec (Figure 3). For rationalization of the experimental observations, use is made of Koopmans' theorem

$$\Delta E_{total} = IE_n^{vertical} \equiv - \varepsilon_j^{SCF} \qquad (2)$$

according to which the state energy differences (cf. Figure 1) measured PE spectroscopically as vertical ionization energies can be correlated with calculated one-electron orbital energies $-\varepsilon_j^{SCF}$ of 'Self Consistent Field'-quality — as long as differences in electron correlation during the ionization of M and the charge redistribution in $M^{\cdot\oplus}$ approximately cancel each other [6].

Ejection of one of the 8 valence electrons of the HSH molecule, which exhibits an almost spherical charge distribution in its ground state, leads to one of the 4 states of the radical cation $HSH^{\cdot\oplus}$ observable in the He(I) and HE(II) regions of the PE spectrum. In these $HSH^{\cdot\oplus}$ states, the respective electron hole distributions can be described by the difference electron densities between the initial and the pertinent final state (Figure 3): evidently, zero site planes result according to the $M^{\cdot\oplus}$ state symmetry. If Koopmans' theorem (2) applies, then the electronic arrangement can be represented by a single configuration i.e. a symmetry-adapted molecular orbital, the square, ψ_i^2 of which can be correlated with the electron density — or in other words, the zero-site planes of the electron hole in the radical cation states apparently are reflected by the orbital nodal planes. On inclusion of qualitative orbital perturbation arguments even the observed and/or calculated structural changes on adiabatic relaxation might be rationalized (Figure 2): if the $M^{\cdot\oplus}$ state is represented by a now only singly occupied molecular orbital exhibiting strongly bonding or antibonding interactions, then structural changes mitigating these interactions are expected. For the $HSH^{\cdot\oplus}$ ground state with a nodal plane through the SH bonds hardly any change at all should occur: the sulfur lone pair

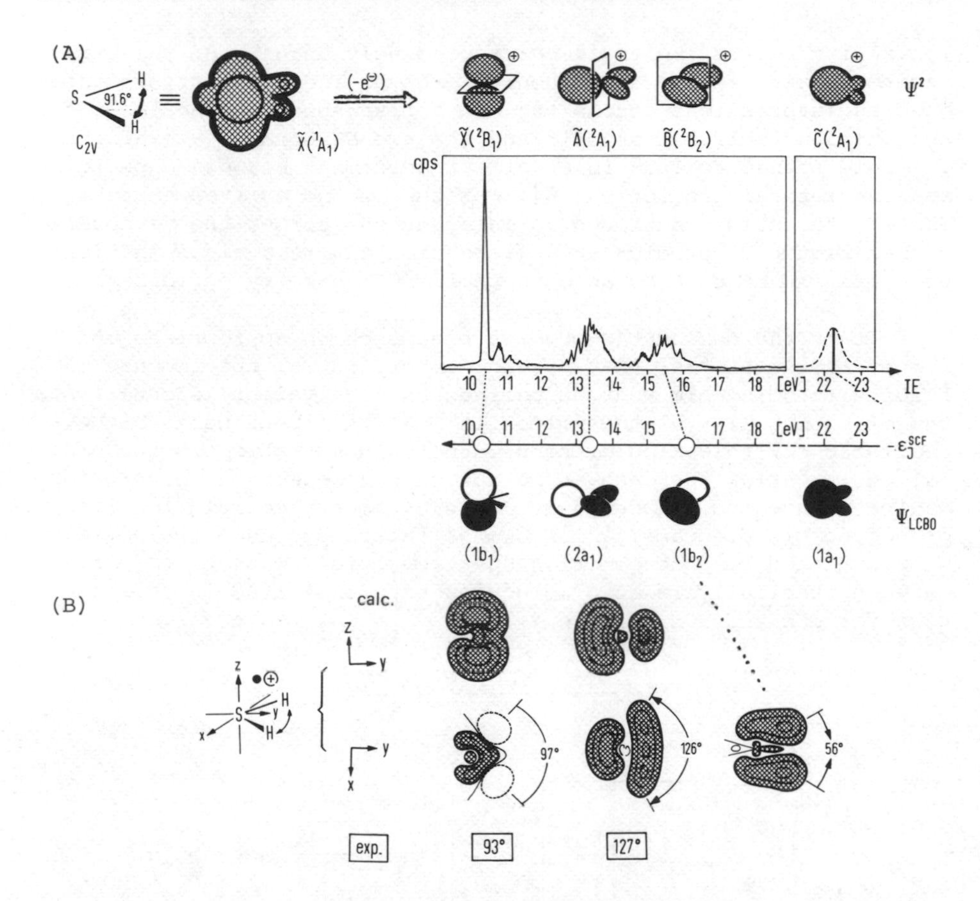

Figure 3. The HSH molecule with C_{2v} structure possesses in its totally symmetric ground state, $\tilde{X}(^1A_1)$, an almost spherical electron density. Ionization of one of its 8 valence electrons leads to the 4 $HSH^{\cdot\oplus}$ states shown. The first three exhibit zero-site planes with respect to the resulting positively charged hole, which are reflected by the squared wavefunction ψ^2 derived e.g. from symmetry adapted Linear Combinations of Bond Orbitals with nodal planes, ψ_{LCBO}. The ionization energies measured correlate satisfactorily via Koopmans theorem, $IE_n^v = -\varepsilon_J^{SCF}$, with SCF-eigenvalues calculated for the ground state geometry (A). The $HSH^{\cdot\oplus}$ radical cation (cf. (B)) on adiabatic relaxation either retains the HSH structure as in its ground state, $\tilde{X}(^2B_1)$, opens the angle to 127° (calc. 126° [5]) in its first and closes it to an estimated 56° in its second excited state. The open shell-SCF spin density plots for the unpaired electron in $HSH^{\cdot\oplus}$ [5] can be rationalized by the LCBO molecular orbitals ψ_{LCBO} as follows: no change with an HSH nodal plane and angle changes for reduced antibonding interaction with respect to the nodal planes.

ionization leaves the angle correspondingly unchanged. The first excited state — the a_1 molecular orbital indicates strong antibonding interactions across the nodal plane between the lobes for the in-plane sulfur lone pair and the two SH bonds — exhibits an angle opened to 127° in excellent agreement between experiment and theoretical prediction [6]. For the second excited state of $HSH^{\cdot\oplus}$, the strong antibonding interactions across the yz nodal plane (Figure 2) predict an angle decrease on removal of an electron, as calculated by an open shell SCF geometry optimization.

It is the demand the chemist places on an appropriate model for molecular states, that he can compare out of the immense manifold surrounding him what he defines as 'equivalent molecular states of chemically related molecules'. And, the best basis hitherto available for this kind of comprehensive description is the orbital perturbation approach. Long before photoelectron spectroscopy had been invented, Walsh [7] guided by symmetry correlation lines presented his diagrams, which in most facets proposed are nowadays substantiated by photoelectron spectroscopic $M^{\cdot\oplus}$ state comparison between chemically related molecules as exemplified in this context for HSH and HOH (Figure 4).

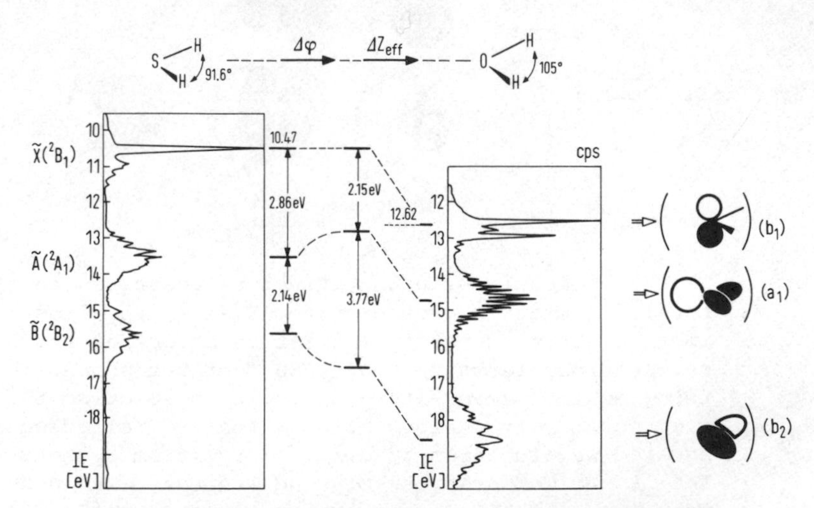

Figure 4. Walsh diagram [7] for the first 3 $M^{\cdot\oplus}$ states of the iso-(valence)electronic molecules HSH and HOH, differing in the effective nuclear charge of the central atom and in bond angle. Assuming that the ionization of the nonbonding lone pair electron may be used as an internal standard, the effect ΔZ_{eff} is compensated by a 2.15eV shift between the two spectra. The angle opening is discussed orbitalwise as follows: decreased bonding and increased antibonding interaction destabilizes a_1, while decreased antibonding interaction stabilizes b_2. This allows to rationalize the observed decrease $\Delta IE_{1,2}$ and increase $\Delta IE_{2,3}$ on comparing the PE spectra of HSH and HOH.

To conclude this introductory chapter on interpretation of photoelectron spectra by assigning them to individual radical cation states, based - if necessary - on the orbital perturbation approach, some larger molecules should also be discussed. Referring, however, to the PE spectra of thousands of molecules published, which are readily accessible via reviews like [2,3] containing hundreds of literature quotations, a few but especially clear-cut examples (Figures 5 and 6) may suffice.

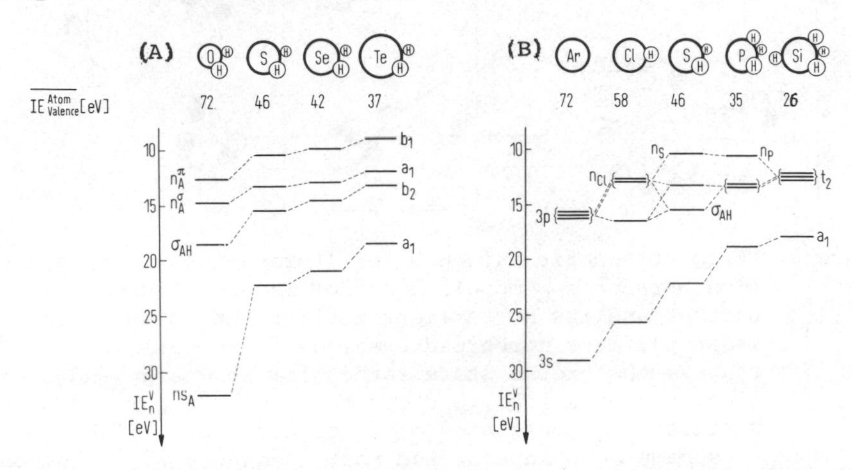

Figure 5. PE spectroscopic comparison of hydrides AH_n within the periodic table of the elements:
(A) Group VIb hydrides AH_2 clearly demonstrate the decrease of the effective nuclear charge Z_{eff}, here represented by the average ionization energy of all valence electrons of the central atom A. Especially the fourth ionization energy, assigned to the ns_A-type radical cation state characterized by an (almost) 'spherical' electron hole distribution, decreases dramatically along the series.
(B) Hydrides along the period with main quantum number n = 3 (so-called 'United Atom' model): to begin with, it is interesting to note, that the sums of all PES ionization energies of HOH (A) and of Ar (B) are approximately equal, demonstrating the dominating influence of Z_{eff}. Pulling out 1 H from the Ar nucleus generates the σ_{HCl} bond, with the corresponding radical cation state stabilized relative to the ones originating from chlorine lone pair ionization (cf. Figure 2). Altogether, nonbonding lone pair n_X electrons are more easily ionized than p-type bonding electrons under the influence of more than one core potential. Highest valence ionization energies are needed for the s-type electrons in closest proximity to the nucleus and exhibiting the strongest influence of Z_{eff} (cf. (A)). Within the p-valence electron band, symmetry-degenerate $M^{\cdot\oplus}$ states are recognizable for Ar, HCl, PH_3 and SiH_4 [8].

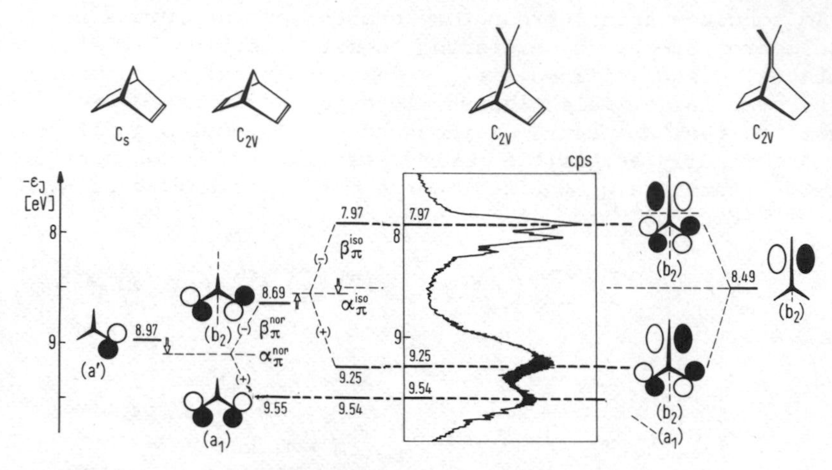

Figure 5. PE spectroscopic evidence for 'through space' interaction between π_{CC} double bonds of isopropylidene norbornadiene and its MO parametrization: The PE spectrum of isopropylidene norbornadiene reveals up to 10 eV 3 $M^{\cdot\oplus}$ states, the ground state rather low in energy and 2 excited states which lie close together. Comparison with π radical cation states of other bicyclo[2.2.1]hydrocarbons have been interpreted most elegantly [9] based on the concept of symmetry-allowed 'through space' π interactions as follows: starting from the π(a') orbital of norbornene after some small inductive correction (⇓), the Coulomb parameter α_π^{nor} is derived, around which the 2 π combinations a_1 and b_2 of norbornadiene split according to the π interaction parameter β_π^{nor} = .43 eV. The π radical cation state of isopropylidene norbornane, on the other hand, is of b_2 irreducible representation, and thus can mix only with the $\pi(b_2)$ of norbornadiene according to the second π interaction parameter β_π^{iso} = .62 eV. The third π orbital belongs to the irreducible representation a_1 and, therefore, cannot interact with any other π orbital —— the corresponding π radical cation state of isopropylidene norbornadiene resembles in ionization energy the second π radical cation state of norbornadiene, 9.54 eV to 9.55 eV, respectively!

The examples selected from thousands of PE spectra recorded and assigned since photoelectron spectroscopy has been introduced in 1962 [3a] —— the main group element hydrides AH_n displaying the influence of the effective nuclear charge (figures 2,3 and 4), and isopropylidene norbornadiene demonstrating the charge delocalization over a molecule with non-conjugate π_{CC} bonds (Figure 5) —— illustrate the wealth of information obtainable. In addition, assignment and discussion of PE spectra by comparing equivalent radical cation states of chemically related molecules provide insight into some general principles. Above all, however, the ionization patterns of molecules containing information on energy and structure of the individual M·⊕ states can be used as 'fingerprints' for their identification and characterization.

2. PHOTOELECTRON SPECTRA FOR REAL-TIME GAS ANALYSIS IN FLOW SYSTEMS [2c,3c]

The illustrative display of some PE spectra (Figure 6) has been chosen as close to reality as possible in such a way that the molecules presented could occur as components of a synthesis gas mixture or a cracking product fraction.

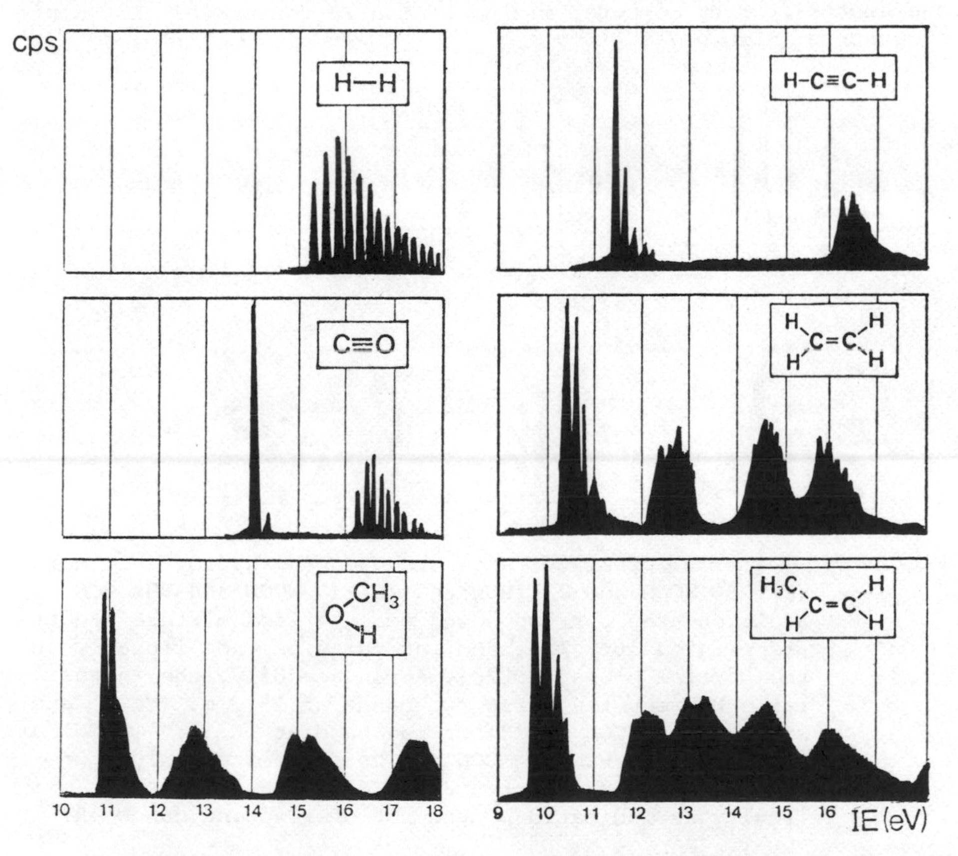

Figure 6. PE spectroscopic comparison of some synthesis gas components $2H_2 + CO \rightarrow H_3COH$ (left-hand side) and of some C_2 + C_3 products which possibly can occur together in a fraction of cracking gas.

Obviously (cf. Figure 6), gas analysis by means of ionization patterns - exhibiting different band numbers, band energies and band shapes ranging from needles to broad ionization hills - can be easily accomplished as long as the PE spectra exhibit separated characteristic bands and do not overlap completely. Even an observer less proficent in the daily use of PE spectra should recognize that there are no particular difficulties involved, just as with infrared vibrational frequencies or with mass spectroscopic fragments. In order to identify molecules by their ionization 'fingerprints', the corresponding PE spectra can be looked up in the vast literature including summaries and collections [2,3], or can be recorded in advance. For hitherto unknown compounds, predictions via Koopmans´theorem (2) are an additional possibility (cf. Figure 10). The feasibility of analyses in gaseous mixtures can be judged by comparison, projection of the PE spectra upon one another, or by computer mixing of the PE spectra of the main components (Figure 7).

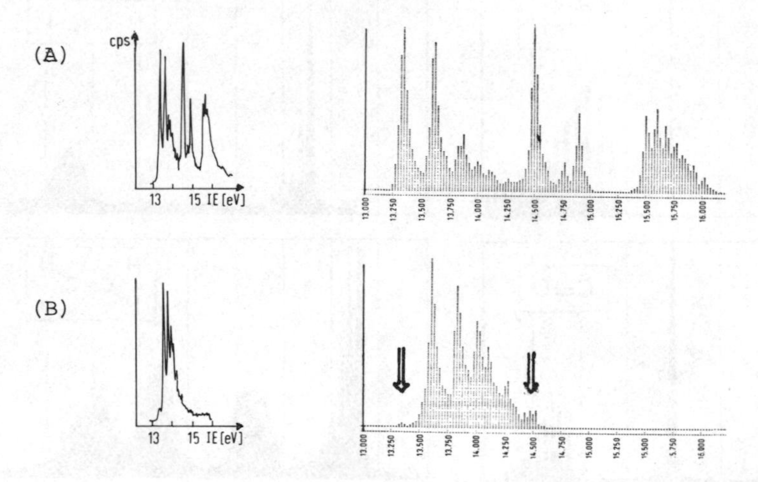

Figure 7. PE spectra of mixtures of cyanogen and hydrocyanic acid: (A) 75% NCCN and 25% HCN and (B) 2% NCCN and 98% HCN. In addition, the corresponding computerized mixture spectra are printed out. The band intensity depends linearly on the partial pressure [2c]. Advantageously, the intensity ratio between the first two bands at 13.3 eV (NCCN) and at 13.6 eV (NCCN + HCN) or the band at 14.5 eV (NCCN) are used: whereas HCN in NCCN can be determined only approximately, one can detect 2% NCCN in HCN (B) without difficulty as indicated by the not overlapping bands (⇓).

The PES analysis method, which can be set up within a day if done by recording the spectra of mixtures according to their partial pressure, or within an hour by computer mixing, permits rapid elucidation of gas phase reactions.

The working principle of a photoelectron spectrometer [3a] is also well-suited for an investigation of gas phase reactions in flow systems: a beam of molecules is pumped from a storage vessel to a cold trap in the instrument at pressures between 10^{-1} to 10^{-5} torr. As a rule-of-thumb, only about 1 millimole of compound is needed for a series of measurements over 6 to 8 hours. Since the helium photons are generated in an open gas discharge, the spectrometer is filled with a protecting inert gas atmosphere. Among the many modifications reported [2c,3c], the following should be mentioned: reactions at atmospheric pressure using a bleeding-valve to connect the spectrometer (Figure 12), or generation and identification of short-lived intermediates using microwave or heating devices in down to 2 cm distance from the ionization chamber (Figure 11).

The gasphase reaction apparatus connected to the photoelectron spectrometer can be constructed from a building-block set [3c,12] depending on the choice of the inlet system, reaction tube or product isolation traps (Figures 8 to 12).

Among the many PE spectroscopically analyzed and optimized gasphase reactions, the following examples are chosen for illustration (cf. [2c,3c] and the literature quoted) from own work:

====> gasphase synthesis (Figure 8: SSO [13])

====> determination of the lowest thermal decomposition channel (Figure 9: RSSR [14])

====> compound separation by forming a solid deposit (Figure 10: $H_2C{=}S$ after removing HCl with NH_3 [15])

====> generation of a presumably short-lived intermediate (Figure 11: silatoluene [16])

====> catalyst screening (Figure 12: cyanation of benzene [11]).

GAS PHASE SYNTHESES:

Reactions between gases or between a vaporized compound and a solid (Figure 8) have been optimized in many cases [2c,3c] using a photoelectron spectrometer to control the stoichiometry — the 'fingerprints' of the reactants have to vanish completely — as well as to achieve the optimum reaction conditions: product bands should dominate the PE spectrum of the mixture as much as possible and all contaminations reduced by scanning the temperature and/or the partial pressures of the starting materials. Under normal spectrometer working conditions, the optimization usually takes about a day and approximately millimole quantities of compounds. The example chosen, SSO, demonstrates nicely that unstable intermediates may also be prepared by PE spectroscopically optimized gas phase synthesis [13].

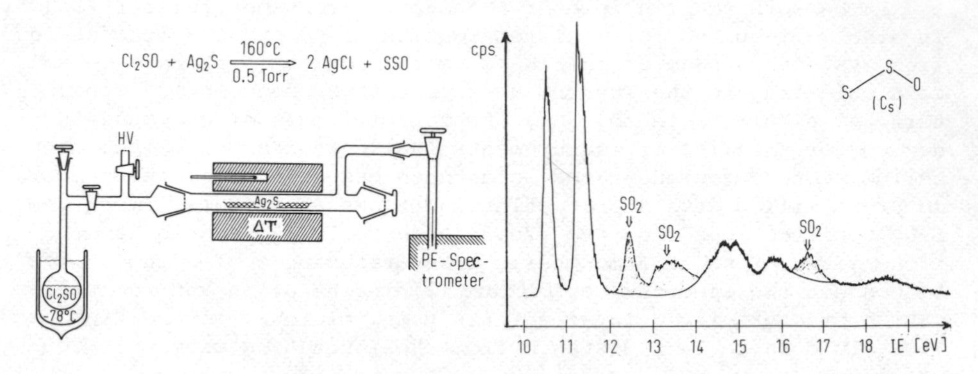

Figure 8. The gas phase synthesis of disulfuroxide SSO, which readily polymerizes at pressures above 1 torr under redox disproportionation into SO_2 and solid S_nO_m, can be optimized PE spectroscopically: thionylchloride is best passed over silver sulfide at 160°C to minimize the unavoidable SO_2 formation.

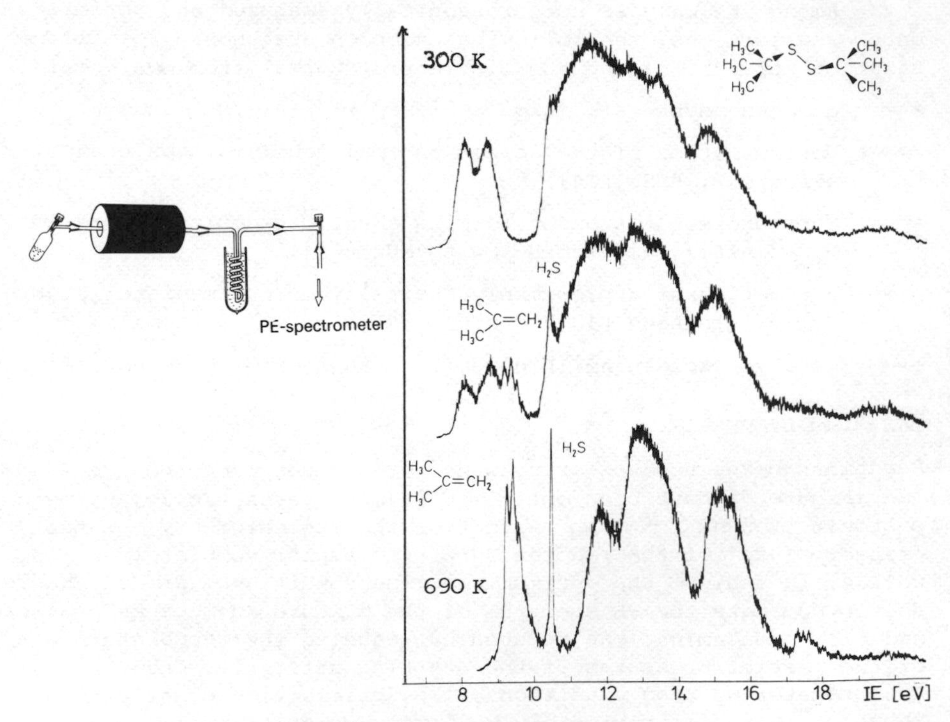

Figure 9. The thermolysis of di(t.butyl)disulfide yields exclusively isobutene, H_2S and S_8, if the temperature is raised carefully in 2o K steps until the lowest thermal decomposition channel has been reached at 690 K.

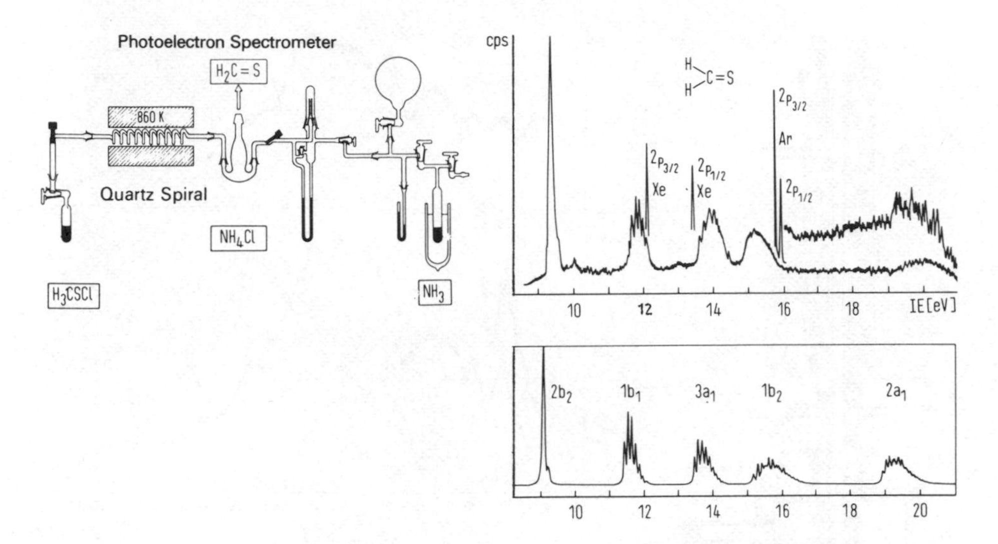

Figure 10. For gas phase preparation of thioformaldehyde, advantageously methanesulfenylchloride is passed through a quartz spiral heated to 860 K. From the resulting mixture HCl can be removed by stoichiometric injection of NH_3 from a storage system and deposited as NH_4Cl on the walls of the mixing bulb. The reaction conditions e.g. the temperature and also the stoichiometry are optimized PE spectroscopically. The unknown compound can be identified by independently calculating the PE spectrum e.g. by the Greens' function procedure (cf. v.Niessen et al., J.Chem.Phys. 1977,66, 4893).

DETECTION OF LOW-TEMPERATURE REACTION CHANNELS:

Thermal decompositions do not have to result in tar production: if the temperature is **carefully raised** step-wise, then the point of specific thermal breakdown of a molecule becomes visible in small changes of the ionization pattern. In the example of dialkyldisulfides (cf. Figure 9), only at the lowest decomposition temperature an exclusive split into olefine, H_2S and sulfur is observed — higher temperatures produce a random mixture containing even CS_2 (no carbon is linked to 2 sulfurs in the starting material!) and also HC≡CH, CH_4, H_2 and others [14].

THERMAL GENERATION OF SHORT-LIVED MOLECULES:

The applicability of photoelectron spectroscopy to prepare and to characterize unstable intermediates of half-live times larger than needed to travel between the zone of their generation and the PES ionization chamber is quite obvious from preceding remarks: as mentioned above, thermal decomposition temperatures are readily opti-

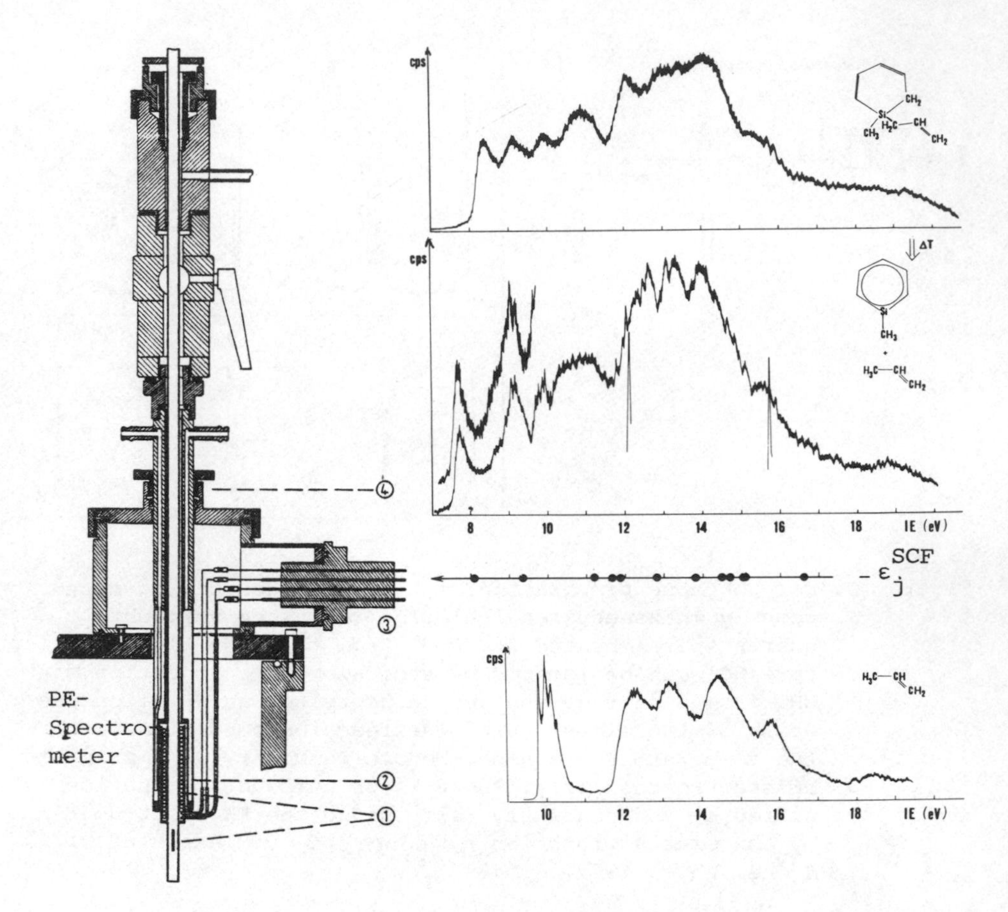

Figure 11. For the PE spectroscopic detection of silatoluene, a short-path pyrolysis apparatus had to be constructed ((1) heating furnace about 2cm above the electron exit slit of a PS 16 ionization chamber, (2) water cooling, (3) current feed and (4) centering device). The precursor chosen, 1-methyl-1-silacyclohexa-2,4-diene, splits off propene quantitatively at about 1000 K, identified by its fine-structured band at 10 eV. The 2 lowest radical cation states of the silatoluene generated can be identified either via a Koopmans' correlation $IE^{v}_{1,2} = \varepsilon^{SCF}_{1,2}$ or by radical cation state comparison with either toluene or silabenzene. In addition, the parent peak, e/m = 109, has been determined mass spectroscopically, and in the case of the analogously generated silabenzene, an argon matrix isolation could be accomplished using the PE spectroscopically optimized pyrolysis conditions (Maier G. et al., Angew. Chem. int. Ed. Engl. 1980, 19, 52).

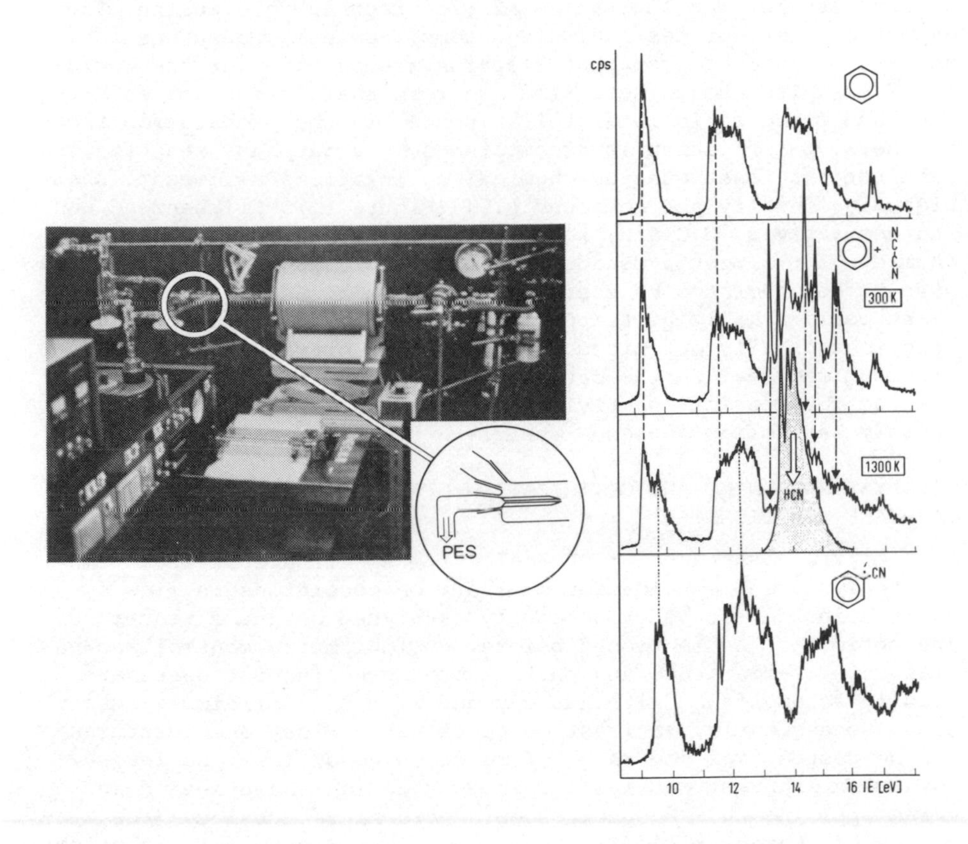

Figure 12. For the screening of catalysts for gasphase reactions at atmospheric pressure, the following standard apparatus is used (from right to left): steel gas containers, heatable if necessary, manometer, precision valves, mixing bulb, temperature-programmable furnace, bleeding capillary to PE spectrometer, cold traps, pump). A mixture of benzene (upper PE spectrum) and cyanogen (second PES) is heated e.g. over a copper catalyst on γ-aluminumoxide to 1300 K (third PES). HCN is evolved in large quantities (for PE spectrum cf. Figure 7). A 40% conversion in one run over the catalyst can be estimated from the low-energy band being composed of the partly overlapping first ionizations of benzene and of benzonitrile (fourth PE spectrum), allowing to judge the quality of the catalyst tested at a certain temperature.

mized, and in section 1 the energy and structure information on radical cation states contained in PE spectra has been dealt with at some length. The literature [2c,3c] reports on a wealth of unexpected intermediates including atoms, radicals, unsaturated σ- and π-type species from photoelectron groups all over the world. The 2 examples chosen here from own work shall demonstrate: $H_2C=S$ has been detected in interstellar space, in the preparation reported here, purification is accomplished by a gasphase reaction, and last but not least quantum chemical calculations allowed to unambigously identify the compound [15] (Figure 10). Silabenzene and its derivatives, H_5C_5SiR, [16] (Figure 11), had been hunted by chemists for almost a decade, are presumably as short-lived as to have to be generated by short-path pyrolysis, and last but not least could finally be trapped in an argon matrix using the PE-spectroscopically optimized conditions for preparation: the UV spectrum recorded in the matrix proves in addition, that silabenzene despite its reactivity is still a 6 π electron perimeter closely resembling benzene itself!

CATALYST SCREENING AND OPTIMIZATION OF HETEROGENEOUSLY-CATALYZED GASPHASE REACTIONS:

Many of the advantages of photoelectron spectroscopic real-time gas analysis and elucidation of reaction conditions in flow systems coincide, if heterogeneously catalyzed gasphase reactions are optimized. As mentioned before, stoichiometry control and variation of temperature and partial pressures in most cases are readily accomplished within a day and with only millimole quantities. Consequently, at least one catalyst per day and instrument can be tested over the range of reaction conditions. So far, two novel heterogeneous catalysts for new reactions have been discovered applying the PE spectroscopic screening technique: the cyanation of benzene [11] (Figure 12, cf. also Figure 8 for the analysis of NCCN/HCN mixtures) and the bromination of trifluormethane:

$$F_3CH + Br_2 \xrightarrow{[NiF_2 \text{ on carbon}]} F_3CBr + HBr \quad [17] \qquad (3)$$

Although the precision of PE spectroscopic gasphase analysis is not sufficient for the final optimization of those heterogeneously catalyzed syntheses, which has to be carried out additionally on a preparative scale using conventional analysis technique, the saving of time in the screening process is a considerable one. Therefore, it may well be that in the future and especially with the development of portable inexpensive photoelectron spectrometers, this application of photoelectron spectroscopy will become one of its important facets.

(Fimis)

Acknowledgement: The gas analysis in flow systems has been investigated by the following members of the Frankfurt Photoelectron Spectroscopic Group: S. Aygen, G. Bert, R. Dammel, Dr. T. Hirabayashi, S. Mohmand, Dr. H. Müller, Frau M. Pohlenz, Dr. P. Rosmus, Dr. A. Semkow, Dr. B. Solouki, Dr. U. Stein and J. Wittmann, whose impetus and dedication should be emphasized here as well. Dozent J. Mintzer as well as Dr. H.J.Arpe and Dr. J. Russow from Hoechst A.G. have contributed advice for the development of catalysts. The University of Frankfurt and the State of Hesse provided the PE spectrometers and supported the investigations; additional support of the Deutsche Forschungsgemeinschaft, the Fonds der Chemischen Industrie, the Hoechst A.G., the Max Buchner and the Hermann Schlosser Foundations is gratefully acknowledged.

REFERENCES

[1] Part 102 of Photoelectron Spectra and Molecular Properties as well as Part 27 of Gasphase Reactions. For the preceding publication cf. Block E., Corey E.R., Penn R.E., Renken T.L. and Bock H., Hirabayashi T., Mohmand S., Solouki B., J. Am. Chem. Soc. 1981, 103, in print.

[2] For preceding summaries on photoelectron spectroscopy cf.
  [a] Bock H. and Ramsey B. G.: Photoelectron Spectra of Nonmetal Compounds and Their Interpretation by MO Models, Angew. Chem. int. Ed. Engl. 1973, 12, pp 734 - 752;
  [b] Bock H.: Molecular States and Molecular Orbitals, Angew. Chem. int. Ed. Engl. 1977, 16, pp 613 - 637;
  [c] Bock H. and Solouki B.:Real-Time Gas Analysis in Flow Systems, Angew. Chem. int. Ed. Engl. 1981, 20, issue 5 (in print)
and the numerous literature references quoted.

[3] Cf. e.g. the books on photoelectron spectroscopy by
  [a] Turner D.W., Baker C., Baker A.D. and Brundle C.R.: Molecular Photoelectron Spectroscopy, Wiley-Interscience, London 1970;
  [b] Rabalais J.B.: Principles of Ultraviolet Photoelectron Spectroscopy, Wiley & Sons, New York 1977;
  [c] the three-volume handbook: Electron Spectroscopy - Theory, Techniques and Applications (ed. Baker A.D. and Brundle C.R.), Academic Press, London 1977 - 1979,
and the thousands of literature quotations given.

[4] Cf. e.g. Bock H. and Mollère P.D.: Photoelectron Spectra, an Experimental Approach to Teach Molecular Orbital Models, J. Chem. Educat. 1974, 51, pp 506 - 514 and lit. cit.

[5] For spectroscopic details as quoted in [3b] cf. Potts A.W. and Price W.C., Proc. Roy. Soc. London 1972, A 326, 181; for the

SCF calculations on HSH cf. Ross B. and Siegbahn P., Theor. Chim. Acta 1971, 21, 368, and for the HSH·⊕ potential surfaces and spin populations cf. Sakai S, Yamabe S., Yamabe T., Fukui K. and Kato H., Chem. Phys. Lett. 1974, 25, 541.

[6] Cf. e.g. Kutzelnigg W.: Einführung in die Theoretische Chemie, Vol. I and II, Verlag Chemie, Weinheim 1975 and 1978.

[7] Walsh A.D., J. Chem. Soc. London 1953, pp 2260 - 2331.

[8] Potts A.W. and Price W. C., Proc. Roy Soc. London 1972, A326, 165 and Lempka H.J., Passmore T. R. and Price W. C., ibid. 1968, A304, 53.

[9] Heilbronner E., Isr. J. Chem. 1972, 10, pp 143 - 156; Heilbronner E. and Martin H.D., Helv. Chim. Acta 1972, 55, 1490. Cf. also Hoffmann R., Acc. Chem. Res. 1971, 4, 1.

[10] Bock H., Solouki B. and Wittmann J., Angew. Chem. int. Ed. Engl. 1978, 17, 932.

[11] Bock H., Solouki B., Wittmann J. and Arpe H.J., Angew. Chem. int. Ed. Engl. 1978, 17, 933.

[12] Bock H., Solouki B., Bert G., Hirabayashi T., Mohmand S. and Rosmus P., Nachr. Chem. Techn. Lab. 1978, 26, 634.

[13] Bock H., Solouki B., Rosmus P. and Steudel R., Angew, Chem. int. Ed. Engl. 1973, 12, 933.

[14] Bock H. and Mohmand S., Angew. Chem. int. Ed. Engl. 1977, 16, 104.

[15] Solouki B., Rosmus P. and Bock H., J. Am. Chem. Soc. 1976, 98, 6054, and Bock H., Solouki B. , Mohmand S., Block E. and Revelle L.K., Chem. Comm. 1977, 287.

[16] Bock H., Bowling A., Solouki B., Barton T.J. and Burns G.T., J. Am. Chem. Soc. 1980, 102, 429. Cf. also Solouki B., Rosmus P., Bock H. and Maier G., Angew. Chem. int Ed. Engl. 1980, 19, 51.

[17] Bock H., Mintzer J., Wittmann J. and Russow J., Angew. Chem. int Ed. Engl. 1980, 19, 147.

XPS APPLIED TO SMALL METAL PARTICLES AND TO METAL-POLYMER ADHESION: PHYSICAL INSIGHT FROM INITIAL AND FINAL-STATE EFFECTS

P. Ascarelli and M. Cini

Assoreni, Monterotondo (Roma) Italy

ABSTRACT

The potentiality of XPS technique as a surface research tool is evidentiated through the description of two applications.

On one side the sensitivity of XPS to final state structure is applied to study the electronic properties of small metal particles. To explain the size dependent broadening effect apparent in our data we propose an extension of previous dynamic response formulations to the case of unhomogeneous systems. The conduction electron relaxation time may have a non neglegible effect in determing line shapes.

On an other side XPS quantitative analysis is applied to the characterization of Al-foils surfaces relevant to the problems of polyethylene-aluminum adhesion.

1. INTRODUCTION

In the last few years the XPS technique attracted a wide interest as a basic research tool, meanwhile its use in applied science and industrial research has continuously proliferated. However the successfull exploitation of the technique in applied research still requires a good deal of ingenuity to select and interpret each time the proper "observable" feature

P. Day (ed.), Emission and Scattering Techniques, 353–373.

that can provide insight on the problem on hand. The aim of this lecture is to evidentiate, through the description of some applications, peculiar qualities and limits of the XPS technique. On one side the well known sensitivity of XPS to final state structure is applied to study the electronic properties of small metal particles. To understand the data we propose an extension of previous dynamic response formulations to the case of unhomogeneous systems and show that the conduction electron relaxation time may have a non neglegible effect in determining line shapes.

On the other side the information that XPS may provide when used as a surface analytical tool is made apparent by studying the problem of metal to polymer adhesion.

2. SIZE EFFECTS ON THE XPS SPECTRUM OF SMALL METAL PARTICLES

The characterization of the physical properties of small metal particles is relevant to the applications of these systems in various fields, recalling among others solar energy collection and heterogeneous catalysis. Many metallic catalysts consist of very small metallic particles (10 to 100 Å) supported on high surface area insulator like Al_2O_3 (alumina) or SiO_2 (silica).

Among small size particle properties one seems to be particularly relevant to its catalytic properties, this is the reduced electron ability to screen electrostatic fields(1, 2).

A consequence of this fact is a longer interaction range which would affect chemisorption properties especially when electron transfer is involved (3).

The application of metallic particle dispersions for solar energy absorption is well known (4), and the characteristic phenomena of plasma resonance absorption occurring in the optical frequency range has also been shown to depend on particle size and shape (5).

2.1 Theory

The core XPS spectrum is proportional (6) to the core hole density of occupied states $N(\omega)$, which may be written in the form

$$N(\omega) = \int_{-\infty}^{\infty} \frac{dt}{2\pi} \, e^{i\omega t} \exp\left(C(t)\right) \qquad (1)$$

where C(t) is the sum of all the distinct linked diagrams that contribute to the deep propagator. Langreth (7) proposed an approximation to C(t) suitable for a translationally invariant metal. By extending the Langreth formulation to an unhomogeneous system like a small metal particle we get

$$C(t) = \frac{1}{\pi\hbar}\int_0^\infty d\omega\left[\frac{1+i\omega t-e^{i\omega t}}{\omega^2}\right]\int d^3x\, V_b(x)\, Im\, \rho_i(x,\omega) \quad (2)$$

where V_b is the bare core hole potential and ρ_i is the induced charge density given in terms of the retarded response function K^R by the usual linear response equation

$$\rho_i(x,\omega) = \frac{2\pi}{\hbar}\int d^3x'\, K^R(x,x',\omega)\, V_e(x',\omega)\,. \quad (3)$$

Here we omit the derivation of equation (2), but the reader can readily check that for translationally invariant systems, when K^R depends on x-x' only, an application of the fluctuation-dissipation theorem (8)

$$S(q,\omega) = -2\ \Theta(\omega)\ Im\, K^R(q,\omega)$$

recovers Langreth's expression of C(t) in terms of the dynamic form factor S of the electron liquid (equation 44 of ref.7). One term of C(t) is linear in time and can be rewritten as $\Delta E\cdot t/h$, where ΔE represents the relaxation energy shift. Since $\rho(t)$ is a causal function we can use the dispersion relations and the oddness of Im $\rho(\omega)$ to cast ΔE in the more familiar form

$$\Delta E = 1/2\int d^3x\ V_e(x)\ \rho_i(x,\omega = 0)\ .$$

The asymptotic behaviour (9) of C(t) for long times yields

$$C(t) = -\alpha \ln(t) \quad \text{as} \quad t \to \infty.$$

Equation (2) yields the following expression for the "asymmetry exponent" α

$$\alpha = -\frac{1}{\pi\hbar}\, Im\int d^3x\ V_e(x)\ \frac{\partial}{\partial\omega}\rho_i(x,\omega)\Big|_{\omega=0} \quad (4)$$

The above results simplify in the limiting case of a point core hole placed at $x=x_o$. Inserting $V_e(x,\omega) = e/|x - x_o|$ in equations (3) and (4) we get

$$\Delta E = \frac{1}{2} e \varphi(x_0, 0) \tag{5}$$

$$\alpha = -\frac{e}{\pi \hbar} \frac{\partial}{\partial \omega} \varphi(x_0, \omega) \Big|_{\omega=0} \tag{6}$$

where $\varphi(x, \omega)$ is the induced potential.

We wish to use this simplified form of the theory in a qualitative study of the effects of size in XPS from small metal particles. The main effects can be demonstrated by an entirely analytical calculation. One obvious consequence of small size is that a particle gets charged when ionized. This should reduce the relaxation shift ΔE by an amount that depends on the dielectric function of the insulator surrounding tha particle. A further well known size effect is the broadening of the plasma optical absorption. The broadening is inversely proportional to particle radius R and can be understood semiclassically (10) in terms of the mean free path limited electron lifetime $\tau \sim R/V_F$ where V_F is the Fermi velocity. Theories based on quantum mechanical dielectric function calculations (5) lead to qualitatively similar results.

The quantum mechanical theories (1, 5) also lead to the prediction that the screening ability of the electron gas is reduced with reducing R, and the long-wavelength limit of the static dielectric function is proportional to R^2.

We should like to see whether such size effects can be observed in photoemission. Any theory that describes dynamic screening leads to an approximation to C(t) (equation 2) and we shall discuss the simplest scheme that embodies the above mentioned features, namely hydrodynamic theory. The hydrodynamic dielectric function

$$\epsilon(q, \omega) = 1 - \frac{\omega_p^2}{\left(\omega + \frac{i}{2\tau}\right)^2 - \beta^2 q^2} \tag{7}$$

is a phenomenological formula that interpolates between high and low frequencies. The velocity β is close to the Fermi velocity V_F ($\beta = V_F\sqrt{\frac{2}{3}}$ gives the correct Thomas-Fermi limit, whereas $\beta = V_F\sqrt{\frac{3}{5}}$ according to the Block electron gas model derivation (11)). Assuming, $\tau \sim R/V_F$, $\epsilon(0, 0)$ has the correct qualitative behaviour. The broadening of the plasma mode also has the correct R^{-1} dependence. Moreover the model can easily be formulated to describe the dynamic screening of a point charge placed at the centre of a spherical metal particle in vacuo. In one possible formulation the polarization potential $\varphi(r,\omega)$ (where r is the distance from the point charge) is the solution of the following boundary value problem:

$$\nabla^2\varphi = \lambda^2\varphi + \frac{\omega_p^2}{\beta^2}\left(\frac{e}{r} + \text{constant}\right) \tag{8}$$

$$\lambda^2 = \frac{\omega_p^2 - (\omega + i/2\tau)^2}{\beta^2}$$

$$\varphi(\infty,\omega) = 0 \quad , \quad \frac{\partial}{\partial r}\varphi(r,\omega)\bigg|_{r=R} = 0$$

The last condition ensures that the integral of the polarization charge density over the particle vanishes, as it should. The solution is readily found to be

$$\varphi(r,\omega) = e\left[\frac{\omega_p^2}{\beta^2\lambda}\,\frac{1-c+zs}{s-zc}\,\frac{\sinh(\lambda r)}{\lambda r} + \frac{\omega_p^2}{\beta^2\lambda}\,\frac{\cosh(\lambda r)-1}{\lambda r} - \frac{K^2}{\lambda}\,\frac{c-1}{s-zc}\right] \tag{9}$$

where $z = \lambda R$, $c = \cosh(z)$, $s = \sinh(z)$.
The relaxation shift is given by

$$\Delta E = \frac{e^2\omega_p^2}{2\beta^2\lambda}\,\frac{zs+2-2c}{s-zc}\bigg|_{\omega=0} \tag{10}$$

For $\omega = 0$, λ is real and even for R ~ a few Ångstroms, $z \gg 1$. Therefore to a very good approximation

$$\Delta E \approx -\frac{e^2\omega_p^2}{2\beta^2\lambda}\left(1 - \frac{1}{z}\right) \tag{11}$$

which shows that there is a R^{-1} correction that is due mostly (but not entirely) to the electrostatic energy of charging a sphere of radius R. We can also use equation (6) to obtain an expression for the asymmetric exponent. Strictly speaking, a finite system with quantized levels should not be described by a continuous lineshape and therefore should have no α. However, as our energy resolution is too poor to see the quantized levels as separate peaks in the XPS spectrum it is reasonable to use a continuous description of the particle response. It has been shown (12) that even for particles containing a few electrons an asymmetry exponent can be defined. For large R ($z \gg 1$) we obtain from equation (6)

$$\alpha_\tau \approx \frac{e^2}{2\pi\tau\beta\hbar\omega_p}\left(1 - \frac{2\beta}{R\omega_p}\right) \tag{12}$$

This should be interpreted as the τ dependent part of α. Usually one computes α without including electron lifetime effects in the response function of the system and obtains values of the order of (13)

$$\alpha_\infty = \frac{e^2}{2\pi\hbar V_F} \frac{4K_F^2}{4K_F^2 + K_{TF}^2} \tag{13}$$

for a point core hole (here K_{TF} is the Thomas-Fermi wave-vector). In order to see more clearly the relationship between the size-dependent contribution α_τ and the bulk contribution α_∞ we need a theory including both the ordinary R.P.A. dynamic screening and electron lifetime effects. Since the two contributions have different origins one expects that to a good aproximation $\alpha \approx \alpha_\infty + \alpha_\tau$, and this can indeed be shown (14) to be the case. For R in the range of a few tens of atomic units, α_τ is an appreciable correction to α_∞. For instance, assuming $r_s = 4$ and $R = 50$ a.u., $\alpha_\infty = 0.199$ while $\alpha_\tau = 0.0017$. We conclude that small particles have a larger asymmetry exponent than their parent metals.

The present idealized model has the merit of being analytically soluble in any detail. The trick for doing so is to expand the polarization potential of equation (9) in a complete orthonormal set of spherically symmetric, electrostatically independent normal modes first introduced in ref. 2,

$$\rho_n(r) = \frac{\lambda_n}{|\sin(x_n)|\sqrt{2\pi R}} \frac{\sin(\lambda_n r)}{\lambda_n r}, \tag{14}$$

where x_n are the roots of $\tan(x_n) = x_n$ and $\lambda_n = x_n/R$. We get

$$\varphi(r,\omega) = \frac{2eR\omega_p^2}{\beta^2} \sum_n \frac{1}{z^2 + x_n^2} \frac{1-\cos(x_n)}{\sin^2(x_n)} \left(\frac{\sin(\lambda_n r)}{\lambda_n r} - \frac{\sin(x_n)}{x_n} \right). \tag{15}$$

If we calculate $\varphi(0, 0)$ we find the same result for the interaction energy that we had obtained previously (2). One advantage of this expansion is that we can take the $\tau \to \infty$ limit explicitly,

$$\mathrm{Im}\,\varphi(0,\omega) = \frac{\pi e \omega_p^2}{R} \sum_n \left(\frac{\cos(x_n) - 1}{\sin(x_n)} \right)^2 \frac{\delta(\omega - \omega_p(n))}{\omega_p(n)} \tag{16}$$

where

$$\omega_p^2(n) = \omega_p^2 + \frac{\beta^2 x_n^2}{R^2} \tag{17}$$

We see by inspection that equation (21) when inserted into equation (2) yields C(t) as a discrete sum of phasefactors in time. Hence, the XPS spectrum becomes a convolution of many spectra of the Langreth (7) type, each corresponding to a size quantized plasmon mode. For finite τ, the spectra are broadened.

However, the main advantage of the normal mode expansion is that it suggests a natural means to introduce a plasmon cutoff by the condition $\lambda_n < \omega_p/v_F$, thereby reducing the overestimate of screening effects inherent in the hydrodynamic model. Thus, the n summations have only a finite number of terms, of the order of $\omega_p R/\pi v_F$. These refinements modify the above results for ΔE and α somewhat, but do not change the qualitative aspects of the predicted size effects.

In summary, we expect a size dependence of ΔE due mainly to metal particle charging, and a size dependence of α due to the mean free path limited electron lifetime.

2.2 Experiment and conclusion

Published XPS core spectra (15) of small Au particles evaporated on teflon show a considerable broadening with reducing the nominal film thickness from 100 Å to about 1 Å. A comparison of 4 f 5/2, 7/2 Au lines from bulk gold and thin films (figure 1) reveals a remarkable, continuous increase of the width with reducing film thickness while the Fluorine 1S line width remains constant. This latter fact indicates that field inhomogeneity due to sample charging plaies only a minor role. On the other hand, the width of lines from 50 ÷ 100 Å thick films was the same as for discontinuous films composed of macroscopic particles. The further broadening by about 1ev (see figure 2) with reducing particle size is not a trivial charging effect, but an intrinsic property of the particles themselves. Our theoretical discussion suggests two different intrinsic broadening mechanisms. A particle size distribution within the sample could produce a distribution of relaxation energies ΔE (equation (11)) and hence an inhomogeneous broadening of core lines. On the other hand, the increase of the asymmetry parameter predicted by equation (12) could easily produce a broadening of the observed magnitude. We conclude that XPS can be used to study screening effects in very small metal particles.

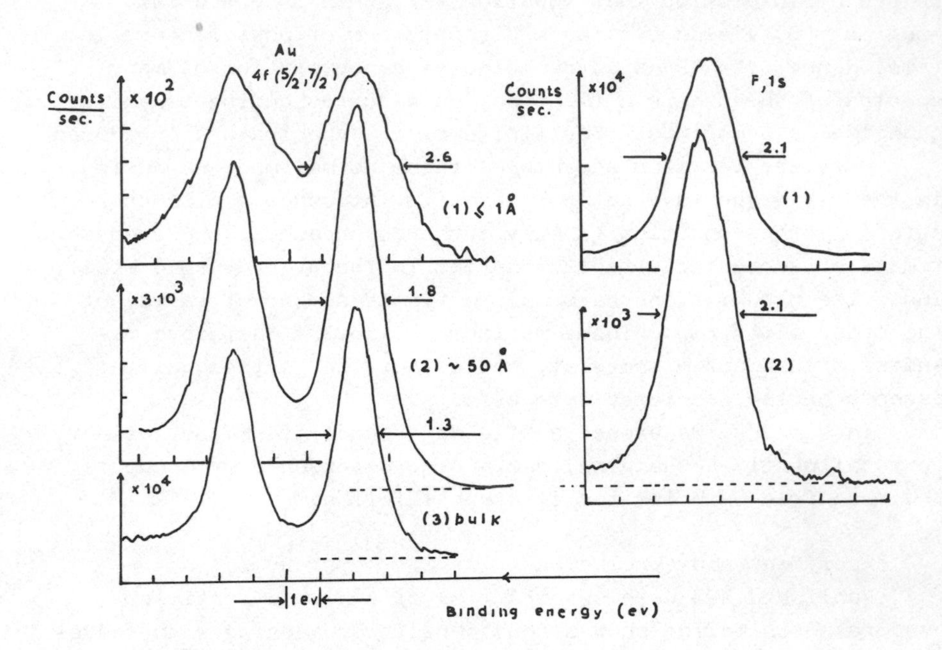

Fig. 1 - Typical X.P.S. spectra of gold films of different thickness evaporated on teflon. Sample (1) had a thickness of about or less than 1 A, sample 2 of about 50 A, and sample (3) was a piece of bulk gold.
- The F.W.H.M. of the lines are reported and are respectively for sample (1) 2.6 (eV), (2) 1.8 (eV), (3) 1.3 (eV), the fluorine for sample (1) and (2) are also reported, they show the same F.W.H.M.

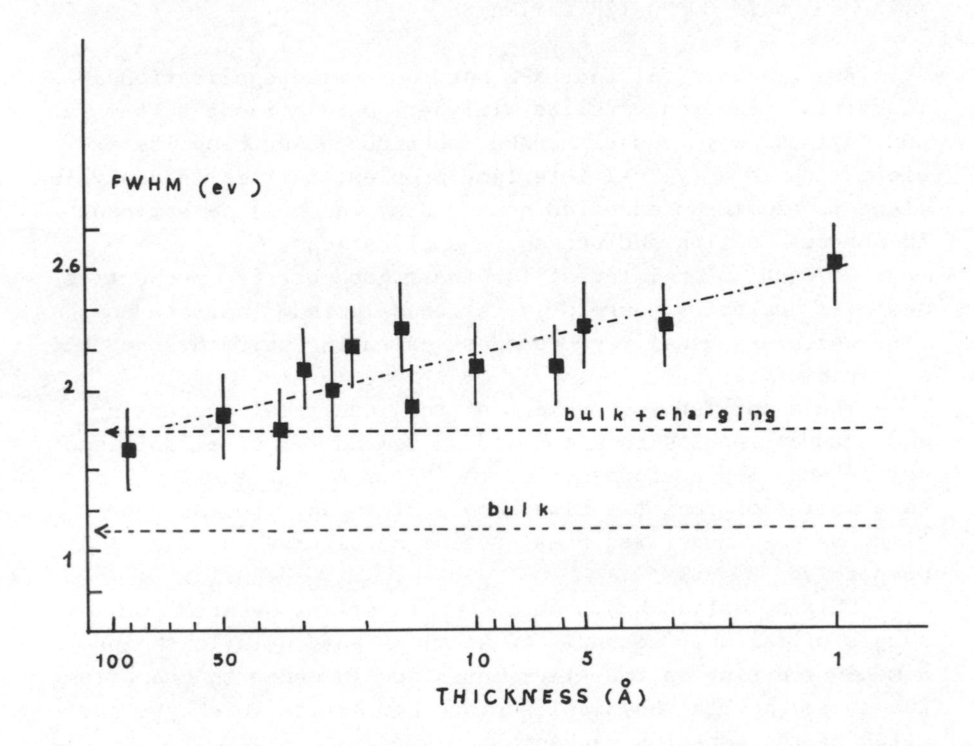

Fig. 2 - F.W.H.M. as a function of film thickness. The F.W.H.M. of bulk gold and that of bulk gold plus charging are indicated with a dotted line. The point dotted line is draw to indicate the trend of the data.

3. XPS APPLICATION TO THE PROBLEM OF POLYETHYLENE TO ALUMINUM ADHESION IN COMPOSITE LAMINATES

As an analytical tool XPS has found wide application in industrial research work. We will here make evident this more conventional way to use the XPS technique describing its exploitation in a typical interface problem, that is the polyethylene to aluminium adhesion phenomenon, which is determinant in various coating and packaging applications.

Of particular interest for their good barrier properties and heat sealability are polyethylene/aluminum laminate composites which are usually produced by extruding polyethylene onto aluminium foils.

The quest for improvement of the industrial processing and product results in a continuous demand for a reliable and better adhesion performance of the PE to Al surfaces.
As a matter of fact the always more stringent process conditions have evidentiated PE-Al adhesion failures on several Al composites.

This situation has prompted us firstly to search for the causes of failure, secondly to study in some details the phenomena occurring on the Al-PE interface in order to characterize those chemical physical surface properties which are specific of the adhesion mechanism.

The work has been focused on several points as follows:
1) characterization of the Al-foil with identification of the Al oxide surfaces properties and of the impurities there segregated by the rolling-cycle.
2) characterization of the PE with identification of the "functional" groups present on the surface or there introduced by oxidation.
3) study of the PE-Al oxide interface chemistry with regard to the interaction between PE and the surface of the Al oxide.
4) effect of the impurities on the PE-Al interface interaction. However being interested in evidentiating XPS analytical capability here we will concentrate ourselves to mostly one aspect of the Al-PE adhesion phenomenon which is better approachable by XPS technique; that is the character of the Al-foil surfaces.

3.1 Characterization of the Al-foil surface

The characteristics of the Al-foil relevant to the PE-Al

adhesion phenomenon are typical of the outmost atomic layers at the foil surface, which consists of Al-oxide usually about 20-30 A thick. The properties of this Al-oxide layer are determined by the past hystory of the Al sample, that is by the rolling cycle conditions used in the factory.

In the first stage of this work we made use of the most evident analytical characteristics of XPS:

1) the capability to identify all elements except hydrogen (atomic composition analysis)

2) the capability to determine the chemical state of the atomic species involved (i.e. the Al ion involved in the Al_2O_3 bond is distinguishable from the Al-atom in the metallic state).

3) but more important the information is derived and concerns the first 4-5 atomic layers at the sample surface.

We examined several Al-foil sample surfaces and we reached the following conclusions:

a) Al foils show frequently the impurities segregated at the surface. These impurities are uniformly distributed in the Al-oxide surface layer. This was established comparing the O, Mg, Al^{+3}, $\overset{\circ}{Al}$ line intensities obtained for various electron emission angles relatively to the sample surface. It is well known that the contribution of the top-most atomic layers is more favorably weighted for electrons emitted grazing than perpendicularly to the surface (19). The costancy of the ratio of the Mg to the (O plus 3/2 Al^{+3}) line intensities when compared to the corresponding one of the Al^{+3} to Al was accordingly interpreted as an evidence that the Mg impurities were uniformly distributed in the Al-oxide surface layer.

b) when the Mg impurity concentration at the surface reaches about 8%-10% (in atomic concentration) faulty adhesion is reported with PE.

c) the Mg surface concentration is produced by an usually very small bulk Mg content which however can be segregated at the surface during the heat treatment consequent to the rolling cycle of Al sheet production.

We have been able to confirm this effect on the Al foil by comparing the Mg line intensity (in particular the more intense MgKLL Auger line) before and after the final thermic treatment (fig. 3) (at 400°C for several hours). The bulk Mg content was initially 50 p.p.m., this was all segregated at the surface during the heat treatment.

However other factors can affect the adhesion properties even if their influence is small in respect to that of Mg. One of these is the thickness of the Al-oxide layer. In fact one can show that a slight increase in the Mg surface concentration can be tolerated if the annealing conditions are such to yield a thicker Al-oxide.

In this respect together with the Mg atomic surface composition percent also the ratio of Al^{+3} to Al^{0} as indicated by an X.P.S. analysis is an interesting datum being related to the oxide-layer thickness and consequently to the length of heat treatment. This can be a meaningful key to the adhesion performance although of less information than the Mg concentration, which we found to be determinant.

As a last point the presence of carbon, which may be related with the surface contamination by Al-rolling oil (or other), has been shown to affect the adhesion performance. In fact rolling-oil can both influence the formation of the oxide layer and be detrimental per se preventing the wettability of Al foil by the PE. The work just summarized was of value in indicating some of the properties of importance in the PE-Al adhesion phenomenon however the microscopic mechanism explaining the Al-PE molecular interaction level and how this was influenced by Mg impurities was not cleared.

Consequently it was apparent to us the need for a characterization of Al-oxide surface layer to the end of relating Al-oxide surface structure with its reactivity toward PE (in particular toward a suitabily functionalized PE). To this end, many questions were unanswered, for example:

1) due to the existence of different types of Al-oxides (like α, γ-Al_2O_3) showing different catalytic properties can we identify which specie corresponds better to Al-oxide formed at the surface of the Al-foil?

2) due to the different catalytic activity of γ-Al_2O_3 as a function of the degree of hydroxilation (20) can we determine the degree of hydroxilation of the Al-oxide on the Al foil surface with respect to common Al_2O_3 powder sample at room temperature?

To answer these questions it is not sufficient to make a more or less standard analytical use of the X.P.S. technique but it is necessary to exploit its quantitative analytical capability.

However in accord with the few works appeared on this subject (21-24) we consider that today X.P.S. analysis cannot be reliably quantified more accurately than about 15% (atomic con.).

On one side this depends on the actual state of the theoretical description of the photoemission phenomena where in the framework of the three steps model (19) use is made of such physical quantities as photoelectron cross sections, electron cross sections, electron mean free paths, intensity losses due to satellite lines, each of which seems to be difficult to be accounted for with a precision of better than about 10%.

On the other side, no standard procedure has been worked out to determine such instrumental factors like the transmission sensitivity of the apparatus, the influence of the geometry of the scattering chamber, the effect of the dimension of the sample and the aging of the electron detector response.

Unfortunately a 15% undeterminacy in the elemental surface composition of the Al-oxide would for example be enough to neglect almost an $Al(OH)_3$ atomic monolayer contamination on the top of the Al_2O_3, which in our case is rather important in defining the sample surface reactivity.

In order to reduce the undeterminacy associated with the influence of the energy dependence of the quantities, previously described, we compared line intensities only for lines taken in a narrow energy range. Furthermore we preferred to refer our data to that of known model compounds (measured under standardized conditions).

Several compounds and Al-foils surfaces have been analysed by X.P.S., the data are reported and compared in Tab. 1.

Our apparatus is an A.E.I. 100. Model compounds have been chosen because of their different stoichiometry, structure and degree of hydroxylation in such a way to allow a comparison with the characteristics of the Al-oxide formed on the surface of the Al-foil.

In tab. 1 are shown twelve columns of data. In col. I we indicate the samples examined. In col. II are reported the ratio between the intensities of the X.P.S. lines associates with oxygen (O_{2s}) and aluminum (Al^{+3}_{2p}) in oxydes. This datum is proportional to the ratio between the total number of Oxygen atoms and the Aluminium atoms in aluminium oxide present at the surface of the sample (Al-oxide of 20-30 A thickness). The $O^{tot}_{2s}/Al^{+3}_{2p}$ intensity ratio has been chosen because

TABLE 1

Characteristics of the Al-oxide formed on the surface of Al-foil.

I	II	III	IV	V	VI	VII	VIII	IX	X	XI	XII
Sample	$\frac{O_{2s}^{tot}}{Al_{2p}^{+3}}$	$\frac{O_{2s}^{tot} - O_{2s}^{CO_3^-}}{Al_{2p}^{+3}}$	$\%Mg = \frac{Mg}{Mg + Al^{+3}}$	Mg p.p.m.	carbon cont. thickness $\frac{Al_{2p}}{C_{1s}}$	$d^{carb} = \frac{5.35}{\frac{Al_{2p}^{+3}}{C_{1s}} + 3}$	$\frac{O_{1s}^{CO_3^-}}{O_{1s}^{t}}$	$Al_{2p} - O_{1s}$	$Mg_{2p} - O_{1s}$	FWHM	FWHM
	stoichiometry in arbitrary unit		on the surface	in the bulk			from $C_{1s}^{CO_3^-}$ %	ΔE_b(ev)	ΔE_b(ev)	O_{1s}	Al_{2p}
α-Al_2O_3	0.33	0.33			3	1.5		456.5		20	18
γ-Al_2O_3	0.41	0.41			9.8	0.5		456.6		29	24
$Al(OH)_3$	0.7	0.7			2.1	2		457.1		22	20
AlO(OH)	0.46	0.46			5	1		547.1		30	20
Al 1	0.41	0.36	3.5	10	1.2	3.5	16	456.7		26	20
Al 8	0.42	0.37	2.5	10	1.1	3.8	17	456.7		27	20
Al 5	0.51	0.42	25	240	0.8	5	18	456.6	480.9	26	20

the line energy positions (i.e. binding energies) of the O_{2s} and Al^{+3}_{2p} lines are similar. Their intensity ratio can thus be shown to be largely independent on the thickness of the surface carbon contamination and (in the case of the Al-foils) on the thickness of the Al-oxide.

Various other factors related to the characteristics of the instrumentation can in this case shown to be neglegible. Returning to the model compounds, we recall that γ-Al_2O_3 is known to be a compact solid, scarcely reactive and poorly hydroxylated on the surface; these properties are in agreement with the results obtained by X.P.S. analysis.

Thus to a good approximation the ratio of O to Al-atoms, at the surface of γ-Al_2O_3, may be considered 3/2 and the reported experimental datum 0.33 should be taken thereafter as indicative of an O to Al ratio of 3/2.

The data obtained for $Al(OH)_3$ and AlO(OH) shown in the first column are in agreement with that of γ-Al_2O_3, in fact their O_{2s}/Al^{+3}_{2p} ratios 0.7 ($Al(OH)_3$), 0.46 (AlO(OH)), 0.33 (γ-Al_2O_3) correspond quite well with their respective stoichiomctrics O/Al of 3, 2, 3/2.

The carbon C_{1s} line is indicative of the surface contamination. This line is splitted in two components, one of which suggests the presence of a carbonate ion CO_3^-. Accordingly it was possible to calculate the percentage of oxygen associated with the carbonate and by substraction to evaluate the amount of oxygen related with the oxides and hydroxides. While on the non metallic compounds the carbon and carbonate contamination were neglegible it was not so on Al-foils (Al 1, Al 8, Al 5).

The percentage of oxygen to be associated with carbonate is displayed in column VIII, when this is substracted from the total oxygen contribution we get the number in column III, which gives a better evaluation of the stoichiometry of the Al-oxide surface of the foils.

From this series of data (column III) we may already conclude that γ-Al_2O_3 showing a O_{2s}/Al^{+3}_{2p} ratio of 0.41 has a surface stoichimetry which is intermediate between that of γ-Al_2O_3 and that of AlO(OH). This evidentiates the partial hydroxylation of γ-Al_2O_3 at room temperature. From the same type of comparison we see that Al foil samples have a surface stoichiometry intermediate between that of γ - Al_2O_3 and α - Al_2O_3. On the basis of ours and in agreement with other published data we shall conclude that the oxide on the surface of Al-foil

samples is a γ - Al_2O_3, less hydroxylated than pure γ - Al_2O_3. In col. IV we report the percentage of Mg found on the surface and we compare this with the p.p.m. of Mg present in the bulk. From these two sets of data it becomes apparent the large amount of Mg which is segregated at the surface during the rolling cycle of the Al.
In practice up to an impurity level of 100 p.p.m. all the Mg present in the bulk is segregated at the surface.
In column VI we report the line intensity ratio Al^{+3}_{2p}/C_{1s} which, when approximately corrected for various element sensitivity factors and instrumental factors (21), can be used to calculate the nominal carbon surface contamination of the samples. To do this, we made the simplest assumption, which consists in considering the carbon contamination as a homogenous layer with a certain nominal thickness. The carbon contamination, which is probably formed by hydrocarbons of various nature, has been considered of an average density comparable to that of graphite. Surface carbon contamination layer thickness are reported in column VII and given in A.

The maximum amount is shown to be of the order of 1 atomic monolayer. This does not mean that the surface is unaccessible, being covered by a homogeneous layer of surface contamination, in fact this is surely not evenly distributed, leaving a large part of the surface easily accessible, as it is shown by the smalless of the total surface contamination (as already described).

In column VIII we reports the CO_3^- contamination as a percentage of the total oxygen content.

In column IX we show the difference in binding energies between the Al_{2p} and O_{1s} levels. The interest of this comparison derives from the observation that oxygen in Al-hydroxide has a different binding energy than oxygen in oxides. It is again confirmed that the oxide on the surface of Al foils consists of a less hydroxylated form of γ -Al_2O_3, when compared with the γ -Al_2O_3 powder solid. This is substantiated by the following XI and XII columns where the lines widths of O and Al are reported. These line-widths are indicative of the oxide and hydroxides contributions that are not resolved due to their small energy splitting.
For the Al-foil surface the O line width appears again as intermediate between those of the γ and α Al_2O_3. Furthermore we notice the large width of the oxygen line in γ -Al_2O_3.

In part this may be also related to the particular configuration of the γ-Al_2O_3 lattice, where two different Al-oxygen first neighbors distances are known to exist, probably associable with a slightly different character of the short-range bonds. Finally in column X is reported the binding energy of Mg relative to that of oxygen.
From all the data shown in the tab.1 we may then conclude that Al-foil surface oxide may be described as a less-hydroxylated form of γ-Al_2O_3 with traces of Al carbonatation contamination.

b) We consider now the question of characterization of the Al-oxide-foil surface when Mg impurities are present.
The questions that we may ask are: is Mg, which segregated at the surface during the rolling process, going to form islands of $Mg(OH)_2$ or MgO, or instead is Mg segregating in the spinel lattice of γ-Al_2O_3 and forming a $MgAl_2O_4$ spinel compound?
We compared samples of Mg-oxides formed on the surface of a Mg-metal giving particular attention to the behaviour of the oxygen line. In Fig. 4(a) is shown the oxygen O_{1s} line of a sample of Mg-metal. Two chemical species are evident the first one due to hydroxide at 952.4 (eV) and the second at 953.9 (eV) for the Mg-oxide, the hydroxide being preminent. By a heating treatment under vacuum (10^{-6} torr) one can observe the trasnformation in "situ" of the Mg-hydroxide in Mg-oxide, Fig. 4(b).
A second Mg metal sample has been heated after a scraping treatment to 400°C and then left 20 days at normal temperature Fig. 5(a). The analysis of the same sample after 20 days in air are shown in Fig.5(b), where the Mg-hydroxide is shown again to be the preminent feature.
These data evidentiate the high reactivity of Mg-metal and of the Mg-oxide formed on a Mg-metal with water vapour contained in air to form hydroxide.
In Fig.6 is instead shown the O_{1s} line of the Al-foil (5), where no Mg hydroxide contribution is apparent.
At this point the following arguments can be given to support the view that Mg at the surface of Al foil consists of $MgAl_2O_4$ instead that MgO.
1) We have shown that MgO on the surface of Mg-metal is usually very quickly transformed and stabilized as $Mg(OH)_2$. This is even more so for MgO powder.

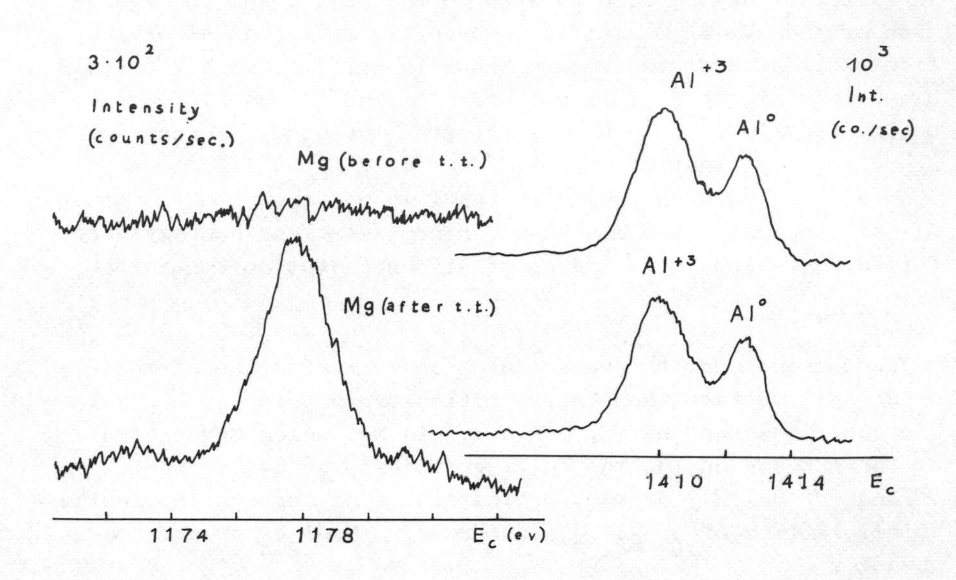

Fig. 3 - Mg on the surface of an Al foil (containing 50 p.p.m.) of Mg before and after final thermic treatment).

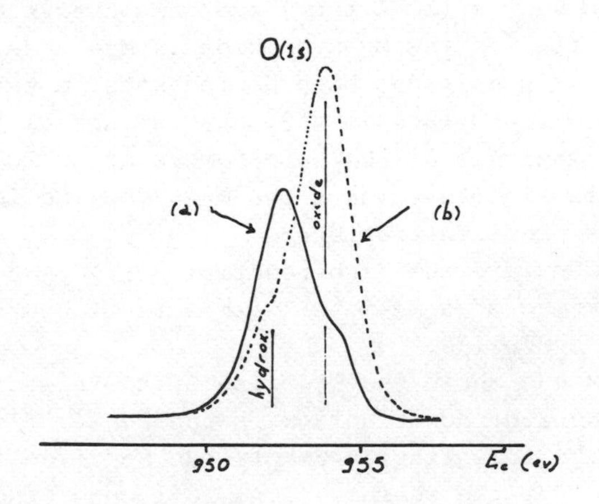

Fig. 4 - The oxygen O_{1s} line of a sample of Mg metal. Two chemical species are evident: hydroxide at 952.4 (eV), oxide 953.9 (eV).

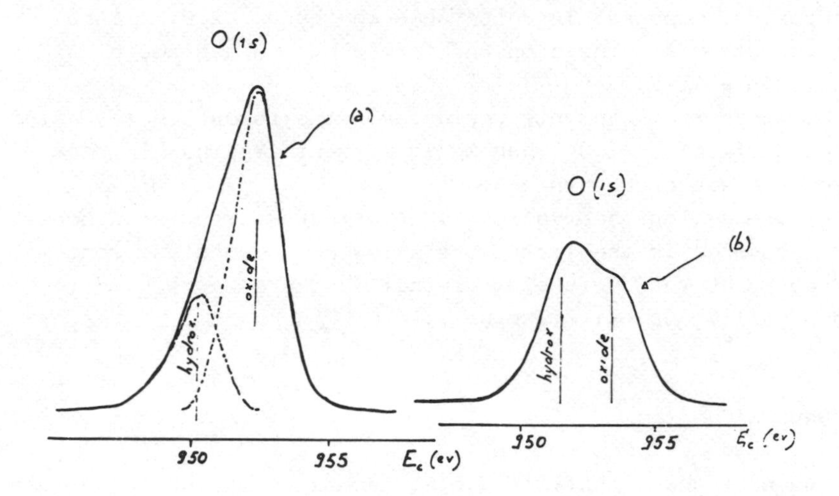

Fig. 5 - a) The oxygen line of Mg sample that after a scraping treatment has been heated to 400°C.
b) The oxygen line of the same sample after 20 days in air.

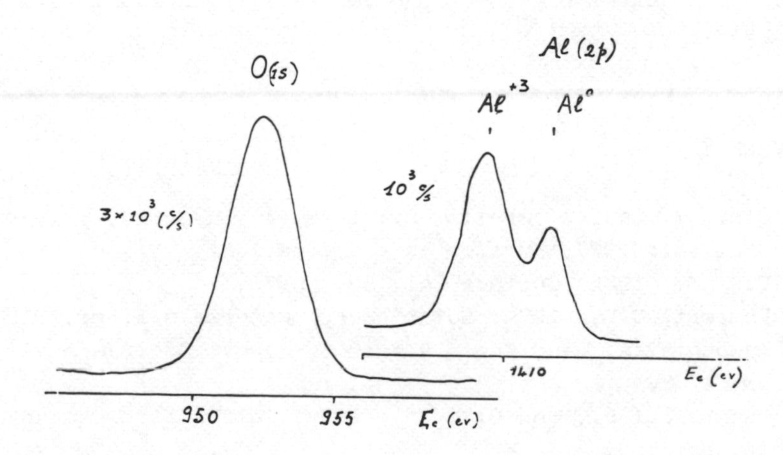

Fig. 6 - The oxygen line of Al-foil of Al foil sample (5) where no Mg hydroxide contribution is apparent. The Aluminium line is also shown.

2) When MgO is mixed to Al_2O_3 in the proportion of one to one a hydroxide component is detectable on the O_{1s} X.P.S. line. This is not so for the compound formed on the surface of Al-foil (sample Al-5).
3) Thermodynamic arguments would favor the formation of $MgAl_2O_4$ on the surface of Al_2O_3 when Mg is present. MgO has in fact larger surface energy than Al_2O_3.
4) The preceeding observations when compared with the discussions appeared in the literature allow us to conclude that Mg segregated at the surface of Al-foil is there aggregated to form a $MgAl_2O_4$ spinel structure.

4. CONCLUSION

We have shown that the X.P.S. technique can be on extremely useful tool in surface research if for each problem some care is taken in selecting the proper "observable" feature to exploit.

In this sense, perhaps for the pleasure of the reasearcher, X.P.S. is still not a technique which allows standard analytical procedures.

Both theoretical and experimental work remains to be done to achieve better precision and reliability in quantitative analysis.

REFERENCES

1) Cini, M., and Ascarelli ,P.: 1978, J. Phys. F4, pp. 1998.
2) Cini, M.: 1977, Surface Sci. 62, pp.148.
3) Cini M.: 1975, Surface Sci. 52, pp.75.
4) Lampert, C.M.: 1979, Solar Energy Materials 1, pp. 319.
5) Kawabata, A., and Kubo, R.: 1966, J.Phy. Soc. Japan 21, pp. 1765.
   Ascarelli, P., and Cini, M.: 1976, Solid State Commun. 18, pp. 385.
   Cini, M.: J. Optical Soc. Amer. (in press).
6) Nozieres, M., and De Dominicis, C.J.: 1969, Phys. Rev. 178, pp. 1097.
7) Langreth, D.C.: 1970, Phys. Rev. B1, pp. 471.

8) Doniach, S., and Sondheimer, E.: 1974, "Green's functions for Solids State Physicists", Benjamin, Reading, Mass., U.S.A.
9) Lighthill, M.J.: 1962, "Fourier Analysis and Generalized Functions", Cambridge University Press.
10) Ganiere, J.D., Rechsteiner, R., and Smithard, M.A.: 1975, Solid State Commun. 16, pp.113.
11) Ritchie, R.H., and Wilems, R.E.: 1969, Phys. Rev. 178, pp. 372.
12) Dow, J.D., and Flynn, C.P.: 1980, J. Phys. C: Solid St. Phys. 13, pp.1341.
13) Ascarelli, P.: 1977, Solid State Commun. 21, pp. 205.
14) Ascarelli, P., and Cini, M.: to be published.
15) Ascarelli, P., Cini, M., Missoni, G., and Nisticò,N.: 1977, J. Physique, Colloque C2, supplément au n°7, 38, pp.C2-125.
16) Internal Report ASSORENI Prot. BASE/188 (unpublished).
17) Ascarelli, P., and Cernia, E.: Spectroscopic Analysis of Interactions between Metallic Surfaces and Organic Polymers. In Proc. Fourth Int. Conf. in Organic Coating Science and Technology, Athens, pp.146, Ed. Parfitt, G.D., and A.V. Patsis. Technomic publ. Westport CT 06880.
18) Internal Report ASSORENI, Prot. BASE.234 (unpublished)
19) Fadley, C.S.: 1976, Progress in Solis State Chemistry, Vol.II, Part 3.
20) Knozinger, H., and Rotnasamy, P.: 1978, Cat. Rev. - Sci. Eng. 17, pp.31.
21) Erans, S., Pritchard; R.G., and Thomas, Y.M.: 1978, J. Electron Spect. Rel. Phenomena 14, pp.341.
22) Powell, C.Y.: 1978, in Quantitative surface Analysis of Materials" ASTM STP 643, ed. Mc. Intyre, N.S., pp.5. American Society for Testing and Materials, Phila.
23) Powell, C.J., and Larson, P.E.: 1978, App. of Surface Science 1, pp.186.
24) Seah, M.P., 1980: Surface and Interface Analysis 222, Vol.2, n.6.
and references there mentioned.

COMPARING VALENCE AUGER LINESHAPES AND PHOTOEMISSION SPECTRA: THE Cd AND CdO CASES

L. Braicovich

Istituto di Fisica del Politecnico - Milano, Italy

We discuss the information on valence states in solids which can be obtained from lineshapes of Auger transitions and we compare Auger lineshapes with photoemission spectra in order to point out the specific aspects of Auger spectroscopy. The discussion is centered on the electron states of Cd and CdO.

Auger electron spectroscopy (AES) is well known as a method to identify chemical elements; taking advantage of the very short electron escape depths the typical application is in surface chemistry and undoubtely AES is the basic method to identify small fractions of a monolayer of adatoms onto surfaces. This diagnostic use relies upon the fact that Auger energies are characteristic of the elements and Auger transitions are very well seen with derivative techniques in the electrons emitted from a solid surface onto which an electron beam is sent. Although some problems are still open when the method is used for quantitative analysis, AES is nowadays a routine technique for chemical identification; for this reason it will be not treated in this lecture which is devoted to an overview of some problems which are encountered when AES is used beyond the scope of chemical analysis.

In this connection a very important point is the interpretation of Auger transition lineshapes when valence electrons of a solid are involved (we use here "valence states" as synonim of extended electron states in a solid and thus we include also conduction electrons in a metal). Of course these transitions are not very convenient in order to get information on the chemical composition. In fact the energies of different transi-

P. Day (ed.), Emission and Scattering Techniques, 375–384.

tions are generally very close with frequent overlaps and the lines are broader. This decrease of analytical chemistry power is compensated by the fact that other important information is contained in the lineshapes which are in some way connected to the nature of the valence states involved in the transition.

Thus the crucial point is: what can be learned from Auger lineshape on the valence electron states in a solid? And the other point connected with the preceding is: is there anything specific to be learnt which cannot be seen with the spectroscopic methods usually employed in electron state studies such as optical and photoemission spectroscopy? These very general problems are receiving increasing attention and have not yet been solved in general. In what follows we want to present a specific case in order to point out qualitatively some aspects of the problem and to stress that AES is not merely an analytical technique and is also growing to become a powerful source of information on the valence electron states of the solids.

When Auger transitions involving valence states are considered it is important to distinguish two cases:
CCV (core-core-valence) transitions in which only one valence electron is involved
CVV transitions in which two valence electrons are involved.
We do not consider VVV transitions which are very difficult to observe for technical reasons connected with the very low energy of the transitions.

In both cases the initial state has a hole in a core level. Thus the matrix element of the transition projects the final state onto this localized hole and the information on the density of states is is someway "local". On the other hand the two final holes interact and this interaction, in general, modifies the lineshape. In the absence of any localization effect and of any two and many body interaction in the final states the Auger lineshape would be connected in a very simple way to the valence electron density of states. In this crude approximation (by assuming constant matrix elements) the lineshape would be proportional respectively to

$\rho_V(E)$ density of valence states in CCV transitions
selfconvolution of $\rho_V(E)$ in CVV transitions (the convolution is an obvious consequence of the energy conservation).

The reason of interest of valence Auger spectroscopy is not in the possibility of obtaining ρ as one might suppose from this rough scheme but in the deviations from this scheme. In fact the valence density of states in itself is best studied with photoemission. On the other hand any deviation from the above scheme in AES is significant in terms of the nature of the valence states which are explored and of the many body interactions in the solid. Thus a very fruitful approach is that of contrasting AES information on the density of states with photoemission results in order to point out the effects which are specific of AES. In this sense AES is complementary to photoemission in understanding the valence states of the solids.

The specific case treated here belongs to CCV transitions. We will focus our attention on CCV transitions in order to avoid the complications due to the deconvolution procedures necessary to get density of states information from CVV transitions. Incidentally we note that CCV transitions have been studied very little up to now due to technical problems since they are very weak; in effect they involve final holes in different states and not onto the same state as in intense Auger transitions. One of the purposes of this lecture is to show the usefulness of CCV spectroscopy which can now be studied more extensively since excellent signal to noise ratios can be obtained in present equipments. Before proceding to the specific case of Cd and CdO we underline also the analogy between valence CVV spectroscopy and ion neutralization spectroscopy where the lineshapes of valence Auger transitions involving the ion neutralization during ion-surface scattering are detected.

The specific case we want to treat is that of MNV transitions in Cd and in CdO by relying upon ref. (1) where a concise summary of the results is given.

The whole spectrum involving M core levels and valence electrons is given in Fig. 1 both for Cd and for CdO obtained by oxidizing in situ to saturation the Cd sample used in the metal spectrum (for technical details see ref. (1) (2)). We note explicitly that the spectra presented here are the Auger intensities vs. energy and not the derivative as is usually done; this is necessary to get the lineshapes which are the object of the discussion.

In Fig. 1 we have two groups of transitions indicated with MNN (intense peaks) and MNV (small structures shown also expanded in the insert). The MNN transitions involve 4d electrons of Cd in the final state; the 4d bands in Cd are known to be very narrow (3) i.e. the d-electrons are considerably localized. For this reason it is expected that MNN transitions are atomic like and the symbol V does not appear in the label. In effect it has been recognized that these transitions are atomic like (4,5,6) and they will not be discussed here.

We discuss the MNV transitions which involve in the final states 4d levels and the conduction band in Cd (and the valence

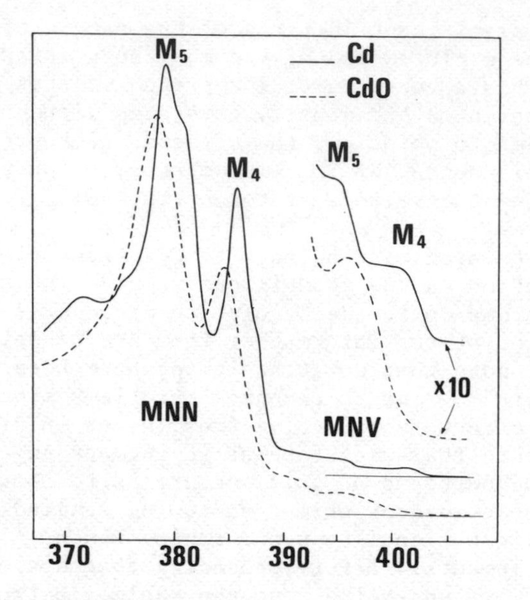

Fig. 1 -
Auger MNN and MNV transitions in Cd and CdO.

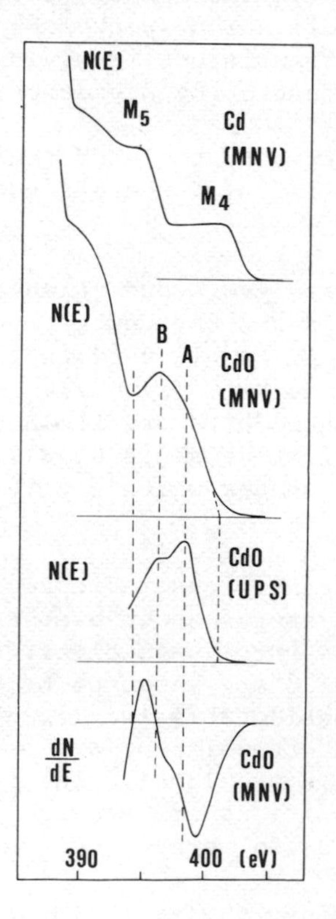

Fig. 2 -
MNV Auger spectra for Cd and CdO and photoemission spectrum of CdO (photon energy 21.2eV).

bands in CdO). In Fig. 1 two distinct groups of MNV transitions appear as in the MNN case; these groups are due to the presence of a hole in a M_4 or in M_5 level in the initial state. The M_4NV transitions appear at higher energies so that they are better seen than M_5VV where the pile up with the background complicates the interpretation; thus in what follows we will refer to M_4NV transition lineshape. These transitions are given in an expanded scale in Fig. 2 both for the metal and CdO. The modification of MNV lineshape is dramatic upon oxidation of the metal and this is one of the key - points of the discussion. Another important point, accordingly to the phylosophy discussed above, is the comparison with photoemission data (2); this is done in Fig. 2 for CdO where the top of the valence band obtained by extrapolation of the photoemission spectrum has been aligned with the extrapolated edge of the MNV Auger spectrum. The photoemission spectrum from the metal is typical of a simple metal and is similar to M_4NV in Cd so that it is not reported here. In Fig. 2 we have given also the $\frac{dN}{dE}$ Auger spectrum for CdO (i.e. what is known as Auger spectrum in standard techniques) in order to point out small structures in CdO valence Auger transitions; in particular the weak shoulder A is well resolved in the derivative.

Before discussing the Auger lineshapes we mention qualitatively a very important result which comes out from ref. (7) (8) (9) in connection with the transition between an atomic like case to a bandlike case in Auger valence emission. We have already seen that a bandlike regime requires that the hole-hole interaction in the final states is negligible. The basic result is that the transition from one regime to the other is determined by the relative weight of the hole-hole interaction U and of the bandwidth a. When $U/a \gg 1$ the hole-hole interaction is so strong that the band effects are small and it is necessary to rely upon an atomiclike interpretation of the spectra. On the contrary, when $U/a \ll 1$ one recovers a bandlike picture. These results proved theoretically in the original references are very intuitive; as a qualitative guideline in the discussion we will assume this point of view also in the case of a CCV transition where U is the hole-hole interaction between the two holes in the final C and V states and a is the width of the valence band V. This heuristic assumption is basically motivated by the physical picture underliing the qualitative interpretation of the above criterion.

If we refer to the lineshapes given in Fig. 2 several considerations can be made.
(1) The shape of MNV transition in the metal is very similar to that expected for the density of states of a free - electron metal with a ruther sharp edge at the heigher energies corresponding to the Fermi edge. This shows clearly that the MNV

transitions in the metal are bandlike; this point can be understood on the basis of the preceeding discussion since the electron gas screening of the interaction between the final holes is very efficient. Another important point deals with the localization effect due to the projection onto the localized hole in the initial state. Undoubtfully an effect of this kind takes place and a kind of local density of states is sampled; nevertheless this local density is not expected to be qualitatively different from the total density of states since the Bloch states are very symilar to plane waves. This discussion shows that in several simple metals a symilar behaviour is expected. It will be interesting in the future to study the deviations from a purely bandlike behaviour introduced by final state interaction which in turn depends on the dielectric properties of the different metals. In this sense CCV (and CVV) transitions in the metals can have a considerable impact in the study of the electron gas properties.

(ii) In the CdO case the MNV lineshape is very different with respect to the UPS spectrum ($h\nu$=21.2 eV)which is known to be connected to the density of states more closely than the Auger lineshape. The difference is evident from Fig. 2 which nevertheless indicates a correlation between the two spectra. In effect a correspondence between the structures is found but the intensities are quite different. Structure B is the strongest in Auger spectrum and is the weakest in photoemission spectrum whereas structure A which is the dominant one in photoemission is very weak in Auger. We can thus say that considerably different information on the electron states in CdO can be obtained with the two methods; the differences in the spectra are primarily due to the different nature of the sampling of the valence states which is done in the two cases and to final state interaction as we will show below. The fact that the differences in Auger valence spectroscopy with respect to photoemission are much higher in CdO ruther than in the metal is very significant. In ionic CdO the localization effects are expected to be much higher due to the different cationic and anionic contributions to the density of states; moreover the metallic screening of the holes is not present. Thus compound semiconductors and ionic crystals are the ideal candidates for valence Auger lineshape spectroscopy and a tremendous work remains to be done in this field.

(iii) In CdO considerable hole-hole interaction effects are expected. It is difficult to assess the point quantitatively owing to the difficulty in evaluating the true hole-hole interaction in the solid but it is possible to give a qualitative argument. By assuming a purely intraatomic picture and by referring to atomic spectroscopy properties it is possible to evaluate the hole-hole interaction as suggested in ref. (10)

$$U = W(YZ) - W(Y) - W(Z)$$

where W are the spectroscopically determined free ion energies and Y, Z and YZ indicate the appropriate one and two holes states. From values of ref. (11) one obtains $U \simeq 10-11$ eV. The valence bandwidth estimated from photoemission data (2) is around 8 eV so that $U/a \simeq 1.3$. For this reason we expect hole-hole effect to be relevant in originating the distorsion of the Auger lineshape with respect to the density of states which would be samples in the absence of this interaction. Since the hole-hole interaction is repulsive we expect an increase of the intensity of the Auger emission near the top of the valence band with respect to what would be found with U=0.

(iv) The localization effects are expected to be strong in CdO as suggested above. In the specific case since the initial hole is in a cationic position the spectrum must reflect the cationic density of states, while the hole-hole interaction originates a distorsion of this local density of states. The comparison between AES and photoemission results shows that at higher binding energies in the valence bands the MNV intensity is higher than at lower binding energies; this is the evidence of a cationic contribution to the electron density of states which is stronger at higher binding energies. Namely structure B has a stronger cationic contribution than structure A. The Auger results are the direct evidence of this point which could be hardly assessed with standard photoemission techniques. The tail in MNV line up to the top of the valence band could be due at least in part to final state interaction but there is no doubt about the evidence of the cationic contribution to the density of states at higher binding energies. This is very important since the evidence of this cationic contribution is not a trivial point. In fact in polar II-VI systems the anionic contribution to the density of states is known to be the most relevant; in some extreme models as in (12), (13) the valence density of states is fully anionic although it has been suggested in (14) that some cationic contribution can be present in polar semiconductor. The correctness of the cationic assignement in the Auger spectra presented here is also confirmed by the comparison of the intensities of MNV lines in the metal and in the oxide. In the figure the lines are given in such a scale that the MNN transitions have the same intensity in Cd and CdO. This means that the intensities of MNV is of the same order of magnitude in CdO and in Cd (more precisely is higher in CdO than in Cd). In Cd there is no doubt that the greatest contribution to MNV comes from the overlap of conduction Bloch states onto the localized hole on Cd atoms; since the intensity is comparable in CdO this means that MNV comes from the overlap between valence states localized in the cationic position with the cationic hole. A possible interatomic Auger effect giving an anionic contribution to the MNV line should be much weaker due to the small overlap accordingly to the general rule which gives much weaker interatomic transition with respect to intra-atomic ones (10). We can thus conclude that cationic contribution is the dominant term in the measured line; in the upper region of the valence band the cationic contribution is known to be very small and the hole-hole interaction is the leading effect in determining Auger lineshape.

The above discussion is qualitative but it is largely sufficient to show the power of valence lineshape analysis in understanding the nature of the chemical bonds in solids. More precise discussions will be possible when detailed theoretical treatments will be available. We want to stress that in such a case a great amount of information onto the dielectric properties of solids will be obtained from realistic discussion of final states interaction. This research line is at the very beginning but has already given excellent results (15) (16) even in the absence of extensive theoretical support. In this connection we want to mention, before concluding, the work by IBM group on Auger lineshapes of Si LVV in transition metal silicides and in Si d-metal interfaces (17). The reader is sent to the original papers and to the references quoted therein. We mention only some basic ideas of this work which is the brightest recent application of Auger lineshape analysis. The phenomenological basis is the fact that LVV Si lineshape is very different in pure Si and in Silicon combined in a silicide--like phase. This result, which has been already found by several other authors, has been applied to the study of the electron states which are formed in an interface (Si(111)+Pd) grown in situ by evaporating increasing amounts of Pd onto the Si substrate from fractions of a monolayer up to several monolayers. In the interface case Auger spectra contain two contributions: one from the uncombined Silicon beneath the interface an the other from the reacted Silicon in the interface. With an accurate subtraction of the unreacted contribution it is possible to point out the interface contribution which gives, after deconvolution, information on the electron states at the interface. The whole procedure is not simple but it is very confortable that an independent research with photoemission (18) has given basically the same picture. Incidentally we note that in this photoemission experiment we have taken advantage of the tunability of synchrotron radiation to measure interface valence photoemission in two distinct conditions

(i) when the photoionization probability from d-states of the metal is very low (19) i.e. at the Cooper minimum for the d shell

(ii) when the total photocurrent from d-states is the dominant term (this situation is basically the standard case).

By contrasting the two cases it is possible to point out respectively the contribution from Si and from the metal atoms to the electron density of states.

Of course this Cooper minimum approach, when it can be used (for example the first period of transition metal does not give Cooper minima for d shells) is very powerful and direct. This does not imply that the Auger technique is not interesting also in the cases in which Cooper minimum approach can be used. In our opinion the two techniques are complementary since they are two distinct methods to look at the same object and each method has its specific character. For example a future account of final state interaction in AES applied to these problems can give very important information on the dielectric properties in the interface region.

In conclusion we have shown that valence Auger lineshape analysis can complement usefully photoemission to give unique possibilities of understanding the nature of the chemical bond in solids and their dielectric properties.

References

1. Braicovich,L., and Powell,R.A.: 1980, Solid State Commun. 33, pp. 377-379.
2. Braicovich, L., Rossi, G., Powell, R.A., and Spicer, W.E.: 1980, Phys. Rev. B21, pp. 3539-3544.
3. Stark, R.W., and Falicov, L.M.: 1967, Phys. Rev. Lett. 19, pp. 795-798.
4. Aksela, S., Aksela, H., Vuontisjarvi, M., Väyrynen, J., and Lähteenkarva, E.: 1977, J.Electron. Spectr. 11, pp. 137-145.
5. Weightman, P., Andrews, P.T., and Hiscott, L.A.: 1975, J. Phys. F5, pp. L220-L224.
6. Weightman, P.: 1976, J.Phys. C9, pp. 1117-1128.
7. Cini, M.: 1976, Solid State Commun. 20, pp. 605-607.
8. Sawatzky, G.A.: 1977, Phys. Rev. Lett. 39, pp.504-507.
9. Cini, M.: 1977, Solid State Commun. 24, pp. 681-684.
10. Citrin, P.H., Rowe, J.E., and Christman, S.B.: 1976, Phys. Rev. B14, pp. 2642-2658.
11. Moore, C.E.: 1958, NBS Circular, pp. 55-56.
12. Pantelides, S.T.: 1975, Phys. Rev. B11, pp. 5082-5093.
13. Pantelides, S.T., and Harrison, W.A.: 1975, Phys. Rev. B11, pp. 3006-3021.
14. Baldereschi, A., Maschke, K.: 1979, Inst. Phys. Conf. Series 43, pp. 1167-1170.
15. Davis, G.D., and Lagally, M.G.: 1978, J.Vac. Sci. Technol. 15, pp. 1311-1316.
16. Salmeron, M., Baro, A.M., Rajo, J.M.: 1975, Surf. Sci. 53, pp. 689-697.

17. Ho, P.S., Rubloff, G.W., Lewis, J.E., Moruzzi, V.L., and Williams, A.R.: 1980, Phys. Rev. B22, pp. 4784-4790.
Rubloff, G.W., Ho, P.S., Freeouf, J.L., and Lewis, J.E.: on print on Phys. Rev.
and references quoted therein.
18. Rossi, G., Abbati, I., Braicovich, L., Lindau, I., and Spicer, W.E.: on print on Solid State Commun..
19. Miller, J.N., Schwartz, S.A., Lindau, I., Spicer, W.E., De Michelis, B., Abbati, I., and Braicovich, L.: 1980, J. Vac. Sci. Techn. 17, pp. 920-926.

INDEX